Sustainable Artificial Intelligence-Powered Applications

IEREK Interdisciplinary Series for Sustainable Development

The interdisciplinary series "Sustainable Artificial intelligence Powered Applications," (SAIPA) aims to cover a wide range of Artificial Intelligence applications with a specific focus on sustainability to further support the sustainable development goals. AI systems are able to perform a wide range of tasks, including high-level image and video recognition, perspective modeling, smart automation, advanced simulation, and complex analytics among many others, and has high learning and improvement potential through machine learning and deep learning. These capabilities have many uses and provide many benefits in all industries applications. SAIPA includes a diverse range of these applications, perspectives, and expertise, where it seeks to advance the understanding and implementation of sustainable AI-powered solutions across various scientific disciplines. Its aim is to become a comprehensive guide to different experimental sciences. The "Sustainable AI-Powered Applications" publishes conference proceedings, monographs, and textbooks that address the intersection of artificial intelligence, sustainability, and advanced technologies. Authors, researchers, contributors, and scientists from all scientific disciplines related to AI and advanced technologies, are all invited to contribute to this series to shape the future of AI in a sustainable and responsible manner. Thus, the series represents an excellent endeavor that promotes a collaborative and interdisciplinary strategy for the development of sustainable AI-driven applications. The scope of this series includes, but is not limited to: Promoting sustainability and human well- being, health and medical use, optimizing smart cities planning, transportation systems management, infrastructure development, creating livable urban environments, renewable energy, agricultural Innovations, precision agriculture, weather agriculture, sustainable food production, crop & animal monitoring, and agricultural robotics, Socio-Economic Impacts, optimizing energy consumption in buildings, factories in energy usage and lower carbon, smart grids applications, smart grids management, climate modelling, environmental monitoring, waste management, Internet of things (IOT) and other domains related to the use of AI applications that improve overall environmental impact.

Series Editor : Mourad Amer **editor@saipa-series.com**

Brahim El Bhiri · Amir Hussain · Yassine Maleh
Editors

AI and Digital Transformation: Innovations in Supply Chain, Education, and Energy Systems

Proceedings of the 6th International Conference on Advanced Technologies for Humanity (ICATH'2024)

Springer

Editors
Brahim El Bhiri
CEIRA
Harmony Technology
Rabat, Morocco

Amir Hussain
School of Computing
Edinburgh Napier University
Edinburgh, UK

Yassine Maleh
École Nationale des Sciences Appliquées de Béni Mellal
Khouribga, Morocco

ISSN 3005-1762 ISSN 3005-1770 (electronic)
Sustainable Artificial Intelligence-Powered Applications
ISBN 978-3-031-86836-8 ISBN 978-3-031-86837-5 (eBook)
https://doi.org/10.1007/978-3-031-86837-5

This Springer imprint is published by the registered company Springer Nature Switzerland AG
The registered company address is: Gewerbestrasse 11, 6330 Cham, Switzerland

Contents

A Comparative Study of HBOS and Isolation Forest Anomaly Detection Models: An Experimental Analysis

Naoual Mouhni, Sana Chakri, Ibtissam Amalou, Mohamedou Cheikh Tourad, and Abdelmounaim Abdali

Abstract

Anomaly detection plays a pivotal role across various domains. This paper introduces a comparative analysis between two prevalent anomaly detection techniques: Histogram-based Outlier Detection (HBOS) and Isolation Forest. A comprehensive elucidation of the operational mechanisms of each model is presented along with empirical investigations conducted on authentic datasets. The efficacy of both approaches was assessed through performance metrics, including accuracy. Our findings demonstrate that both HBOS and Isolation Forest attained a commendable accuracy rate of 95% in detecting normal instances, besides encountering challenges in flagging anomalies. Notwithstanding their limitations in anomaly detection, these models furnish valuable insights for academics and professionals in the anomaly detection domain, facilitating sound decision making in selecting the optimal model for their requirements.

Keywords

Histogram-based outlier detection · Isolation forest · Anomaly detection · Outliers · Deep learning

1 Introduction

Detecting anomalies is a crucial task in various fields, where identifying unusual patterns or outliers is highly important. It plays a vital role in maintaining system security, identifying fraudulent activities, and monitoring network behavior. Numerous anomaly detection models have been developed over the years (Time-Series Neural Network Predictions Using LSTM 2024), each having its strengths and limitations. In this article, we present a detailed comparative study between two commonly used anomaly detection models: Detection (HBOS) (Goldstein and Dengel 2012) and Isolation Forest (Anomaly Detection Using Isolation Forest in Python 2024). We also include a comparison with the Local Outlier Factor (LOF) (Breunig et al. 2000) to better understand the results.

Detecting anomalies can be challenging due to the diverse nature of data and the complex patterns that anomalies can exhibit. Traditional statistical methods often struggle to capture these anomalies effectively. Therefore, novel approaches, such as HBOS and Isolation Forest, have emerged to overcome these limitations and enhance anomaly detection accuracy. HBOS is an un-supervised anomaly detection algorithm that utilizes a histogram-based approach. It estimates the probability of an observation being an anomaly by calculating the density of instances within predefined bins. By leveraging the distribution of normal data, HBOS assigns lower scores to anomalies, enabling their efficient detection. This model is particularly suitable for high- dimensional data and is known for its simplicity and computational efficiency. On the other hand, Isolation Forest is an ensemble-based

N. Mouhni (✉)
GL-ISI Team, Department of Computer Science, FSTE, UMI of Meknès Morocco, Meknes, Morocco
e-mail: n.mouhni@umi.ac.ma

N. Mouhni · I. Amalou
LAMIGEP EMSI Marrakech, Marrakech, Morocco

S. Chakri
LIM Laboratory, Faculty of Sciences and Techniques of Mohammedia, University Hassan II, Casablanca, Morocco

I. Amalou · A. Abdali
Department of Physics, CISIEV, FSTG, UCA, Marrakech, Morocco

M. C. Tourad
Scientific Computing, Computer Science and Data Science Research Unit, Universityof Nouakchott, Nouakchott, Mauritania

B. El Bhiri et al. (eds.), *AI and Digital Transformation: Innovations in Supply Chain, Education, and Energy Systems*, Sustainable Artificial Intelligence-Powered Applications, https://doi.org/10.1007/978-3-031-86837-5_1

anomaly detection model that employs random decision trees to isolate anomalies from normal data instances. It exploits the principle that anomalies are often easier to separate from the rest of the data points. By constructing isolation trees, this model measures the average number of partitions required to isolate an anomaly, thus assigning higher scores to anomalies and lower scores to normal instances. Isolation Forest is effective in handling large datasets and is robust against outliers and irrelevant features.

To evaluate the performance of the study, we conducted an experimental analysis using a real-world dataset. We measured their effectiveness in detecting various types of anomalies, including point anomalies, contextual anomalies, and collective anomalies. Performance evaluation metrics, such as the Area Under the Receiver Operating Characteristic curve (AUC-ROC) (AUC ROC 2024), were used to quantify the models' accuracy and ability to distinguish anomalies from normal instances. The results of our study revealed significant differences between HBOS and Isolation Forest in terms of their detection capabilities and overall performance. While both models demonstrated promising results, HBOS showed higher accuracy in detecting point anomalies, whereas Isolation Forest excelled in identifying contextual and collective anomalies (Ali et al. 2012). These findings highlight the importance of considering the specific types of anomalies when selecting an appropriate detection model. In conclusion, this article provides a comprehensive comparative study between HBOS and Isolation Forest, two popular anomaly detection models. We have discussed the functioning of each model, its advantages, and limitations. Additionally, we presented the results of our experimentation, highlighting the differences in their performance for different types of anomalies. To offer a systematic analysis.

2 Materials and Methods

One reason to choose the HBOS (Histogram-based Outlier Score) and Isolation Forest models for comparison in anomaly detection is their popularity and wide usage in the field. HBOS is a lightweight and computationally efficient model that has been widely used for anomaly detection tasks. It leverages the density estimation approach, assuming that normal instances are densely clustered and anomalies are located in sparse regions. HBOS is particularly suitable for large datasets and can handle noisy or imbalanced data effectively. Isolation Forest, on the other hand, is a tree-based ensemble model specifically designed for anomaly detection. It works by isolating anomalies in a random and recursive partitioning process. Isolation Forest has the advantage of being able to handle high-dimensional datasets and is less sensitive to the presence of outliers. By comparing these two models, we can gain insights into their performance and limitations in detecting anomalies. This helps in understanding the strengths and weaknesses of different approaches and provides a basis for selecting the most suitable model for specific anomaly detection tasks. It is worth mentioning that the choice of models for comparison depends on various factors, including the characteristics of the dataset, the specific requirements of the anomaly detection task, and the availability of resources. Other models like Support Vector Machines (SVM) ((PDF) Machine Learning Approach Using MLP 2024) or Gaussian Mixture Models (GMM) (Support vector machines using GMM supervectors for speaker verification 2024) could also be considered, but the selection of HBOS and Isolation Forest for comparison is justified by their popularity, efficiency, and suitability for different types of datasets. In addition to the comparison between HBOS and Isolation Forest models for anomaly detection, we consider including the Local Outlier Factor (LOF) algorithm (Breunig et al. 2000) to broaden the scope of our analysis. LOF is a well-established unsupervised outlier detection algorithm known for its capability to measure the local density deviation of a data point about its neighboring points. Through a comparison of the point's density with its local neighborhood, LOF can effectively identify outliers exhibiting significantly lower density. This algorithm proves valuable in detecting anomalies within datasets characterized by varying densities and complex distributions, offering a complementary perspective to the existing comparison. The incorporation of LOF enriches the analysis by considering a diverse array of outlier detection techniques and providing a comprehensive assessment of their performance across different data scenarios.

The outlier detection algorithm integrates advanced techniques, including Histogram-based Outlier Detection (HBOS), Isolation Forest, and Cluster-based Local Outlier Factor (CBLOF), to detect anomalies within a dataset. The process begins by importing necessary libraries and classifiers for outlier detection methods and defining classifiers with specific parameters. Training data from a CSV file is then read and key features are visualized through scatter plots to understand the underlying data distribution. Normalization of data using MinMaxScaler improves model performance, followed by preparing input data for outlier detection by concatenating scaled features (Chakri et al. 2017). By setting a random seed and contamination fraction, the algorithm facilitates outlier detection, with the creation of a mesh grid enabling visualization of algorithm outputs. The algorithm iterates over each classifier, involving model fitting, calculation of raw anomaly scores using decision functions, and prediction of class labels for data points. By distinguishing between inliers and outliers based on predicted labels, the algorithm identifies anomalous instances. Visualization elements include generating heatmaps of anomaly scores and plotting inliers and outliers in distinct colors to aid resulting interpretation. Evaluationmetrics, such as classification reports (Classification: courbe ROC et AUC 2024) and AUC- ROC scores, offer a quantitative assessment of model performance across various outlier detection techniques.

2.1 Data Description

Figure 1 illustrates the data used for experimentation in this project which consists of a dataset of sales transactions for products in various stores (E-Commerce Sales Dataset 2024). This dataset comprises transaction records with details such as the product sold, store location, transaction timestamp, sales value, and additional features like customer demographics, product category, and promotional activities. Before applying machine learning algorithms, the data undergoes preprocessing steps to ensure quality and suitability for analysis, including handling missing values, removing duplicates, and feature engineering. The dataset is split into training and test sets to evaluate the anomaly detection model's performance on unseen data, maintaining a consistent proportion of normal and anomalous transactions in each subset. Additionally, this dataset offers a comprehensive overview of e-commerce sales data from various channels, enabling retailers and digital marketers to measure campaign performance accurately. Users can analyze sales trends by month, category, currency, stock level, and customer, assess profitability through different fulfillment methods like Shiprocket and INCREF, and compare prices across channels to optimize profit margins while maintaining product quality. One research idea involves analyzing the profitability difference between sales via Shiprocket and INCREFF to identify areas with the highest profit margins and develop strategic decisions based on the findings. The removal of outliers from the dataset played a crucial role in enhancing the quality and reliability of the analysis. By eliminating these extreme values, the statistical models were able to capture the underlying patterns more accurately, leading to improved performance and better insights. The decision to remove outliers not only improved the robustness of the models but also ensured compliance with key assumptions, thereby strengthening the validity of the findings. The performance of the anomaly detection model is assessed using two main metrics: (AUC ROC 2024) and precision. AUC-ROC (Area Under the Receiver Operating Characteristic curve) measures the model's ability to distinguish between normal and anomalous transactions. The ROC curve plots the true positive rate (sensitivity) against the false positive rate (1-specificity) at various classification thresholds. AUC-ROC provides an overall measure of how well the model separates anomalies from normal instances, with higher values indicating better performance. (Precision and Recall 2024), also known as the positive predictive value, quantifies the accuracy of the model in identifying anomalies. It is calculated as the ratio of true positives (correctly identified anomalies) to the sum of true positives and false positives (instances incorrectly identified as anomalies). Precision reflects the model's ability to minimize false alarms and is particularly useful when the cost of false positives is high. Both AUC-ROC and precision provide valuable insights into the performance of the anomaly detection model. AUC-ROC assesses the overall quality of the model's predictions, while precision focuses on the accuracy of anomaly identification specifically. In summary, the data used for experimentation in this program consists of sales transactions for products in different stores. The data undergoes preprocessing and is divided into training and test sets. The performance of the anomaly detection model is evaluated using AUC- ROC and precision metrics, which assess its ability to distinguish anomalies from normal instances and its accuracy in identifying anomalies, respectively.

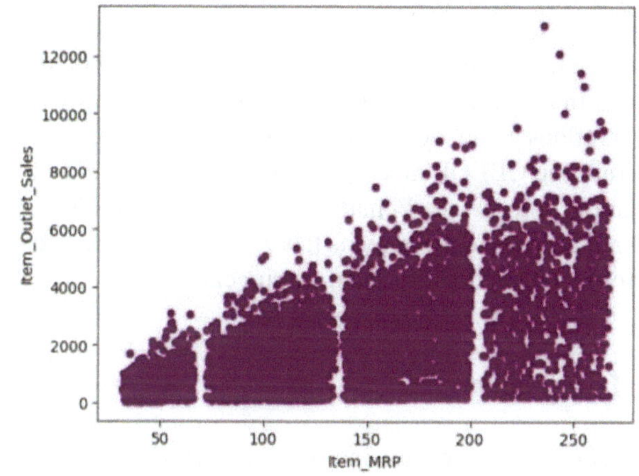

Fig. 1 Sales transactions for products in various stores

3 Results

Figure 2 illustrates the results of the HBOS approach. The accuracy of the HBOS model is 0.95, which means it correctly classified 95% of the instances in the dataset. However, it failed to identify any anomalies (class 1) correctly. The macro average of precision, recall (Precision and Recall 2024), and f1-score is 0.48 (Understanding and Applying F1 Score 2024), indicating poor performance. The weighted average of these metrics is 0.90, suggesting that the model performs well in the majority class (class 0) but poorly in the minority class (class 1). The AUC-ROC score is 0.5, which implies that the HBOS model cannot distinguish between normal and anomalous instances effectively. An AUC-ROC score of 0.5 indicates random guessing, meaning the model's performance is

Histogram-base Outlier Detection (HBOS)

	precision	recall	f1-score	support
0	0.95	1.00	0.98	5108
1	0.00	0.00	0.00	251
accuracy			0.95	5359
macro avg	0.48	0.50	0.49	5359
weighted avg	0.91	0.95	0.93	5359

AUC-ROC: 0.5

OUTLIERS : 251 INLIERS : 5108 Histogram-base Outlier Detection (HBOS)

Fig. 2 HBOS results

similar to a random classifier. Similar to the HBOS model, the results presented in Fig. 3 which belong to the Isolation Forest model achieved an accuracy of 0.95, correctly classifying the majority class (class 0). However, it failed to identify any anomalies correctly (class 1). The macro average of precision, recall, and f1-score is 0.47, indicating poor performance. The weighted average is 0.90, suggesting that the model performs well on the majority class but poorly on the minority class. The AUC-ROC score for the Isolation Forest model is also 0.5, indicating that it is not effective in distinguishing between normal and anomalous instances. Comparison of Performance: Based on the AUC-ROC scores and other relevant metrics, both the HBOS and Isolation Forest models show poor performance in identifying anomalies. They have an AUC-ROC score of 0.5, suggesting that their ability to separate normal and anomalous instances is similar to random guessing. Moreover, both models have a precision of 0.00 for class 1, indicating that they failed to correctly identify any anomalies. The recall for class 1 is also 0.00, meaning that the models missed all the true anomalies in the dataset. The f1-scores for class 1 are also 0.00, further highlighting the poor performance. In terms of the majority class (class 0), both models achieved high precision (0.95) and recall (1.00), indicating that they performed well in correctly classifying normal instances. The f1-scores for class 0 are also high (0.97), suggesting good overall performance for the majority class. In summary, both the HBOS and Isolation Forest models exhibit poor performance in identifying anomalies, as indicated by their low AUC-ROC scores, precision, recall, and f1-scores for class 1. The models perform well on the majority class but fail to detect anomalies effectively.

To add another algorithm to the comparison, we consider including the Local Outlier Factor (LOF) algorithm. LOF is a popular unsupervised outlier detection algorithm that measures the local density deviation of a data point with respect to its neighbors. By comparing the density of a point to the density of its neighbors, LOF can identify outliers that have significantly lower density compared to their local neighborhood. This algorithm is effective in detecting outliers in datasets with varying densities and complex distributions.

Isolation Forest

	precision	recall	f1-score	support
0	0.95	1.00	0.97	5091
1	0.00	0.00	0.00	268
accuracy			0.95	5359
macro avg	0.47	0.50	0.49	5359
weighted avg	0.90	0.95	0.93	5359

AUC-ROC: 0.5

OUTLIERS : 268 INLIERS : 5091 Isolation Forest

Fig. 3 Isolation forest results

Figure 4 illustrates the results of The CBLOF model which achieved an overall accuracy of 0.95, correctly classifying the majority class (class 0). However, similar to the HBOS and Isolation Forest models, the CBLOF model had difficulty identifying anomalies correctly, with a precision, recall, and f1-score of 0.00 for the minority class (class 1). The AUC-ROC score for the CBLOF model was 0.5, indicating that it also struggled to distinguish between normal and anomalous instances effectively. The precision, recall, and f1-score for class 1 all being 0.00 demonstrate the model's inability to detect true anomalies within the dataset. In contrast, the model performed well in terms of precision (0.95), recall (1.00), and f1-score (0.97) for the majority class (class 0), indicating its proficiency in correctly identifying normal instances. The high scores for class 0 highlight the CBLOF model's strength in classifying normal data points but its limitations in detecting anomalies effectively, aligning with the observations made for the HBOS and Isolation Forest models. Overall, the CBLOF model, like its counterparts HBOS and Isolation Forest, faced challenges in accurately identifying anomalies, as evidenced by its low AUC-ROC score and poor performance metrics for the minority class.

4 Discussion

The comparative evaluation of the HBOS and Isolation Forest models, along with the Cluster-based Local Outlier Factor (CBLOF) model, on the dataset provides valuable

Fig. 4 Cluster-based local outlier factor

```
Cluster-based Local Outlier Factor (CBLOF)
              precision    recall  f1-score   support

           0       0.95      1.00      0.97      5091
           1       0.00      0.00      0.00       268

    accuracy                           0.95      5359
   macro avg       0.47      0.50      0.49      5359
weighted avg       0.90      0.95      0.93      5359

AUC-ROC: 0.5

OUTLIERS :  268 INLIERS :  5091 Cluster-based Local Outlier Factor (CBLOF)
```

insights into their performance characteristics and suitability for various applications. The experiment results highlight the strengths, weaknesses, and potential applications of each model. Both the HBOS and Isolation Forest models exhibited suboptimal performance in anomaly detection, as indicated by their AUC-ROC scores of 0.5, signifying a limited ability to differentiate between normal and anomalous instances accurately. Specifically, both models displayed a precision, recall, and F1-score of 0.00 for class 1 (anomalies), indicating a failure to correctly classify any anomalies in the dataset. Furthermore, the CBLOF model showed similar trends in performance, achieving an accuracy of 0.95 in classifying the majority class (class 0) but encountering challenges in identifying anomalies, with a precision, recall, and F1-score of 0.00 for class 1. The AUC-ROC score of 0.5 underscores the difficulty of accurately distinguishing between normal and anomalous instances, highlighting the models' limitations in effective anomaly detection tasks. While the HBOS and Isolation Forest models excel in identifying normal instances, making them suitable for applications prioritizing the detection of regular patterns, such as fraud detection or network intrusion detection, their shortcomings in accurate anomaly detection should be acknowledged. The HBOS model's computational efficiency and robustness in handling outliers make it a good fit for large, noisy datasets, while the Isolation Forest model's proficiency in high-dimensional data scenarios suits complex data structures. However, both models have inherent limitations. HBOS relies on the assumption of dense clusters for normal instances and feature independence, while Isolation Forest may face challenges with clustered anomalies and overfitting small anomalous datasets. Considering these findings, it is evident that the HBOS, Isolation Forest, and CBLOF models are better suited for tasks emphasizing the identification of normal instances rather than precise anomaly detection. For applications requiring accurate anomaly detection, such as rare disease identification or detecting unusual events in complex systems, alternative models like Support Vector Machines (SVM), Gaussian Mixture Models (GMM) (Support vector machines using GMM supervectors for speaker verification 2024), or Deep Learning-based approaches like Autoencoders or Variational Autoencoders (Zhang et al. 2023) may offer improved performance. Selecting the appropriate anomaly detection model tailored to specific dataset characteristics and anomaly patterns is crucial for achieving optimal results in real-world applications.

5 Perspectives

To further advance anomaly detection capabilities, it is essential to explore strategies for enhancing the performance of models like HBOS and Isolation Forest. One approach could involve integrating these models with complementary techniques or leveraging domain-specific expertise to optimize their anomaly detection accuracy. By combining these models with other methodologies or incorporating specialized knowledge from the domain, it may be possible to overcome their current limitations and achieve more robust anomaly detection results. Additionally, addressing the challenge of handling imbalanced datasets is crucial for improving model performance. Techniques such as oversampling the minority class or utilizing anomaly detection-specific evaluation metrics can help mitigate imbalanced data issues and enhance the overall effectiveness of anomaly detection algorithms (Alimohammadi and Nancy Chen 2022). Moreover, expanding the scope of the study to include a wider array of anomaly detection models would provide a more comprehensive evaluation framework for assessing the capabilities of HBOS and Isolation Forest. By incorporating additional models into the comparison, researchers can better understand the relative strengths and weaknesses of each approach and gain deeper insights into the optimal choice of model for specific anomaly detection tasks (Alghushairy et al. 2021). This broader perspective can offer valuable guidance for selecting the most suitable anomaly detection model based on the unique characteristics of the dataset and the nature of anomalies present. By embracing these perspectives and exploring opportunities for improvement, the field of anomaly detection can continue to evolve and advance towards more effective and reliable anomaly detection solutions.

6 Conclusion

In this comparative study, we evaluated the performance of the HBOS and Isolation Forest models for anomaly detection. The results obtained revealed that both models have limitations in accurately identifying anomalies. Both models achieved AUC-ROC scores of 0.5, indicating limited ability to effectively distinguish normal instances from anomalous instances. Additionally, the precision, recall, and F1-score measures for class 1 (anomalies) were all 0.00, suggesting that the models failed to correctly identify anomalies in the dataset. However, the models showed good performance in the majority class (class 0) with high precision, perfect recall, and a high F1-score. This means that they were successful in correctly classifying normal instances. Therefore, for applications where identifying normal instances is the primary goal, such as fraud detection in financial transactions or network intrusion detection, these models could be useful. Incorporating the Local Outlier Factor (LOF) algorithm into the comparative study brought additional insights into anomaly detection. LOF, known for its ability to measure local density deviations and compare them with neighboring points, offers a unique perspective on identifying outliers in datasets with varying densities and complex distributions. The inclusion of LOF alongside HBOS and Isolation Forest provides a more comprehensive evaluation of outlier detection techniques across different data scenarios. Despite the poor performance of the HBOS and Isolation Forest models in accurately detecting anomalies, they possess certain advantages. The HBOS model's computational efficiency and robustness in handling large datasets, as well as its resilience to outliers, make it a valuable tool for scenarios with noisy or imbalanced data. Similarly, the Isolation Forest model's capability to detect anomalies in high-dimensional datasets proves beneficial for applications involving complex data structures. While acknowledging the limitations of the HBOS and Isolation Forest models, the introduction of LOF underscores the importance of exploring a diverse set of anomaly detection algorithms to address specific data challenges effectively. Future research opportunities may focus on enhancing anomaly detection capabilities through the integration of novel algorithms, feature engineering techniques, or ensemble methods, ultimately improving the overall performance of anomaly detection models. In conclusion, the experimental results of the HBOS, Isolation Forest, and LOF models shed light on their varying strengths and weaknesses in anomaly detection tasks. Thorough evaluation and comparison of different models are essential to select the most suitable approach based on the specific requirements of the anomaly detection task and characteristics of the dataset. The inclusion of LOF in the comparative study expands the understanding of anomaly detection techniques and emphasizes the need to consider a diverse range of algorithms to achieve accurate and effective anomaly detection outcomes.

References

'(12) (PDF) Machine Learning Approach Using MLP and SVM Algorithms for the Fault Prediction of a Centrifugal Pump in the Oil and Gas Industry'. Accessed: Feb. 02, 2024. Available: https://www.researchgate.net/publication/342173662_Machine_Learning_Approach_Using_MLP_and_SVM_Algorithms_for_the_Fault_Prediction_of_a_Centrifugal_Pump_in_the_Oil_and_Gas_Industry

'Anomaly Detection Using Isolation Forest in Python', Paperspace Blog. Accessed: Feb. 02, 2024. Available: https://blog.paperspace.com/anomaly-detection-isolation-forest/

'AUC ROC Curve in Machine Learning—GeeksforGeeks'. Accessed: Feb. 02, 2024. Available: https://www.geeksforgeeks.org/auc-roc-curve/

Alghushairy, O., Alsini, R., Soule, T., Ma, X.: A review of local outlier factor algorithms for outlier detection in big data streams. Big Data Cogn. Comput., **5**(1), (2021). https://doi.org/10.3390/bdcc5010001.

Ali, J., Khan, R., Ahmad, N., Maqsood, I.: Random forests and decision trees. Sep. 2012. Accessed: Feb. 04, 2024. Available: https://www.semanticscholar.org/paper/Random-Forests-and-Decision-Trees-Ali-Khan/959a8e906ee26b940374b719253c8e188ed78fd3

Alimohammadi, H., Nancy Chen, S.: Performance evaluation of outlier detection techniques in production timeseries: A systematic review and meta-analysis. Expert. Syst. Appl., **191**, p. 116371, (2022). https://doi.org/10.1016/j.eswa.2021.116371

Breunig, M.M., Kriegel, H.-P., Ng, R.T., Sander, J.: LOF: identifying density-based local outliers. SIGMOD Rec. **29**(2), 93–104 (2000). https://doi.org/10.1145/335191.335388

Chakri, S., Raghay, S., El Hadaj, S.: Semantic trajectory knowledge discovery: a promising way to extract meaningful patterns from spatiotemporal data. Int. J. Soft. Eng. Knowl. Eng. **27**(03), 399–421 (2017). https://doi.org/10.1142/S0218194017500140

'Classification: courbe ROC et AUC. Machine Learning', Google for Developers. Accessed: Feb. 02, 2024. Available: https://developers.google.com/machine-learning/crash-course/classification/roc-and-auc?hl=fr

'E-Commerce Sales Dataset'. Accessed: Feb. 02, 2024. Available: https://www.kaggle.com/datasets/thedevastator/unlock-profits-with-e-commerce-sales-data

Goldstein, M., Dengel, A.: Histogram-based outlier score (HBOS): A fast unsupervised anomaly detection algorithm (2012)

'Precision and Recall. Essential Metrics for Machine Learning'. Accessed: Feb. 02, 2024. Available: https://www.analyticsvidhya.com/blog/2020/09/precision-recall-machine-learning/

'Support vector machines using GMM supervectors for speaker verification. IEEE J.S & Mag.. IEEE Xplore'. Accessed: Feb. 02, 2024. Available: https://ieeexplore.ieee.org/abstract/document/1618704

Time-Series neural network predictions using LSTM for clustering and forecasting GPS data of bird immigration. IEEE Conference Publication. IEEE Xplore. Accessed: Jun. 20, 2024. Available: https://ieeexplore.ieee.org/document/10520764

'Understanding and Applying F1 Score: A Deep Dive with Hands-On Coding', Arize AI. Accessed: Feb. 02, 2024. Available: https://arize.com/blog-course/f1-score/

Zhang, L., Lu, W., Xue, F., Chang, Y.: A trajectory outlier detection method based on variational auto-encoder. Math. Biosci. Eng. **20**(8), 15075–15093 (2023). https://doi.org/10.3934/mbe.2023675

Recurrent Neural Models for Retail Supply Chain Stock Prices Forecasting

Doha Haidar, Salma Mouatassim, Rajaa Benabbou, and Jamal Benhra

Abstract

Owing to their ability to capture complex non-linear patterns and to memorize long-term dependencies, LSTM and GRU models are widely used to forecast time series. However, stock price data presents another layer of complexity due to their volatility. In this paper, a comparative study between LSTM and GRU is held using Hyperband (HB) as the Hyperparameter Optimization (HPO) technique for 4 different-sized datasets of stock prices of retail supply chains. The paper aims to provide an intuition about the accessibility and the performance of Automated Machine Learning (AutoML) methods in tuning powerful Deep Learning (DL) models such as LSTM and GRU. To this end, a basic forecasting framework is provided to outline the tuning steps of different hyperparameters. After model building, the performance evaluation using RMSE, MAE, RMSPE, MAPE and R^2 leads to a discussion of the models' accuracy in validation and test sets. Finally, an overall conclusion paves the way for the main improvements to implement as well as perspectives for future work.

Keywords

LSTM · GRU · Hyperband · Stock prices · Retail supply chain

1 Introduction

The literature encourages the search for promising forecasting techniques which yield optimal performance and accurately project the extracted patterns into the future is a prominently ongoing task (Parmezan et al. 2019). Thus, a variety of forecasting methods is available to cover different fields and suit different time series properties (non-linearity vs preliminary distributional knowledge) and user needs (accuracy vs parsimony). Time series forecasting techniques belong to one of the following categories (Makridakis et al. 2020; Januschowski et al. 2020): statistical methods (SARIMA, Exponential Smoothing…) (Marzak et al. 2023) requiring pre-distributional knowledge of the dataset before looking for the fit model (Parmezan 2019; Hyndman 2008; Pankratz 1983), ML methods (SVR, Random Forest, Deep Learning (DL), XGBoost…) which implicitly learn the dataset's properties (Parmezan 2019; Januschowski 2020), combinations which aggregate final forecasts from different methods to enhance the forecasting performance (Pankratz 1983) and hybrid models which alter pure statistical or ML forecasting approaches to model both linear and non-linear properties embedded in a given time series (Mattos Neto et al. 2020) (Goodfellow et al. 2016).

In light of the emergence of highly sophisticated models, there's a more pressing need for designing reliable forecasting frameworks which enable efficient model building at different levels of user expertise. Some studies stated their interest in flexible forecasting models which balance the accuracy with the computational cost and can be updated to overcome misfitting issues (Mattos Neto et al. 2020). Dealing with inconsistent data patterns inhibits the global forecasting framework from adapting to dynamic data variations, thereby entailing overfitting or misfitting issues (Mattos Neto et al. 2020). Our paper intends to build a reproducible and flexible data-driven forecasting framework enabling the model selection through Hyperparameter Optimization (HPO) to fit complex data properties (variations, volatility). Thus, this paper represents a stepping stone in designing a reliable forecasting process rather than building an accurate model in a given timeframe. To deal with volatile and inconsistent patterns, we adopted a basic forecasting framework using

D. Haidar (✉) · S. Mouatassim · R. Benabbou · J. Benhra
LARILE Laboratory, OSIL Team, GIL Department, ENSEM, Hassan II University, Casablanca, Morocco
e-mail: doha.haidar@ensem.ac.ma

B. El Bhiri et al. (eds.), *AI and Digital Transformation: Innovations in Supply Chain, Education, and Energy Systems*, Sustainable Artificial Intelligence-Powered Applications, https://doi.org/10.1007/978-3-031-86837-5_2

stacked Long Short-Term Memory (LSTM) and Gated Recurrent Networks (GRU) models tuned using hyperband (HB). We applied the framework to different-sized financial time series to form an intuition about the limitations and opportunities of stacked windowed LSTM and GRU models using efficient and accessible HPO methods.

Recurrent Neural Networks (RNN) is a versatile DL model category designed to deal with sequential data through memory updates (Goodfellow et al. 2016). RNN models are adopted across different tasks including Computer Vision (CV) (Gregor et al. 2015; Ba et al. 2015), Natural Language Processing (NLP) (Köksal and Akgül 2022), speech and human activity recognition (Lalapura et al. 2024; Han et al. 2017), language modelling (Liu et al. 2023) and time series forecasting (Investigation of Load 2024; Rokhsatyazdi et al. 2020) (Kong et al. 2022) amongst other applications (Goodfellow et al. 2016). In time series forecasting, various interesting applications make use of DL models and RNN models to build predictive models for smart agriculture (Mhaned et al. 2023), weather forecasting (Mung and Phyu 2023), stock prices forecasting (Zaheer et al. 2023; Moghar and Hamiche 2020), preventive maintenance (Li et al. 2022) to cite a few applications.

Owing to their embedded recursive connections, RNN models view data sequentially since they process new inputs in light of the recently updated hidden states (Goodfellow et al. 2016). The hidden state of a given RNN cell acts like a memory cell encapsulating a non-linear function of the linear combination of the previous inputs and feeding it the following cells to preserve previous sequential dependencies. Thus, RNN architectures maintain a memory using hidden states and process time series within their context, unlike traditional ANN models which capture data patterns by updating parameters independently (Goodfellow et al. 2016; 'Chapter 2019). This alteration of ANN to form RNN structures resulted in modifying the learning flow from backward propagation to Backpropagation Through Time (BPTT). The latter mechanism describes the sequence in which the RNN parameters are updated (Werbos 1990). The literature offers pseudo-algorithms (Werbos 1990; Sutskever 2013) and computational graphs to model BPTT (Goodfellow 2016; Amini 2023; Salehinejad 2018). Essentially, BPTT unfolds RNN structures by modelling each timestep by a layer. And so, at a given timestamp t, a new input ($x^{<t>}$) is fed to the RNN structure which processes it in light of the hidden state ($h_{t-1}\ or\ a^{<t-1>}$). The forward propagation outputs a prediction ($\hat{y}^{<t>}$) which recapitulates a weighted form of the updated hidden state h_t. After consuming all the unfolded layers, the predictions ($\hat{y}^{<t>}$) across timesteps amount to an estimated cost function $\mathcal{L}_t$ which is backpropagated through each layer/ timestep t to update the shared RNN weights (W_{xh}, W_{hh}, W_{hy}) (Werbos 1990).

Theoretically, BPTT enables RNN models to track long-range dependencies and the weights to incorporate the gradients through timesteps (Schmidt 2019). In practice, for long data sequences, the RNN gradients tend to either exponentially increase (Exploding Gradient) or decrease (Vanishing Gradient), which hinders capturing long-term dependencies (Goodfellow et al. 2016; Sutskever 2013; Hochreiter and Schmidhuber 1997). To monitor exploding gradients, gradient clipping can be implemented. Other variants of BPTT such as Truncated BPTT can also be efficient as they perform BPTT for a limited part of the input sequence, thus enabling RNN to capture long enough dependencies by tuning the size of data sequences while partially mitigating the occurrence of vanishing or exploding gradients in RNN variants (Bird and Polivoda 2021).

LSTM and GRU were introduced to address the vanishing gradient issues in RNN using gates which compute non-linear activation functions on the inputs ($x^{<t>}$) and the hidden states ($h_{t-1}\ or\ a^{<t-1>}$) generating a prediction $\hat{y}^{<t>}$ (Goodfellow et al. 2016; Schmidt 2019). Hence, LSTM and GRU models are a better choice for forecasting time series with long-term dependencies (Chung 2014; Gers 2024). LSTM are more robust to complex patterns (Gers et al. 1999) and the most used RNN variants models for financial time series (Singh and Malhotra 2024). On the other hand, GRU provides a less computationally expensive model with a competing performance with LSTM (Chung 2014; Jia 2020).

This paper represents a basic comparative study of LSTM and GRU models, tuned using an accessible HPO method, for predicting noisy time series with inconsistent patterns throughout time. In the methods section, we introduce the adopted forecasting framework, the HPO process and the performance metrics used to compare the stacked HB-LSTM and HB-GRU models for forecasting stock prices of different-sized retail supply chains. After evaluating the models' performance, the results section shares the fundamental findings including the hyperparameter combinations and the performance evaluation for each model and dataset. Then, a result analysis is conducted to comment on the hyperparameter selection and compare the performance per dataset's size before selecting the fit recurrent model per dataset. Finally, the conclusion recapitulates the main findings and shares perspectives of future works aiming to improve the predictive model selection using recurrent models.

2 Methods

In this section, we present the adopted DL methods, the HPO methods and the forecasting process describing the sequence of steps to fine-tune and build the models. Because of their volatile nature and non-linear patterns, statistical methods are mostly limited to providing reliable and accurate forecasting

Table 1 LSTM and GRU cell structures and equations

	LSTM		GRU	
Structure	Figure 1		Figure 2	
Equations	$f_t = \sigma(W_f[h_{t-1}, x_t] + b_f)$ $i_t = \sigma(W_i[h_{t-1}, x_t] + b_i)$ $\tilde{C}_t = \tanh(W_c[h_{t-1}, x_t] + b_c)$ $C_t = f_t * C_{t-1} + i_t * \tilde{C}_t$[1] $o_t = \sigma(W_o[h_{t-1}, x_t] + b_o)$ $h_t = o_t * \tanh(C_t)$	(1)	$z_t = \sigma(W_z[h_{t-1}, x_t])$ $r_t = \sigma(W_r[h_{t-1}, x_t])$ $\tilde{h}_t = \tanh(W[r_t * h_{t-1}, x_t])$ $h_t = (1 - z_t) * h_{t-1} + z_t * \tilde{h}_t$	(2)

[1] *denotes an element-wise multiplication
[2] *denotes an element-wise multiplication

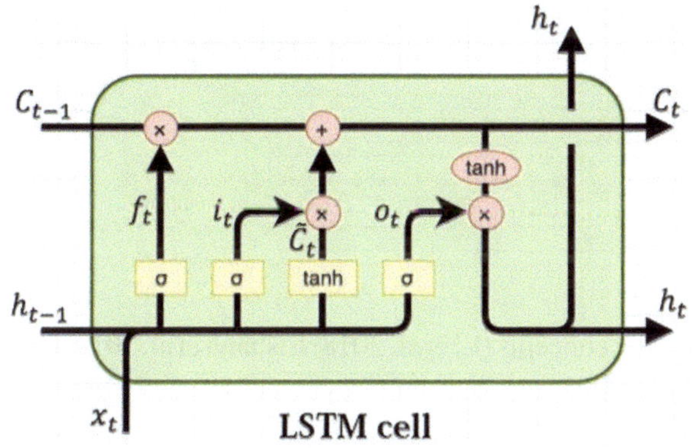

Fig. 1 LSTM cell structure (Varsamopoulos et al. 2018; Olah 2024)

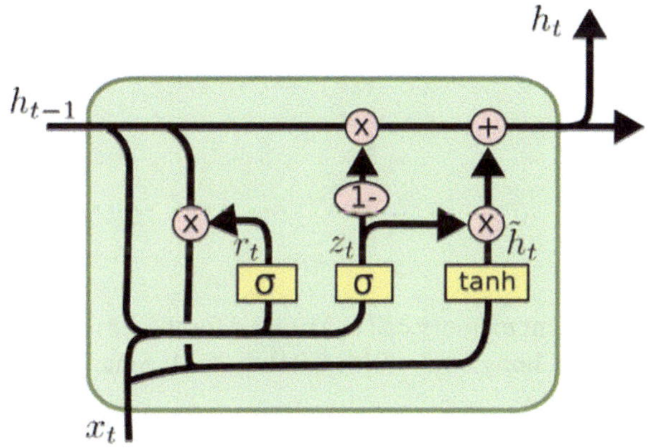

Fig. 2 GRU cell structure (Olah 2024; Daoulas et al. 2021)

distributions in comparison with ML and DL methods (Sezer et al. 2020). Thus, we opted for DL models, specifically LSTM and GRU models to capture the non-linear patterns and long-term dependencies (Sezer et al. 2020).

2.1 LSTM Versus GRU

According to the systematic reviews in references (Singh and Malhotra 2024; Sezer et al. 2020; Lin and Lobo Marques 2024), LSTM and GRU are consistently the most popular and among the most effective AI models used for stock prices forecasting throughout the years. As outlined in the reference (Sezer et al. 2020) systematically reviewing papers from 2005 to 2019, around 51% of the ML models used for forecasting stock prices are RNN-based, 60% of which are LSTM and almost 10% are GRU. Systematic reviews conducted very recently still report the adoption of LSTM specifically among other well-reputed ML models such as Support Vector Machines (SVM), ANN, Convolutional Neural Networks (CNN) and other RNN models (Singh and Malhotra 2024; Lin and Lobo Marques 2024). Moreover, LSTM and GRU are RNN-based models that were introduced to deal with the practical failure of RNN in tracking long-term dependencies (Chung et al. 2014; stanford 2024; Smagulova and James 2020). Table 1 shows the structures and the equations of LSTM and GRU cells.

The gates in LSTM and GRU cells enable selective processing of the hidden states and the inputs (Mung and Phyu 2023; Moghar and Hamiche 2020; GitHub 2024). In LSTM structures, the memory is preserved through the update of the hidden state (h_t or $a^{<t>}$) which represents the short-term memory and the memory cell (C_t or $C^{<t>}$) which encapsulates long-term information (Smagulova and James 2020; Olah 2024). Owing to the forget gate f_t, the input gate i_t, the candidate memory cell ($\tilde{C}_t$ or $\tilde{C}^{<t>}$) and the output gate, both long and short-term memories are updated in light of the previously saved information and the new input (Smagulova and James 2020; Olah 2024). GRU cells are a simpler and less computationally expensive version of LSTM which, among other changes, combine the concepts of f_t and i_t into a single update gate (z_t) while also preserving one form of the hidden state h_t and candidate hidden state $\left(\tilde{h}_t\right)$ (Chung et al. 2014; Olah 2024). The gates act like information filters determining to which extent past hidden states and new inputs x_t should be updated ($i_t and\ z_t$), dropped (f_t and z_t), considered relevant

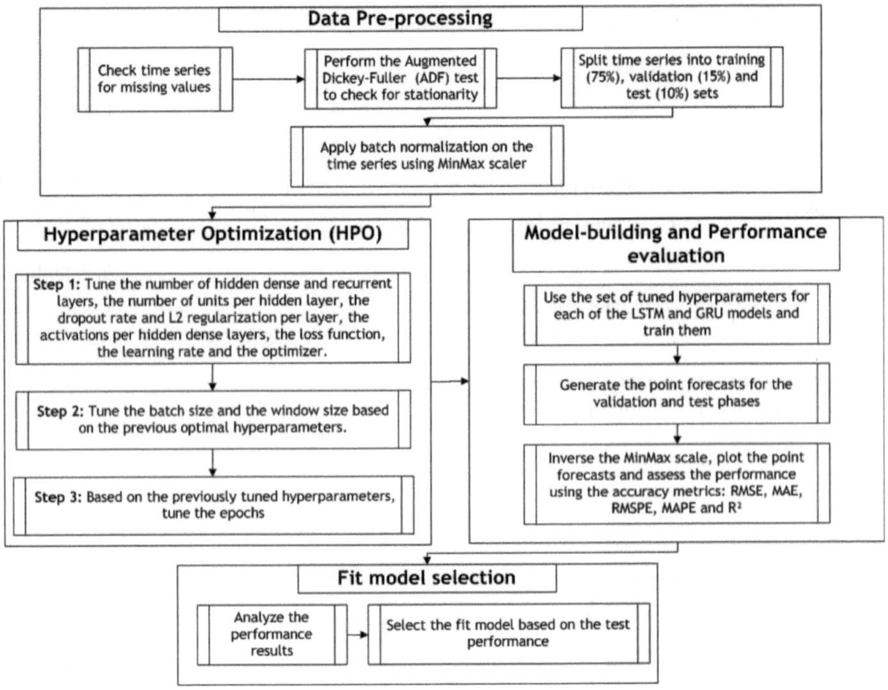

Fig. 3 Forecasting framework using HB-LSTM and HB-GRU

for the current memory cell (r_t) or used for the current hidden state (o_t) (Chung et al. 2014; GitHub 2024; Smagulova and James 2020).

2.2 Forecasting Framework

Figure 3 outlines the adopted steps to forecast our time series using HB-LSTM to HB-GRU. We used the following hyperparameters: number of hidden dense and recurrent layers, number of hidden units per dense and recurrent layer, dropout, L2-regularization, activations per hidden dense layers, optimizer, learning rate, loss function, batch size and window size.

As we couldn't tune all hyperparameters simultaneously using HB, we gathered the ones related to the model's structure first, then we tuned the remaining hyperparameters (batch size, window size and epochs) after shaping the model's main architecture using grid search (Fernández-Delgado et al. 2019). The order in which the hyperparameters are tuned is up for discussion in the next section.

To limit overfitting issues, we used the dropout rate and the L_2-regularization. We included the Huber loss function to balance the models' sensitivity to large errors and outliers (Jadon et al. 2022). We implemented HB using TensorFlow's available and commonly used library 'Keras-tuner'(Sezer et al. May 2020). Table 2 contains the search space per hyperparameter for our daily stock price data.

We opted for HB as our HPO technique in this application as it's an efficient and robust Automated Machine Learning (AutoML) tuning method which consists of combining the following concepts (Li et al. 2018; Elshawi et al. 2019; Falkner et al. 2018):

- Random Search (Bergstra and Bengio 2012) as the base exploratory model for probable and efficient hyperparameter combinations (Li et al. 2018),
- Iterative successive halving which acts like a resource allocation strategy fostering the chances of promising hyperparameter configurations to be tested while discarding poor ones (Li et al. 2018),
- Principled Early Stopping which limits the computational cost of unnecessary explorations by stopping further testing of a given hyperparameter set beyond a certain point or condition (number of consecutive iterations where the loss remains the same) (Li et al. 2018).

While grid search and random search algorithms fully evaluate all hyperparameter combinations, they come with a high computational cost (Elshawi et al. 2019). Multi-fidelity algorithms strike the balance between the accuracy and the computational cost particularly when used for larger datasets, with the limitation of discarding potentially good combinations as they exhibit poor quality in early rounds (Li et al. 2018; Elshawi et al. 2019).

HB is a state-of-the art multi-fidelity HPO technique that proved its robust and efficient performance in optimizing hyperparameters compared to Bayesian Optimization (BO) and random search (Elshawi et al. 2019; Falkner et al. 2018). This multi-armed bandit HPO algorithm is an extended version of successive halving introduced to deal with some of the issues that the latter has (Li et al. 2018; Elshawi et al. 2019;

Table 2 Hyperparameter space

Hyperparameter	Space of values
Number of hidden recurrent layers	$[1, 3]$, $step = 1$
Number of units per hidden recurrent/ dense layer	$\{8, 512\}$, $step = 8$
Number of hidden dense layers	$[1, 2]$
L_2-regularization	$\{10^{-3}, 10^{-2}, 10^{-1}\}$
Dropout rate	$[0, 0.5]$, $step = 0.1$
Activation per dense hidden layer	$[ReLU, tanh, sigmoid]$
Batch size	$\{16, 32, 64, 128\}$
Window size	$\{1, 5, 10, 15, 20, 25, 30\}$
Learning rate	$[10^{-4}, 10^{-2}]$
Optimizer	$\{Adam, SGD, SGD(momentum = 0.9), Adadelta, RMSprop\}$
Loss function	$\{MeanSquaredError(MSE), MeanAbsoluteError(MAE), Huberloss\}$
Epochs	$\{10, 25, 50, 75, 100, 125, 150, 175, 200\}$

Jamieson and Talwalkar 2015). Essentially, HB uses successive halving in dynamic brackets where different randomly selected initial pools of hyperparameter configurations are explored (Li et al. 2018). As the number of successive halving brackets increases, HB decreases the initial pool of hyperparameter configurations and assigns more budget to promising configurations, thereby exploiting the computations to evaluate the best hyperparameter combinations (Li et al. 2018; Elshawi et al. 2019).

Since the first HPO step deals with many hyperparameters at once, for each combination deemed worthy of exploration by HB, we reserved 50 epochs, conditioned with early stopping over the validation loss. The remaining HPO steps use grid search (Bergstra and Bengio 2012) over the given hyperparameter spaces and explore each possibility within 30 epochs. If successive tuning combinations with hyperparameters in common perform poorly (after 5 tuning cycles), the tuning sets will be studied and the search space will be narrowed down to prevent HB from exploring hyperparameters linked to poor performance for 50 epochs. As the remaining hyperparameters lie within the commonly adopted values, we'll restrain the search on common optimizers or activation associated with poor performance.

2.3 Performance Metrics

To assess the accuracy of LSTM and GRU models, we opted for the Root Mean Squared Error (RMSE), the Mean Absolute Error (MAE), the Root Mean Squared Percentage Error (RMSPE), the Mean Absolute Percentage Error (MAPE) and R^2 (Amini 2023; Salehinejad et al. 2018; Singh and Malhotra 2024). The metrics' equations are provided below. The fit model selection is based on the test RMSE. We used RMSPE and MAPE to compare normalised errors of the validation and the test sets.

$$RMSE = \sqrt{\frac{1}{n} \cdot \sum_{i=1}^{n} \left(y_i - \hat{y}_i\right)^2} \tag{3}$$

$$MAE = \frac{1}{n} \sum_{i=1}^{n} \left|y_i - \hat{y}_i\right| \tag{4}$$

$$RMSPE = \sqrt{\frac{1}{n} \cdot \sum_{i=1}^{n} \left(\frac{y_i - \hat{y}_i}{y_i}\right)^2} \tag{5}$$

$$MAPE = \frac{1}{n} \cdot \sum_{i=1}^{n} \left|\frac{y_i - \hat{y}_i}{y_i}\right| \tag{6}$$

$$R^2 = 1 - \frac{\sum_{i=1}^{n}(y_i - \hat{y}_i)^2}{\sum_{i=1}^{n}(y_i - \overline{y})^2} \tag{7}$$

3 Results and Discussion

As indicated by the forecasting process, the HB-LSTM and HB-GRU models were tuned, built and evaluated for four different-sized stock price datasets of financial equities. This section shares the main results. The following plots depict the real observations plotted against the prediction for the entire time series (Figs. 4, 5, 6, 7, 8, 9, 10, 11).

The plots show that point forecasts (PF) of the LSTM and GRU models are close. Though measures were taken to prevent overfitting by fine-tuning the dropout rate, the L_2 regularization and the epochs, the PF are prone to overfitting in the validation area and underperforming to an extent in the test set. We suspect that, in part, this is due to cramming many hyperparameters to be simultaneously optimized.

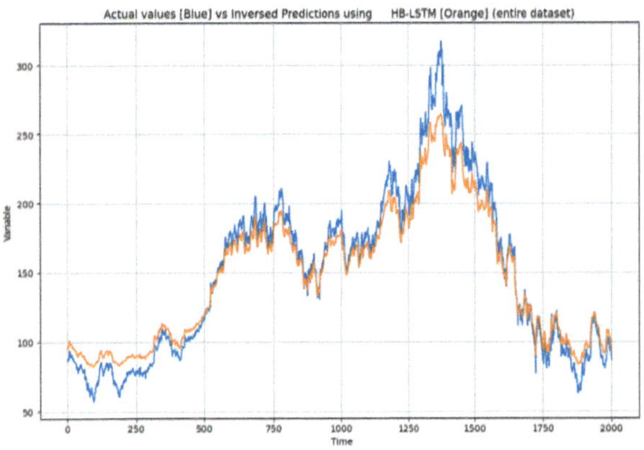

Fig. 4 Point forecasts using HB-LSTM for Alibaba's stock prices

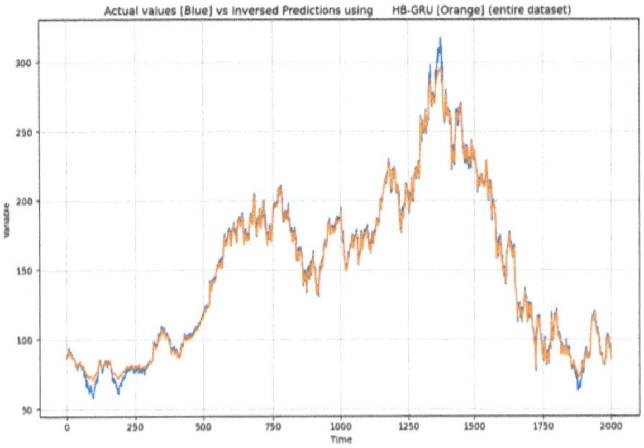

Fig. 5 Point forecasts using HB-GRU for Alibaba's stock prices

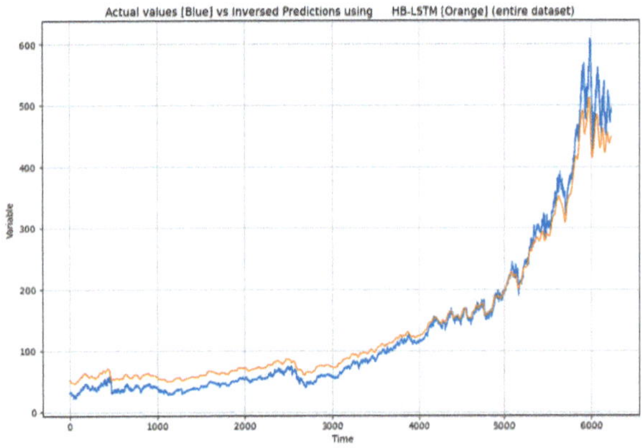

Fig. 6 Point forecasts using HB-LSTM for Costco's stock prices

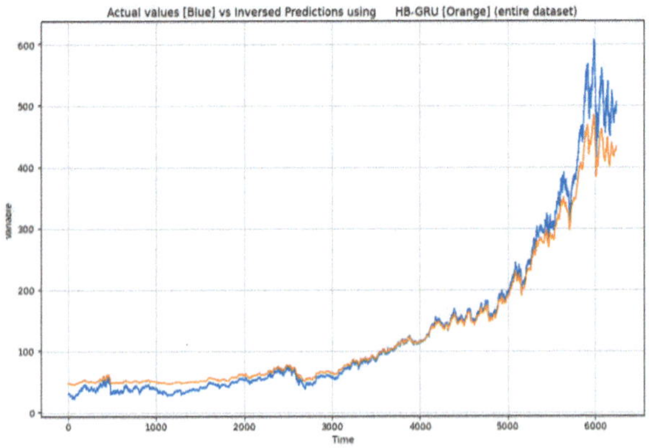

Fig. 7 Point forecasts using HB-GRU for Costco's stock prices

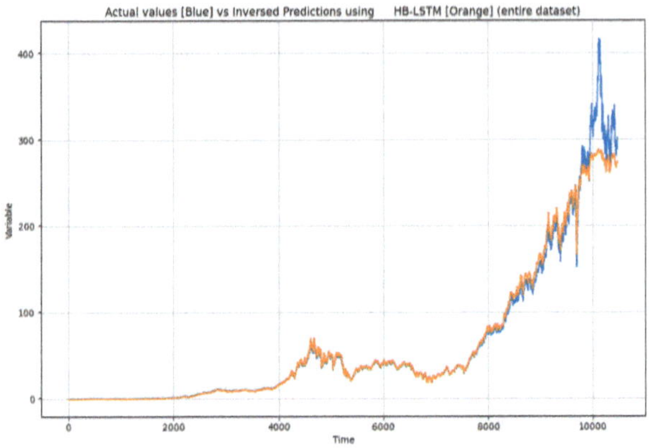

Fig. 8 Point forecasts using HB-LSTM for Home Depot's stock prices

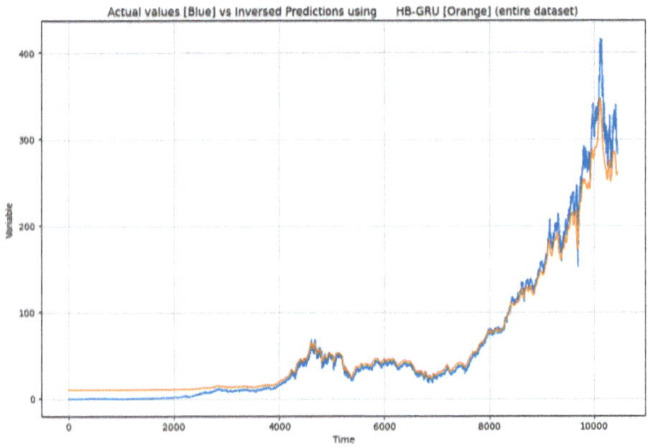

Fig. 9 Point forecasts using HB-GRU for Home Depot's stock prices

HB is suitable for efficient simultaneous tuning particularly in high-dimensional spaces owing to wide hyperparameter space exploration and exploitation of the configurations deemed promising (Li et al. 2018; Elshawi et al. 2019). Nonetheless, if the hyperparameter space contains values that could lead to poor performance, the brackets evaluating these non-reliable configurations or values with other configurations containing reliable but slowly converging values could hinder the selection of better hyperparameter sets. Therefore, switching from complete simultaneous HPO to semi-simultaneous HPO could be interesting. When few hyperparameters are optimized at a time, it's possible that the algorithm has more opportunity to gauge different hyperparameter sets properly (Li et al. 2018; Elshawi et al. 2019). This entails that overall, less dysfunctional hyperparameters would

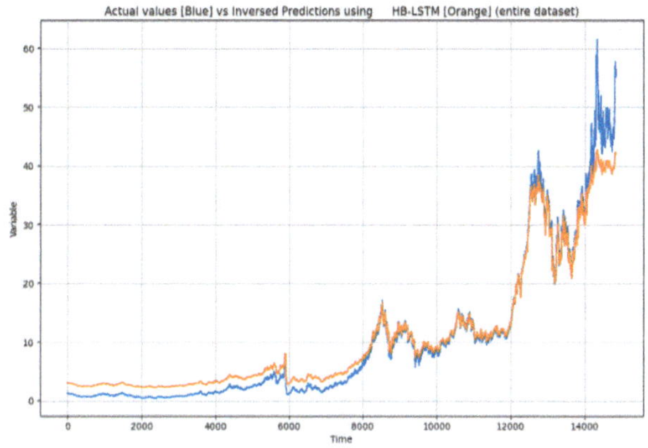

Fig. 10 Point forecasts of HB-LSTM for Kroger's stock prices

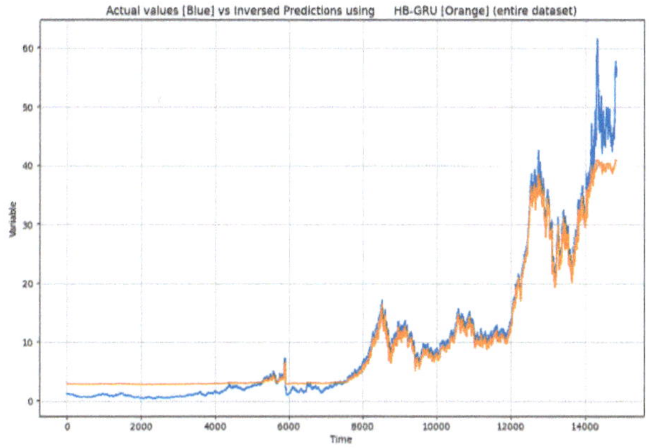

Fig. 11 Point forecasts of HB-GRU for Kroger's stock prices

be selected and the following tuning sections would be shaped to fit the previously optimized hyperparameters.

Moreover, HB relies on successive halving, it prioritizes the exploration of hyperparameter combinations with potentially high accuracy for the given number of epochs or budget (Elshawi et al. 2019). Thus, the algorithm selects the promising combination based on an average of errors (MSE, MAE or Huber loss in our case) for the given 50 epochs. Consequently, the algorithm could discard other promising tunings which could outperform in the model-building phase where the number of epochs is considerably higher.

As the dataset's size increases, the PF in the test area tend to get flattened or tamed. The explanation behind such behaviour is two-fold. From a point of view, the studied datasets are increasingly noisy over time. On the other hand, financial datasets are known for their volatility which explains why the models fail to track inconsistent patterns to what they've previously memorized. This entails the need to update the forecasting framework and monitor the amount of training data fed into recurrent models to prevent them from capturing unnecessary long-term dependencies. However, this should be balanced by a less time-consuming HPO process, either by updating specific hyperparameters such as the learning rate, the optimizer, the batch size and / or the window size, or through exploration of other efficient HPO techniques. Indeed, the systematic over systematic reviews of stock market prediction prompted the pivotal role of adjusting tuning sets as different HPO techniques come with advantages and shortcomings and as the literature has no unanimous hyperparameter set for all stock market datasets (Lin and Lobo Marques 2024).

Table 3 recapitulates the tuning sets of the LSTM and GRU models optimised using HB.

The batch size is 16 across different-sized datasets. Two hidden recurrent layers followed by one hidden dense layer is a consistently optimal choice as per the tuning results as well as popular state-of-the art recommendations (Parmezan et al. 2019; Abumohsen et al. 2023). Some papers argue that a single-layer LSTM outperforms a multilayer-LSTM (Paygude et al. 2023), while other papers recommend a double-layered-GRU over LSTM (Mateus et al. 2021). As the HPO process consists of other fundamental hyperparameters such as the optimizer and the learning, we recommend comparing a single-layered to a double-layer structure before selecting the fit recurrent model. The results show that the Huber loss was selected for the small dataset and yields a satisfying performance as per Figs. 3 and 4. Huber loss was also consistently selected for GRU models. As the dataset's size increases, LSTM models switch between MSE and MAE as the loss function in our case. Generally, the optimal number of epochs hovers around 200 in this study. The optimal activation function for the dense layer of the stacked GRU models is ReLU while LSTM models equally opt for ReLU or tanh in our study.

The tuning sets in Table 3 are a result of conducting the HPO process repeatedly for at least 3 times. Datasets such as Costco and Home Depot presented a particular difficulty in tuning HB-GRU. Overall, LSTM and GRU models performed better with optimizers such as Adam, SGD and Adadelta over RMSprop, with 2 stacked recurrent layers with 1 hidden dense layer over larger hidden recurrent and dense layers.

Overall, the hyperparameter combination selected by our HPO process can be improved if we'd like it to configure recurrent models capable of accurately track time series patterns over longer periods. While it's true that the noise proportions embedded in these datasets increase over time, our HPO process should definitely be able to fairly explore hyperparameter combinations. Thus, if the user is adamant on using accessible AutoML techniques, the sequence in which hyperparameters ought to be tuned should be extended over more steps to allow tuning algorithms such as HB to evaluate the potential of controllable hyperparameter sets. In fact,

Table 3 Hyperparameter combinations of HB-LSTM and HB-GRU models

Dataset_ID	Kroger		Home depot		Costco		Alibaba	
Dataset size	14850		10485		6225		2007	
DL model	HB-LSTM	HB-GRU	HB-LSTM	HB-GRU	HB-LSTM	HB-GRU	HB-LSTM	HB-GRU
Epochs	100	150	175	200	200	200	200	175
Learning rate	3,58491174E-04	2,50874860E-04	5,51164619E-04	8,08141283E-03	1,32942604E-04	3,01013903E-03	8,45085190E-04	1,07480806E-03
Optimizer	'adam'	'adam'	'adam'	'adadelta'	'sgd'	'sgd'	'rmsprop'	'adam'
Loss	MSE	Huber	MAE	Huber	MSE	MSE	Huber	Huber
Batch size	16	16	16	16	16	16	16	16
Window size	10	5	5	30	25	20	5	5
Hidden recurrent layers	2	2	2	3	2	2	2	2
Hidden recurrent units	[208, 96]	[440, 248]	[472, 384]	[304, 136, 328]	[56, 280]	[328, 64]	[376, 360]	[352, 104]
(recurrent) L_2-regularization	[0.001, 0.1]	[0.001, 0.001]	[0.01, 0.01]	[0.001, 0.001, 0.001]	[0.1, 0.001]	[0.1, 0.001]	[0.001, 0,001]	[0.01, 0.01]
Dropout	[0.0, 0.30000000000000004]	[0.30000000000000004, 0.2]	[0.1, 0.0]	[0.2, 0.1, 0.1]	[0.4, 0.0]	[0.2, 0.0]	[0.30000000000000004, 0.4]	[0.4, 0.1]
Hidden dense layers	1	1	1	1	1	1	1	1
Hidden dense units	[128]	[16]	[32]	[16]	[16]	[8]	[8]	[8]
Activations	['tanh']	['relu']	['tanh']	['relu']	['relu']	['relu']	['relu']	['sigmoid']
(dense) L_2-regularization	[0.01]	[0.01]	[0.001]	[0.001]	[0.01]	[0.1]	[0.1]	[0.001]
Dropout	[0.4]	[0.0]	[0.0]	[0.4]	[0.0]	[0.1]	[0.2]	[0.0]

Table 4 Performance assessment of HB-LSTM and HB-GRU models

		Kroger		Home Depot		Costco		Alibaba	
		HB-LSTM	HB-GRU	HB-LSTM	HB-GRU	HB-LSTM	HB-GRU	HB-LSTM	HB-GRU
Validation	RMSE	0,7847	1,1502	7,3460	7,7251	11,3647	13,9611	9,3635	2,5800
	RMSPE	2,731%	5,897%	5,346%	4,812%	4,227%	4,974%	6,082%	1,871%
	MAE	0,5089	1,0575	6,7826	5,4437	8,5684	11,7969	7,3817	1,9988
	MAPE	2,199%	5,529%	5,247%	3,866%	3,4%	4,560%	4,931%	1,444%
	R^2	0,9939	0,9868	0,9679	0,9639	0,9605	0,9445	0,9550	0,9966
Test	RMSE	5,2852	5,6899	37,3076	32,6189	48,4067	63,6701	9,4473	2,6921
	RMSPE	10,598%	11,444%	10,667%	10,611%	9,644%	12,795%	12,159%	3,743%
	MAE	3,6393	4,0932	26,0917	28,1741	41,8550	59,3656	8,5483	1,8935
	MAPE	7,898%	9,163%	8,272%	9,595%	8,709%	12,374%	10,256%	2,323%
	R^2	0,6916	0,6457	0,5060	0,6308	0,5074	0,1315	0,4719	0,9571

during our tuning process, HB repeatedly opted for models with RMSprop or Adam over the other options since they had better chances of outperforming in 50 epochs conditioned by early stopping. In training and test phases, some of these choices which fit the general recommendations performed poorly. Table 4 contains the performance assessment using the accuracy metrics.

The results show that as the dataset's size increases, HB-LSTM tends to outperform HB-GRU considering all metrics. The gap between validation values and test values of RMSPE and MAE show that larger datasets tend to be more prone to overfitting. The poor test values of R2 can be supported by the largest gaps between the validation and test values of RMSE.

For Alibaba and Home Depot's stock prices, HB-GRU outperformed, while HB-LSTM outperformed HB-GRU for Costco and Kroger. Because of the large data size and the resulting variations in patterns and noise, Kroger stock prices predictions exhibited a larger gap in the test set. This implies that beyond significant dataset' sizes, we'd expect LSTM and GRU models to yield poor predictions as they still carry long-term dependencies that no longer fit the dataset's variations.

Moreover, some models exhibited a better quantitative performance contrasted with a widening visual error gap in the test set, such as Kroger's LSTM compared to Kroger's GRU. The disparity between quantitative and visual performance can be explained by the metrics we used for model selection. All the metrics in this study are based on some form of an average between the PF and the actual values, thereby promoting the selection of models with overall best accuracy metrics over models with PF better extrapolating the previous patterns.

As the size of the dataset increases, the performance of both models tends to lie within the same range of values. Since the performance metrics are based on an average of the gap between observations and predictions, some models are quantitatively deemed accurate but don't replicate the exhibited patters as accurately as other candidate models. The R2 values show that the robustness drastically drops in test set in comparison to the high performance in the validation set.

Overall, the performance evaluation shows that analyzing both plots and metrics should go hand in hand to identify the accuracy's variance throughout time and intervene as needed.

While we adopted HB as the HPO technique for its accuracy and efficiency compared to other commonly AutoML methods (random search, BO) (Elshawi et al. 2019) (Falkner et al. 2018), the study underlines some limitations. The random initial selection and successive halving associated with HB might favor some poor hyperparameter configurations for a given dataset while discarding potentially better ones either because they performed poorly in the early HB rounds or they weren't initially selected. Thus, the search space for a given dataset type should be thoroughly studied before starting the tuning process to improve the forecasting pipeline for our financial forecasting application. Otherwise, other HPO techniques such as metaheuristics should be confronted to HB for a smoother and richer forecasting experience.

The results presented in Table 4 can't be generalized since we didn't test enough datasets with similar properties. Hence, as a future step, a deeper comparative study consisting of comparing 2 widely-used AutoML HPO techniques with LSTM, GRU and their bidirectional counterparts is planned to enable a fair analysis of the hyperparameter combination selection, and the performance achieved by common specific hyperparameters.

4 Conclusion

The study represents a preliminary step to building a data-driven framework using recurrent models. The paper describes the forecasting process and the order in which the hyperparameters were tuned to deal with noisy stock price time series of retail supply chains with different sizes. Judging by the results, HB-LSTM tends to outperform HB-GRU as the dataset's size increases. Both the compared recurrent models used 16 as batch size and mostly opted for 2 recurrent hidden layers followed by a single dense hidden layer. The results show that the performance exhibits some variation from validation to test sets considering RMSPE and MAPE, which is partly explained by the inconsistent and noisy patterns, and partly by the heavy tuning process simultaneously exploring too many hyperparameters in one step.

Since the conducted comparison didn't consider enough datasets to elaborate the relationship between recurrent model performance, size of datasets and the hyperparameter selection, we aim to conduct quantitative analysis to measure the correlation between the aforementioned factors.

References

Abumohsen, M., Owda, A.Y., Owda, M.: Electrical load forecasting using LSTM, GRU, and RNN Algorithms. Energ., **16**(5), (2023), https://doi.org/10.3390/en16052283

Amini, A.: Deep sequence modelling. MIT, (2023). Available: http://introtodeeplearning.com/2023/slides/6S191_MIT_DeepLearning_L2.pdf

Ba, J., Mnih, V., Kavukcuoglu, K.: Multiple object recognition with visual attention. arXiv, (2015). https://doi.org/10.48550/arXiv.1412.7755

Bergstra, J., Bengio, Y.: Random search for hyper-parameter optimization. J. Mach. Learn. Res. **13**(10), 281–305 (2012)

Bird, G., Polivoda, M.E.: Backpropagation through time for networks with long-term dependencies. arXiv, (2021). https://doi.org/10.48550/arXiv.2103.15589

'Chapter 12_ Notes_ Recurrent Neural Networks'. MIT, Dec. 18, 2019. Available: https://openlearninglibrary.mit.edu/assets/courseware/v1/0de27572f5d771b35ad094df49a8e200/asset-v1:MITx+6.036+1T2019+type@asset+block/notes_chapter_Recurrent_Neural_Networks.pdf

Chung, J., Gulcehre, C., Cho, K., Bengio, Y.: Empirical evaluation of gated recurrent neural networks on sequence modeling. (2014). https://doi.org/10.48550/arXiv.1412.3555

de Mattos Neto, P.S.G., Cavalcanti, G.D.C., Firmino, P.R.A., Silva, E.G., Vila Nova Filho, S.R.P.: 'A temporal-window framework for modelling and forecasting time series. Knowl.-Based Syst., **193**, p. 105476. (2020), https://doi.org/10.1016/j.knosys.2020.105476

Daoulas, V., Tampouratzis, N., Mousouliotis, P., Papaefstathiou, I.: An Open-source implementation of LSTM and GRU in the ptolemy simulation framework. In: 2021 IEEE/ACM 25th International Symposium on Distributed Simulation and Real Time Applications (DS-RT), Valencia, Spain: IEEE, pp. 1–8. (2021) https://doi.org/10.1109/DS-RT52167.2021.9576137

Elshawi, R., Maher, M., Sakr, S.: Automated machine learning: state-of-the-art and open challenges. arXiv, (2019). Accessed: May 23, 2024. Available: http://arxiv.org/abs/1906.02287

Falkner, S.: Klein, A.: Hutter, F.: BOHB: robust and efficient hyper-parameter optimization at scale. arXiv, (2018). https://doi.org/10.48550/arXiv.1807.01774

Fernández-Delgado, M., Sirsat, M.S., Cernadas, E., Alawadi, S., Barro, S., Febrero-Bande, M.: An extensive experimental survey of regression methods. Neural Netw. **111**, 11–34 (Mar.2019). https://doi.org/10.1016/j.neunet.2018.12.010

Gers, F.A., Schmidhuber, J., Cummins, F.: Learning to forget: continual prediction with LSTM. In: 1999 Ninth International Conference on Artificial Neural Networks ICANN 99. (Conf. Publ. No. 470), Accessed: May 03, 2024. Available: https://www.researchgate.net/publication/12292425_Learning_to_Forget_Continual_Prediction_with_LSTM

Goodfellow, I., Bengio, Y., Courville, A.: Deep Learning. MIT PRess, (2016). Available: http://www.deeplearningbook.org

Gregor, K., Danihelka, I., Graves, A., Rezende, D.J., Wierstra, D.: DRAW: a recurrent neural network for image generation. arXiv, May 20, 2015. Accessed: May 24, 2024. Available: http://arxiv.org/abs/1502.04623

Han, S., et al.: ESE: efficient speech recognition engine with sparse LSTM on FPGA. arXiv, (2017). https://doi.org/10.48550/arXiv.1612.00694

Hochreiter, S., Schmidhuber, J.: Long Short-term Memory. Neural Comput. **9**, 1735–1780 (Dec.1997). https://doi.org/10.1162/neco.1997.9.8.1735

Hyndman, R., Koehler, A., Ord, K., Snyder, R.: Forecasting with exponential smoothing the state space approach. in Springer Series in Statistics. Berlin, Heidelberg: Springer, (2008). https://doi.org/10.1007/978-3-540-71918-2.

Hyndman, R.J., Koehler, A.B.: Another look at measures of forecast accuracy. Int. J. Forecast. **22**(4), 679–688 (Oct.2006). https://doi.org/10.1016/j.ijforecast.2006.03.001

Investigation of load, solar and wind generation as target variables in LSTM time series forecasting, using exogenous weather variables. Accessed: May 24, 2024. Available: https://www.mdpi.com/1996-1073/17/8/1827

Jadon, A., Patil, A., Jadon, S.: A comprehensive survey of regression based loss functions for time series forecasting'. arXiv, (2022). https://doi.org/10.48550/arXiv.2211.02989

Jamieson, K., Talwalkar, A.: Non-stochastic best arm identification and hyperparameter optimization. arXiv, (2015). Accessed: Jun. 20, 2024. Available: http://arxiv.org/abs/1502.07943

Januschowski, T., et al.: Criteria for classifying forecasting methods. Int. J. Forecast. **36**(1), 167–177 (Jan.2020). https://doi.org/10.1016/j.ijforecast.2019.05.008

Jia, P., Miao, Y.: Research on short-term forecast of power load based on adam-gru', J. Phys.: Conf. Ser., **1693**(1), p. 012116, (2020). https://doi.org/10.1088/1742-6596/1693/1/012116

Köksal, Ö., Akgül, Ö.: A comparative text classification study with deep learning-based algorithms. In: 2022 9th International Conference on Electrical and Electronics Engineering (ICEEE), pp. 387–391. (2022). https://doi.org/10.1109/ICEEE55327.2022.9772587

Kong, D., Wang, S., Ping, P.: State-of-health estimation and remaining useful life for lithium-ion battery based on deep learning with Bayesian hyperparameter optimization. Int. J. Energy Res. **46**(5), 6081–6098 (2022). https://doi.org/10.1002/er.7548

Lalapura, V.S., Bhimavarapu, V.R., Amudha, J., Satheesh, H.S.: A systematic evaluation of recurrent neural network models for edge intelligence and human activity recognition applications. Algorithms, **17**(3), (2024), https://doi.org/10.3390/a17030104

Li, C., et al.: Prediction of length of stay on the intensive care unit based on least absolute shrinkage and selection operator. IEEE Access **7**, 110710–110721 (2019). https://doi.org/10.1109/ACCESS.2019.2934166

Li, X., Qin, X., Wu, J., Yang, J., Huang, Z.: Tool wear prediction based on convolutional bidirectional LSTM model with improved particle swarm optimization. Int. J. Adv. Manuf. Technol. **123**(11–12), 4025–4039 (2022). https://doi.org/10.1007/s00170-022-10455-1

Li, L., Jamieson, K., DeSalvo, G., Rostamizadeh, A., Talwalkar, A.: Hyperband: a novel bandit-based approach to hyperparameter optimization'. arXiv, (2018). https://doi.org/10.48550/arXiv.1603.06560

Lin, C.Y., Lobo Marques, J.A.: Stock market prediction using artificial intelligence: A systematic review of systematic reviews. Soc. Sci. & Hum. Open, **9**, p. 100864, (2024). https://doi.org/10.1016/j.ssaho.2024.100864

Liu, S., Fang, Y., Cheng, H., Pan, Y., Liu, Y., Gao, C.: Large language models guided generative prompt for dialogue generation. In: 2023 International Conference on Cyber-Enabled Distributed Computing and Knowledge Discovery (CyberC), pp. 10–17. (2023) https://doi.org/10.1109/CyberC58899.2023.00013

Makridakis, S., Spiliotis, E., Assimakopoulos, V.: The M4 Competition: 100,000 time series and 61 forecasting methods. Int. J. Forecast. **36**(1), 54–74 (Jan.2020). https://doi.org/10.1016/j.ijforecast.2019.04.014

Marzak, Z., Benabbou, R., Mouatassim, S., Benhra, J.: Forecasting seasonal and trend-driven data: a comparative analysis of classical techniques, **2**, (2023)

Mateus, B., Mendes, M., Farinha, J., Cardoso, A.: Comparing LSTM and GRU models to predict the condition of a pulp paper press, Energies, (2021). https://doi.org/10.3390/en14216958

Mhaned, A., Salma, M., Mounia, E.H., Jamal, B.: Contribution to smart irrigation based on internet of things and artificial intelligence. In: Proceedings of the 6th International Conference on Big Data and Internet of Things, Lazaar, M., En-Naimi, E.M., Zouhair, A., Al Achhab, M., Mahboub, O., Eds., Cham: Springer International Publishing, pp. 537–549. (2023) https://doi.org/10.1007/978-3-031-28387-1_45

Moghar, A., Hamiche, M.: Stock market prediction using LSTM recurrent neural network. Procedia Comput. Sci. **170**, 1168–1173 (2020). https://doi.org/10.1016/j.procs.2020.03.049

Mung P.S., Phyu, S.: Time series weather data forecasting using deep learning. In: 2023 IEEE Conference on Computer Applications (ICCA), pp. 254–259. (2023) https://doi.org/10.1109/ICCA51723.2023.10182058

Olah, C.: Understanding LSTM networks. Accessed: Jun. 19, 2024. Available: https://colah.github.io/posts/2015-08-Understanding-LSTMs/

Pankratz, A.: Forecasting with univariate box-jenkins models: concepts and cases. In: Wiley Series in Probability and Statistics. John Wiley & Sons, (1983)

Parmezan, A.R.S., Souza, V.M.A., Batista, G.E.A.P.A.: Evaluation of statistical and machine learning models for time series prediction: Identifying the state-of-the-art and the best conditions for the use of each model. Inf. Sci. **484**, 302–337 (May2019). https://doi.org/10.1016/j.ins.2019.01.076

Paygude, P., et al.: Optimizing hyperparameters for enhanced LSTM-Based prediction system performance. Int. J. Recent. Innov. Trends Comput. Commun. **11**, 203–213 (2023). https://doi.org/10.17762/ijritcc.v11i10s.7620

Rokhsatyazdi, E., Rahnamayan, S., Amirinia, H., Ahmed, S.: Optimizing LSTM based network for forecasting stock market. In: 2020 IEEE Congress on Evolutionary Computation (CEC), pp. 1–7. (2020) https://doi.org/10.1109/CEC48606.2020.9185545

'stanford-cs-230-deep-learning/en/cheatsheet-recurrent-neural-networks.pdf at master · afshinea/stanford-cs-230-deep-learning', GitHub. Accessed: May 03, 2024. Available:https://github.com/afshinea/stanford-cs-230-deep-learning/blob/master/en/cheatsheet-recurrent-neural-networks.pdf

Salehinejad, H., Sankar, S., Barfett, J., Colak, E., Valaee, S.: Recent advances in recurrent neural networks. arXiv, (2018). Accessed: May 03, 2024. Available: http://arxiv.org/abs/1801.01078

Schmidt, R.M.: Recurrent neural networks (RNNs): a gentle introduction and overview. arXiv, (2019). https://doi.org/10.48550/arXiv.1912.05911

Sezer, O.B., Gudelek, M.U., Ozbayoglu, A.M.: Financial time series forecasting with deep learning : A systematic literature review: 2005–2019. Appl. Soft Comput. **90**, 106181 (2020). https://doi.org/10.1016/j.asoc.2020.106181

Singh, H., Malhotra, M.: Stock market and securities index prediction using artificial intelligence: A systematic review. Multidiscip. Rev. **7**, 2024060 (Jan.2024). https://doi.org/10.31893/multirev.2024060

Smagulova, K., James, A.P.: Overview of long short-term memory neural networks. In: Deep Learning Classifiers with Memristive Networks: Theory and Applications, A. P. James, Ed., Cham: Springer International Publishing, pp. 139–153. (2020). https://doi.org/10.1007/978-3-030-14524-8_11

Sutskever, I.: Training recurrent neural networks', graduate department of computer science university of Toronto. (2013). Available: https://www.cs.utoronto.ca/~ilya/pubs/ilya_sutskever_phd_thesis.pdf

Varsamopoulos, S., Bertels, K., Almudever, C.: Designing neural network based decoders for surface codes. (2018)

Werbos, P.: Backpropagation through time: what it does and how to do it. Proc. IEEE **78**, 1550–1560 (1990). https://doi.org/10.1109/5.58337

Zaheer, S. et al.: A multi parameter forecasting for stock time series data using LSTM and deep learning model. Math., **11**(3), (2023). https://doi.org/10.3390/math11030590

The Trend of Integration of Artificial Intelligence in Supply Chain: A Bibliometric Review

Ouahbi Younesse and Ziti Soumia

Abstract

This study delves into the impact of artificial intelligence on the supply chain by examining various employment-related factors. These factors include publication volume, authorship, country of origin, affiliations, and relevant journals. Through a comprehensive analysis of bibliographic data obtained from Scopus, Springer Link, Web of Science, Science Direct, and Google Scholar, the study revealed a significant increase in relevant publications since 2010. Additionally, it uncovers key authorities in the field, source countries, institutions, and influential journals. Furthermore, it examines common keywords used in search queries and recurring themes in research endeavors. Citation analysis aims to identify renowned authors and significant documents that are highly influential in this literature. The purpose of this investigation was to identify key areas of interest by analyzing co-citations among authors. Biased assessments are excluded, and the language used is objective and neutral in terms of value. Technical terms are explained the first time they are used, and common academic elements are included.

Keywords

Bibliometric study · Sustainability · Artificial intelligence · Logistics · Supply chain · Supply chain management

O. Younesse (✉) · Z. Soumia
IPSS Intelligent Processing and Security of Systems Laboratory
Faculty of Sciences, Mohammed 5 University in Rabat, Rabat, Morocco
e-mail: Younesse_ouahbi@um5.ac.ma

Z. Soumia
e-mail: s.ziti@um5r.ac.ma

1 Introduction

The field of logistics and supply chain management (SCM) is witnessing the emergence of a new and promising concept known as artificial intelligence (AI). AI in SCM combines various technological advancements and research domains, including the Internet of Things (IoT), big data, and cloud computing. This comprehensive approach aims to optimize logistics and supply chain processes, ultimately creating more efficient, effective, and sustainable supply chains. The goal of AI in SCM is to replicate the organization of data flows in the digital Internet but applied to the physical flow of goods.

Optimizing resource utilization, waste reduction, and cost reduction in supply chains are significant challenges that AI aims to address. AI has been specifically developed to tackle existing challenges in the logistics sector and the supply chain as a whole, including lack of standardization, transport inefficiency, and negative environmental consequences associated with logistics operations.

The purpose of this systematic mapping study is to provide a comprehensive summary of research on artificial intelligence and supply chain management (SCM). Our study aims to examine publication years, channels, authors, countries, and keywords used in AI, logistics, SC, and SCM-related publications. Subsequently, we will analyze existing literature to uncover potential opportunities and perspectives that will provide valuable insights into future research directions. Additionally, we will establish a comprehensive framework that effectively organizes the field of physical Internet and SCM, ultimately defining future research directions for logistics, SC, and SCM researchers.

B. El Bhiri et al. (eds.), *AI and Digital Transformation: Innovations in Supply Chain, Education, and Energy Systems*, Sustainable Artificial Intelligence-Powered Applications, https://doi.org/10.1007/978-3-031-86837-5_3

2 Background

Artificial Intelligence (AI) is a concept that represents a transformative action across all aspects of the supply chain, drawing inspiration from the digital world. It emerged in the early twenty-first century intending to revolutionize traditional supply chain methods by leveraging ideas from data networks, the Internet, and task automation. The goal of this innovative endeavor is to establish a highly interconnected and efficient infrastructure featuring standardized containers and intelligent routing, facilitating the seamless movement of goods. This new concept aims to reduce waste, enhance resource utilization, bolster sustainability, and increase supply chain resilience. The increasing complexity and global nature of modern supply chains have contributed to its development.

Motivated also by the growth of e-commerce, urbanization, and environmental concerns, this paradigm shift is anticipated to transform the way goods are transported, stored, and distributed, ultimately leading to more efficient and sustainable supply chain management in the future.

3 Related Work

Artificial Intelligence (AI) represents a cutting-edge idea within the realm of supply chain management and logistics. Its primary objective is to enhance the efficiency, effectiveness, and sustainability of supply chains (Yogesh upadhyay faisal talib1 syed ali zaheen mohd 2023; Paul Jayender 2022). AI draws upon various technologies and research domains such as the Internet of Things, big data, and data analytics (Zakharkina et al. 2022; Dubey 2022). This concept introduces a revolutionary framework for operational networks, production industries, and logistics. It is characterized by its open and global nature, providing a deeply rooted multimodal system within universal physical, digital, and operational connectivity. This is made possible through encapsulation, standard protocols, and interfaces (Yogesh upadhyay faisal talib1 syed ali zaheen mohd 2023). The benefits of AI may vary across different types of businesses and may also present potential advantages for customers (Fasso et al. 2023; Dubey 2022).

Numerous studies have explored the application of artificial intelligence in supply chain management. For instance, studies have demonstrated how to enhance innovation and performance in manufacturing supply chains through datafication (Fosso et al. 2024), the role of AI's dynamic capability on environmental performance in France and the United States (Madancian et al. 2024) by Samuel Fosso, as well as the impact of AI on supply chain management in modern enterprises (Essghaier et al. 2023). According to Milenkovic (2022), AI is a disruptive innovation enabling the creation of sustainable supply chains (Fasso et al. 2023).

Although AI is increasingly being utilized and tested by forward-thinking companies through pilot projects, it remains a relatively new concept.

Table 1 Research questions

IDs	RQ	Rational
RQ1	What publication venues were utilized for the selected studies and during which years were they conducted?	To determine if there are any designated publication channels and to ascertain the extent of research undertaken in this field
RQ2	What are the commonly used keywords in this field?	To identify the principal domain of research interest concerning AI, SC and SCM
RQ3	Which authors are showing an interest in this area	To identify the experts in that subject. And to determine where a particular country dominates this field

4 Research Questions

This survey aims to provide an overview of studies concerning AI and SCM that have been published since 2010. To achieve a comprehensive understanding of the topic, we have divided the overall objective into three research questions (RQs), as detailed in Table 1, along with their rationale.

5 Research Methodology

5.1 Data Collection

The initial stage of this bibliometric analysis involves gathering relevant data from five widely recognized databases between [2010–2023]: Scopus, Web of Science, SpringerLink, ScienceDirect, and Google Scholar, all known for their comprehensive cataloging of scientific literature (see Fig. 1). To assemble a comprehensive dataset, a thorough search was conducted using carefully chosen keywords closely related to artificial intelligence (AI) and supply chain management (SCM). This search strategy was meticulously designed to capture a wide range of documents covering the breadth of research in this field, encompassing all published studies.

The search was carried out using the following search string: ("Artificial Intelligence") AND ("Supply Chain" OR "Logistics" OR "Supply Chain Management").

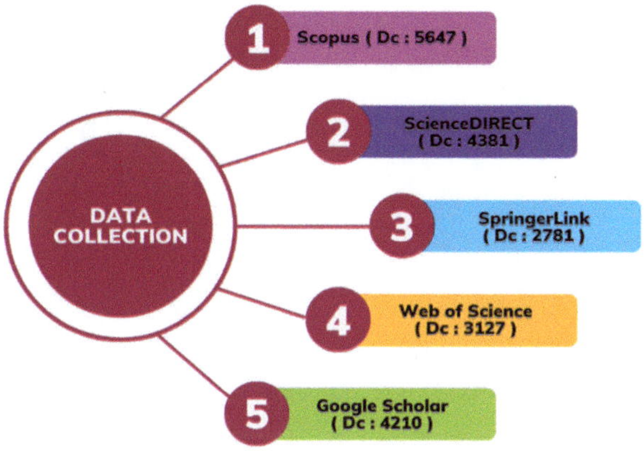

Fig. 1 Organisation structure chart, infographic graph

5.2 Inclusion and Exclusion Criteria

The main goal of the selection procedure is to identify the candidate papers that, based on their title, abstract, and keywords, are most relevant to the goal of this mapping project. A set of inclusion and exclusion criteria were developed and applied to the candidate articles to choose the best papers to address the research questions (RQs). As a result, we determined the following inclusion criteria (IC), which are connected by the Boolean operator OR (Fig. 2).

- **IC1:** Articles published in the last decade.
- **IC2:** The documents must be journal articles or conference papers.

The candidate studies that fulfilled any of the following exclusion criteria (EC) were eliminated (Fig. 3):

Fig. 3 Pyramid of hierarchy of information

- **EC1:** Studies not relating to the subject matter of IA, SC, Logistics, and SCM.
- **EC2:** Papers that were available in the form of abstracts and/or PowerPoint presentations.

5.3 Data Analysis

Figures 1, 2, and 3 enabled us to collect 20,146 potential papers by searching through the five data sources. After

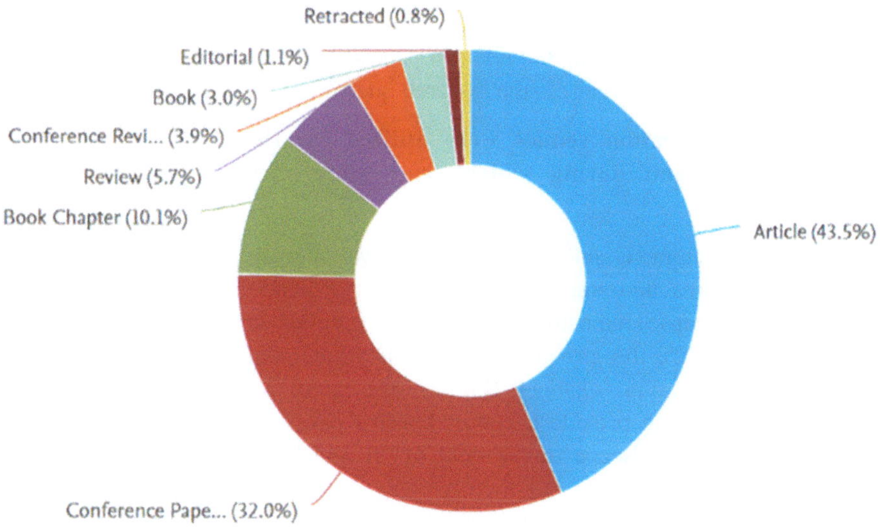

Fig. 2 Results of data collection

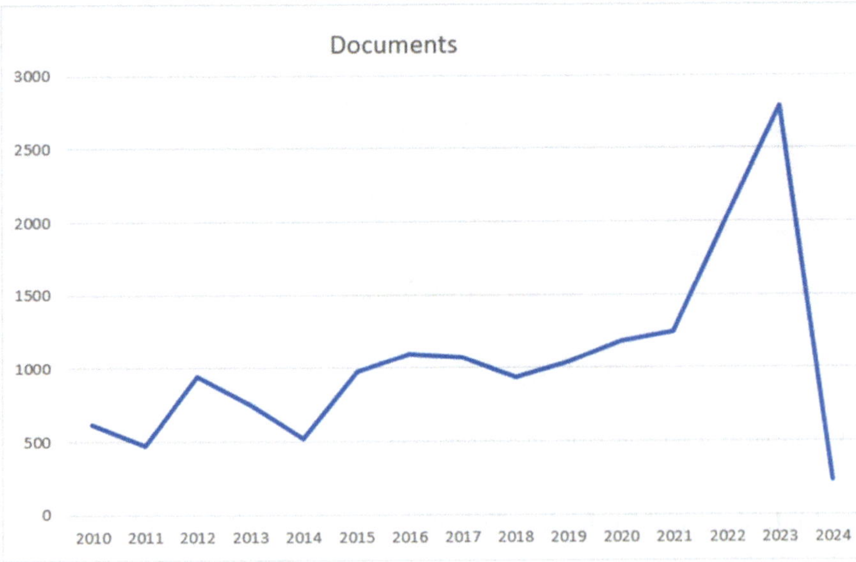

Fig. 4 Number of publications by year between 2010–2023

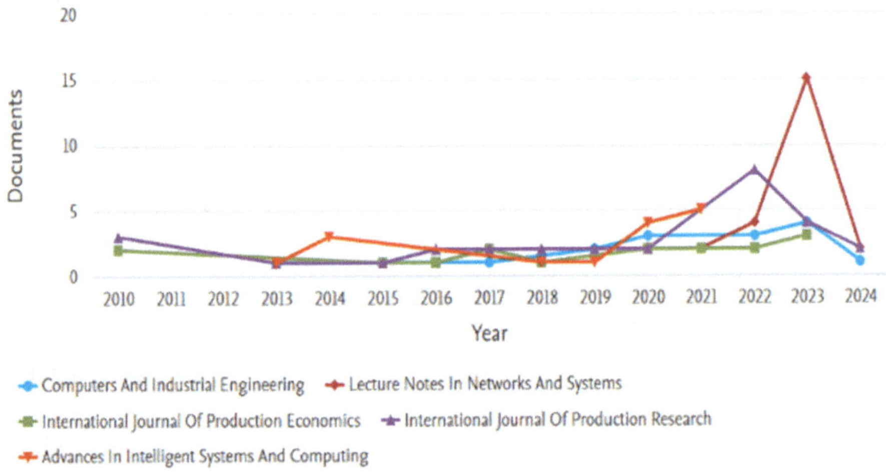

Fig. 5 Journal-wise number of published articles

evaluation, 65.5% of the potential results met the inclusion criteria, resulting in only 13,094 correctly responding papers. Subsequently, we applied the exclusion criteria to the 13,094 papers, resulting in 832 relevant papers that were subsequently analyzed to address the research questions.

RQ1: What publication venues were utilized for the selected studies and during which years were they conducted?

The selected documents underwent quantitative analysis to identify patterns and trends in research performance. Different bibliometric indicators such as the number of publications (Fig. 2), the growth rate of publications over time, and the publication's channels (Fig. 3) were examined. The analysis aimed to provide a comprehensive overview of the publication landscape in the field of AI and SCM (Fig. 4).

Our results document a slight increase in publications between 2010 and 2012, followed by a moderate development of publications between 2014 and 2018, and this was followed by a significant expansion of research in the following years, starting from 2019 (Fig. 5).

RQ2: What are the commonly used keywords in this field?

The objective of this investigation was to identify the fundamental concepts and research subjects that have garnered significant attention within the field. The frequency and distribution of the keywords were investigated to gain insight into current research interests and priorities (Fig. 6).

Six unique clusters were found after keyword analysis, which represents the primary areas of interest for further study on the relationship between AI and SCM. These domains cover important topics like Artificial Intelligence, which has 641 links across 2 clusters, logistics, which has 356 links across 3 clusters, supply chain management, which has 481 links, Warehouse, which is mentioned in 152 papers, as well as Machine Learning, decision-making, and optimization, among other things.

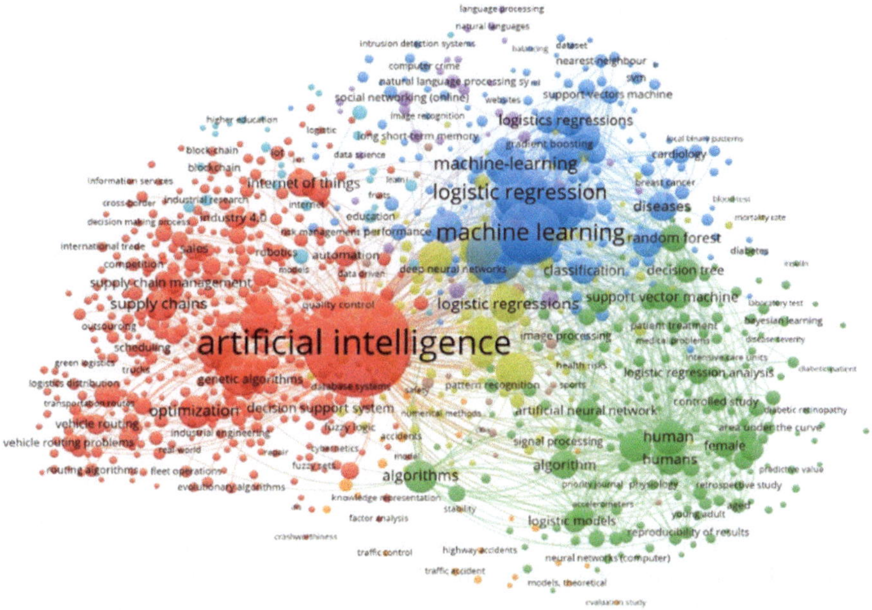

Fig. 6 Keyword co-occurrence map, 6 clusters and 8420 links

RQ3: Which Authors Are Showing an Interest in This Area?

Analyzing the most frequently referenced writers in relevant research publications might discover subject matter experts. Authors who have received a large number of citations are frequently regarded as experts in their professions. Identifying specialists and measuring individual nations' supremacy in a given field can provide significant insights into research activities. The most influential writers were identified using a co-authorship analysis (see Fig. 7). To determine the influence of specific publications and writers on the field, the number of citations obtained by each was evaluated. Co-citation analysis revealed clusters of linked studies and notable research themes in the literature on the durability of construction materials.

The graphical depiction shows 14 separate clusters of writers connected by a total of 522 linkages. Two key individuals stand out: Huang, G.P., who connects with 55 other writers and has contributed to 121 documents, and Zhong, R.Y., who has created 66 documents while maintaining ties with 41 other authors. Identifying specialists and measuring nations' supremacy in a certain field (refer to Fig. 8) might help target cooperation and funding in research areas with specific knowledge needs.

The nation analysis revealed the presence of 9 clusters connected by 745 linkages. The aforementioned groups

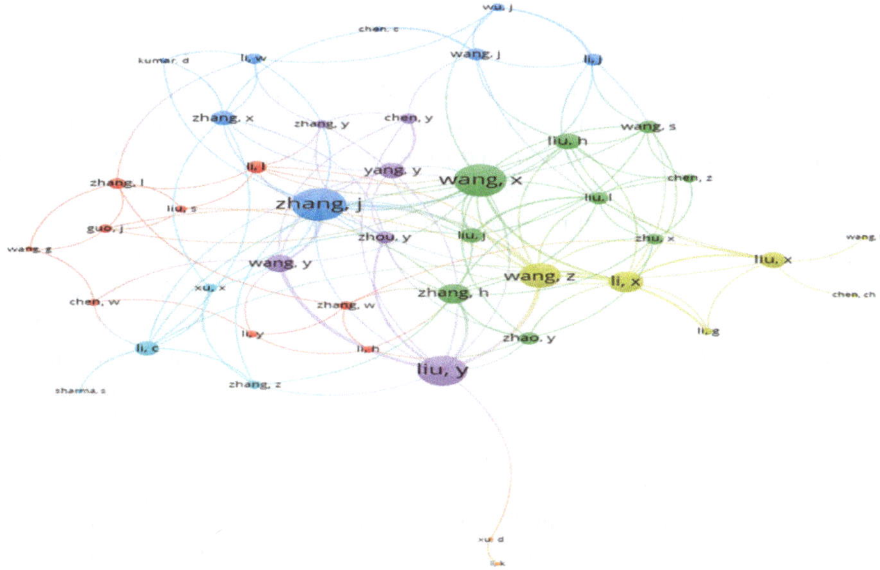

Fig. 7 Co-authorship map, 14 clusters and 522 links

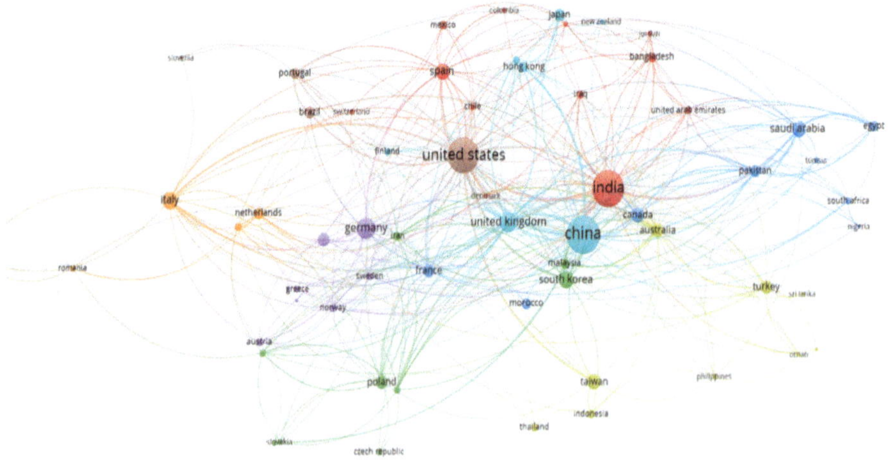

Fig. 8 Country co-occurrence map, 8 clusters and 396 links

exhibit significant differences in terms of socioeconomic situations and legal frameworks. China, the United States, and the United Kingdom are the countries with the most interest in artificial intelligence (AI). Morocco is related to France, Canada, and China, which all contribute to the genesis of 37 texts.

6 Research Trends and Future Directions

This study conducted a comprehensive examination of recent research trends to expand our understanding of the field. The primary objective was to provide a comprehensive perspective that integrates insights gained from both quantitative and qualitative assessments. By integrating various analytical methodologies, we aim to achieve several crucial objectives.

The goal of this analysis is to compile and summarize the extensive knowledge accumulated over the years. Through this synthesis, we aim to identify the most significant developments in the fields of Artificial Intelligence (AI) and supply chain management (SCM). This synthesis serves as a roadmap for future research initiatives.

Notably, the results of our research trends analysis revealed the emergence of new fields within AI and SCM. These emerging fields present a fertile ground for groundbreaking research and innovation. We intend to support this exploration as it represents the cutting-edge of knowledge in these domains.

7 Conclusion

This comprehensive bibliometric analysis delves deeply into the evolving landscape of research concerning the intricate relationship between Artificial Intelligence (AI) and Supply Chain Management (SCM) from 2010 to 2023. The significance of this study lies in its commitment to highlighting the growing of The trend of integration of artificial intelligence in Supply Chain.

Our findings reveal a significant increase in relevant publications since 2010, with notable contributions from key authorities, countries, and institutions. The study identifies emerging research trends, influential authors, and seminal publications, providing a roadmap for future research initiatives. By analyzing common keywords and recurring themes, we have highlighted critical research areas and potential opportunities for further exploration.

AI's integration into SCM is shown to optimize logistics and supply chain processes, aiming to create more efficient, effective, and sustainable supply chains. The identified benefits include enhanced resource utilization, waste reduction, and cost reduction. The study also underscores the potential of AI to address existing challenges such as lack of standardization, transport inefficiency, and negative environmental impacts.

This research serves as a guide for both academics and industry professionals, offering valuable insights into the interplay between AI and SCM. By embracing AI principles, supply chain operations can be transformed and optimized, leading to significant advancements in the field.

The comprehensive framework established in this study effectively organizes the field of AI and SCM, defining future research directions for logistics and SCM researchers. The synthesis of extensive knowledge accumulated over the years highlights the most significant developments and emerging fields within AI and SCM. This exploration represents the cutting-edge of knowledge, supporting groundbreaking research and innovation in these domains.

In conclusion, this bibliometric analysis provides a thorough understanding of the intersection between AI and SCM, guiding future research and practical applications. The study's findings emphasize the transformative potential of AI in optimizing supply chain operations, enhancing sustainability, and driving innovation in the field.

Acknowledgements I would like to express my deepest gratitude to all the individuals and institutions that contributed to the completion of this study. First and foremost, I would like to extend my sincere thanks to the Faculty of Sciences at Mohammed V University of Rabat for their continuous support and for providing me with the necessary resources for this research. I am also grateful to my supervisor, Mrs. Ziti Soumia, whose valuable advice, scientific precision, and unwavering support have significantly contributed to the advancement of this work. Lastly, I wish to thank everyone who has contributed to the realization of this study in one way or another.

References

Bhattacharya, S., Govindan, K., Ghosh, S.: Dastidar, Preeti Sharma. Applications of Artificial Intelligence in Closed-Loop Supply Chain Management: Systematic Literature Review and Future Research Agenda (2024). https://doi.org/10.1016/j.tre.2024.103455

Chen, Chun-Hsi Vivian; Chen, Yu-Cheng; Influence of intellectual capital and integration on operational performance: big data analytical capability perspectives (2022). https://doi.org/10.1108/CMS-02-2021-0037

Debnath, Rimi, Majumder, Pinki, Tarafdar, Anirbanb, Bhattacharya, Baby, Bera, Uttam Kumar.: Artificial intelligence based supply chain management strategy during COVID-19 situation (2024). https://doi.org/10.1080/16258312.2024.2303307

Dubey, Rameshwar, Bryde, David, Dwivedi, Yogesh K, Graham, Gary, Foropon, Cyril.: Impact of artificial intelligence-driven big data analytics culture on agility and resilience in the humanitarian supply chain: A practice-based view (2022). https://doi.org/10.1016/j.ijpe.2022.108618

Essghaier, F., Chargui, T., Hsu, T., Bekrar A., Allaoui H., Trentesaux, D., Goncalves, G.: Fuzzy multiobjective truck scheduling in multi-modal rail–road Physical Internet hubs (2023). https://doi.org/10.1016/j.cie.2023.109404

Fasso, W., Queiroz, M., Trinchera, L.: The role of artificial intelligence-enabled dynamic capability on environmental performance: The mediation effect of a data-driven culture in France and the USA (2023). https://doi.org/10.1016/j.ijpe.2023.109131

Joäo, R.: Exploring applications and practical examples by streamlining material requirements planning (MRP) with Python (2023). https://doi.org/10.3390/logistics7040091

Khan, Pathan Afthab; Ravada, Vinay Teja; Tumna, Manohar; Vidiyala, Sai Suresh Tatavarthy, Srinivas Rao.: A supply chain study of managing multiple routes thru ant colony optimization (2022). https://doi.org/10.1007/978-981-99-4270-1_60

Lin, H, Lin, Ji, Wang, F.: An innovative machine learning model for supply chain management (2022). https://doi.org/10.1016/j.jik.2022.100276

Menzies Jane, Sabert Bianka, Hassan Rohail, Mensah Prince Kofi.: Artificial intelligence for international business: Its use, challenges, and suggestions for future research and practice (2024). https://doi.org/10.1002/tie.22370

Madacancian, Mitra, Taherdoost, Hamed, Javadi, Maassoumeh, Khan, Inam Ullah, Kalantari, Alaeddin; Kumar, Dinesh.: The impact of artificial intelligence on supply chain management in modern business (2023). https://doi.org/10.1007/978-3-031-48573-2_82

Mitra Madancian, Hamed Taherdoost, Maassoumeh Javadi, Inam Ullah Khan, Alaeddin Kalantari, Dinesh Kumar.: L'impact de l'intelligence artificielle sur la gestion de la chaîne d'approvisionnement dans les entreprises modernes (2024). https://doi.org/10.1007/978-3-031-48573-2_82

Nebri Mohamed-Amine, Moussaid Abdellatif, Bouikhalene Belaid.: Artificial intelligence for forecasting sales of agricultural products: A case study of a Moroccan agricultural company (2024). https://doi.org/10.1016/j.joitmc.2023.100189

Paul Jayender, R..: Dibyajyoti Gosh intelligent decision support system of big data and IOT analytics interoperability with ERP promoting SCM sustainability in automotive (2022). https://doi.org/10.1007/978-3-031-25344-7_47

Saad, Sameh, M., Khamkham, Mohamed, The applications of AI in GSCM-A systematic literature review (2022). https://doi.org/10.3233/ATDE220615

Samuel Fosso, Maciel M. Queiroz, Laura Trinchera.: The role of AI's dynamic capability on environmental performance in France and the United States (2024). https://doi.org/10.1016/j.ijpe.2023.109131

Shuang Tiana, Lin Wu, Maria Pia, Marco Ardolino, Kulwant, S.: Pawar. how to enhance innovation and performance in manufacturing supply chains through datafication (2024). https://doi.org/10.1016/j.cie.2023.109841

Wang, Maoyang, Wu, Peng, Luo, Qin, Construction of software supply chain threat portrait based on chain perspective (2023). https://doi.org/10.3390/math11234856

Xinyue Hao, Emrah Demir.: Artificial intelligence in supply chain management: enablers and constraints in pre-development, deployment, and post-development stages (2023). https://doi.org/10.1080/09537287.2024.2302482

Yogesh upadhyay faisal talib1 syed ali zaheen mohd. saifuddin ansari, industry 4.0 adoption framework in msmes using a hybrid fuzzy ahp-topsis approach (2023). https://doi.org/10.24874/pes05.03.010

Zakharkina, L., Rubanov, P., Alibekova, B., Zakharkin, O., Moldashbayeva, L.: The impact of digital transformation in the accounting system of fuel and energy complex enterprises (international experience). Int. J. Energy Econ. Policy **12**(5), 152–161 (2022). https://doi.org/10.32479/ijeep.13317

Artificial Intelligence Approach for Enhancing Vehicle Routing Problem Solutions: Integrating Smart Lockers for Dynamic Delivery and Pickup Operations

Mohamed-Ali Ejjanfi and Jamal Benhra

Abstract

The This paper presents an innovative approach to the Vehicle Routing Problem (VRP) by integrating smart locker technology and employing advanced AI algorithms—Genctic Algorithm (GA), Ant Colony Optimization (ACO), and Particle Swarm Optimization (PSO). This model enhances route flexibility and efficiency in urban logistics, accommodating dynamic route changes and real-time parcel rerouting to smart lockers. Our findings show that these algorithms significantly improve logistical operations, offering scalable solutions for contemporary delivery challenges.

Keywords

Artificial intelligence · Vehicle routing problem · Genetic algorithm · Ant colony optimization · Particle swarm optimization · Smart lockers · Real-time rerouting · Flexible delivery · Pickup systems

1 Introduction

The rapid growth of e-commerce has dramatically transformed the logistics and supply chain landscape, emphasizing the vital importance of last-mile delivery. This final phase of the delivery process involves the complex task of transferring goods from distribution centers to the end consumer, presenting numerous logistical challenges. At the heart of solving these challenges is the Vehicle Routing Problem (VRP), a crucial optimization task designed to enhance the efficiency of delivery routes under various constraints.

This research tackles the complexities of VRP using proven heuristic methods such as Genetic Algorithm (GA), Particle Swarm Optimization (PSO), and Ant Colony Optimization (ACO). These algorithms are chosen for their robustness in addressing complex optimization scenarios. For instance, GA is highly effective in exploring large problem spaces and optimizing routes, as demonstrated by Chunhua et al. (2018). ACO excels in managing multiple constraints like vehicle route allocation, optimizing resource use as described by Negulescu et al. (2008). Similarly, PSO offers significant advantages in constrained environments, enhancing the flexibility of heuristic solutions, as discussed by Akhtar et al. (2015).

By evaluating the performance of these algorithms on key performance indicators such as total tour distance and execution time, this study provides essential insights for logistics practitioners and researchers. It aids in selecting and applying the most effective heuristic methods to meet the operational challenges of modern logistics.

Furthermore, this analysis advances our understanding of heuristic algorithms in the context of last-mile delivery and explores state-of-the-art solutions to the evolving challenges of logistics in the digital era. Incorporating innovative AI algorithms and smart locker technology (Ejjanfi et al. 2024), this research aims to significantly enhance the operational efficiency and adaptability of urban logistics systems, presenting a scalable and effective solution for contemporary logistical challenges.

2 State of Art

The state of the art in flexible routing optimization for problems like Vehicle Routing Problem (VRP) plays a crucial role in tackling complex challenges in logistics, distribution, and smart cities. By employing various methods including the integration of production with transportation, refining genetic algorithm crossover techniques for TSP, developing

M.-A. Ejjanfi (✉) · J. Benhra
LARILE, ENSEM, University Hassan II of Casablanca, Casablanca, Morocco
e-mail: ejjanfimed@gmail.com

B. El Bhiri et al. (eds.), *AI and Digital Transformation: Innovations in Supply Chain, Education, and Energy Systems*, Sustainable Artificial Intelligence-Powered Applications, https://doi.org/10.1007/978-3-031-86837-5_4

Particle Swarm Optimization (PSO) variants for specific routing issues such as the Vehicle Routing Problem with Time Windows (VRPTW) and Time-Dependent Vehicle Routing Problem (TDVRP), and utilizing drones for package delivery, innovative and effective solutions are crafted. These solutions meet both current and future needs in logistics, distribution, and transportation for businesses and smart cities. The following details highlight significant research in this field.

A study focuses on integrating production task planning with vehicle routing optimization using a hybrid algorithm combining the Ant Colony System and local search methods, proving to be effective and efficient (Torres-Tapia et al. 2022).

Another research utilizes a genetic algorithm to enhance daily distribution routes for employees in a Portuguese firm, significantly reducing travel distances, fostering supplier and customer relationships, lowering fuel costs, and promoting environmental sustainability (Gomes et al. 2021).

A novel crossover method for solving the TSP via genetic algorithms is presented, outperforming previous techniques (Akter et al. 2019).

A new PSO variant designed for the VRPTW incorporates three adaptive strategies, demonstrating superiority over other PSO versions and leading algorithms on standard benchmark tests (Marinakis et al. 2019).

Research on TDVRP utilizing PSO is summarized, focusing on route optimization based on cost implications of travel time and distance (Diah et al. 2022).

The application of drones in delivery, combined with trucks, is explored through an Ant Colony Optimization algorithm for the Vehicle Routing Problem with Drones (VRPD), achieving near-optimal results on varied test instances (Huang et al. 2022).

Studies also investigate vehicle routing optimization without time constraints, critical for e-commerce logistics, comparing various algorithms like Intra-Route and Inter-Route Local Search, and Tabu Search, aiming to enhance last-mile delivery efficiency and sustainability (Veer Tiwari and Kumar Sharma 2023).

A routing model coordinating trucks and drones operations is introduced, minimizing total arrival times at depots through effective heuristic algorithms (Moshref-Javadi et al. 2021).

A new TSP with time windows for online purchase deliveries is proposed, focusing on customer pickups at shared facilities, with results supporting the competitiveness of the developed heuristic (Jiang and Dhiaf 2019).

This summary encapsulates the advanced methodologies and results from recent research in flexible routing optimization for the Vehicle Routing Problem (VRP), illustrating diverse approaches to solving complex logistical, distributional, and urban challenges. Furthermore, a comprehensive analysis and critical evaluation are applied across multiple VRP dimensions, introducing a new variant named 'VRP with Real-Time Customer Relocation and Smart Locker Adjustments'. This enhancement incorporates realistic considerations, such as smart locker adjustments, refining the approach to addressing real-life logistic challenges.

3 Methods

In our exploration of routing optimization, we focused on the Vehicle Routing Problem (VRP) and its extensions that incorporate both delivery and pickup operations, employing a range of population-based algorithms to address the challenges and additional constraints presented by this complex logistics scenario. These extensions included not only the standard VRP but also a novel adaptation involving 'smart lockers', which play a critical role in flexible delivery and pickup systems in urban environments.

For the standard VRP, which involves managing vehicle capacity, pickup and delivery points, and varying demands, we utilized Genetic Algorithms (GA), Ant Colony Optimization (ACO), and Particle Swarm Optimization (PSO). These population-based algorithms were selected for their effectiveness in handling large search spaces and optimizing routes while considering multiple operational constraints.

The VRP was further extended to accommodate the innovative use of 'smart lockers'. This adaptation introduced a new layer of complexity, necessitating real-time route modifications and the strategic placement of lockers within the delivery network. For this extended scenario, we applied GA, ACO, and PSO to effectively manage the dynamic and evolving requirements of logistics, highlighting the enhanced flexibility and reliability of our solutions.

The objective behind employing these algorithms across various VRP scenarios was to develop scalable solutions that are adaptable to the intricate and shifting demands of urban logistics. By focusing on these variations of the VRP, we aimed to deepen our understanding of route optimization amidst the growing challenges of delivery and pickup operations.

3.1 A Subsection Sample Data Preparation

In our investigation, we focused on comparing the performance of three distinct algorithms—Genetic Algorithm (GA), Ant Colony Optimization (ACO), and Particle Swarm Optimization (PSO) in solving Vehicle Routing Problem (VRP) scenarios within the city of Casablanca. Casablanca, a major urban center in Morocco, presents unique challenges due to its traffic conditions and urban density, making the resolution of VRP especially pertinent and challenging in this context.

Fig. 1 Spatial Distribution of parcel lockers: a geographic representation

The goal of this comparison was to ascertain the effectiveness and efficiency of each algorithm in terms of the quality of solutions achieved, computation time required, and adaptability under varying problem conditions. We utilized a dataset comprised of 40 points in Casablanca, loaded from a CSV file, to represent the complexities and realities of logistics and routing within a densely populated urban setting.

This analysis allowed us to highlight the strengths and weaknesses of each method, contributing to the identification of the most suitable approaches to solve VRP in similar urban environments. As such, we provided valuable insights into the comparative performance of each algorithm, thereby aiding in the decision-making process for routing optimization in similar contexts (Fig. 1).

3.2 Optimization Algorithms

Genetic Algorithms (GA): is a search heuristic modeled after natural selection. This iterative method enhances solutions by mimicking biological evolution. In the Genetic Algorithm, each potential solution is depicted as a chromosome within a population. These solutions experience operations like genetic mutation and crossover (or reproduction), producing new possible solutions. In our study, we applied the Genetic Algorithm to refine last-mile delivery routes. This approach proved beneficial due to its capacity to navigate extensive search spaces and deliver nearly optimal solutions, even with the problem's inherent complexity (Baker and Ayechew 2003).

The pseudocode can be seen below:

```
Initialize a population of individuals
While not termination condition do
        Calculate fitness for each individual in the population
        Select individuals for crossover
        Perform crossover and create offspring
        Perform mutation on offspring
        Select individuals for the next generation
Replace the current population with the new generation
End while
```

Selection: individuals are selected from the population for reproduction. this is often done probabilistically, where the probability of selection is based on the fitness of an individual.

Crossover: Two individuals (parents) are combined to form one or more offspring. This is often done by swapping parts of the parents' representations. A simple representation could be a binary string, in which case the crossover could be a single point crossover, where the string is cut at a point, and the ends are swapped.

Mutation: small random changes are made to an individual. in the case of a binary string representation, this could be flipping a bit from 0 to 1 or vice versa.

Ant Colony Optimization (ACO): is a probabilistic technique for solving optimization problems that mimics the foraging behavior of ants. In the algorithm, ants traverse a graph that represents the problem space, depositing and following pheromone trails that help guide the colony toward promising solutions. We utilized ACO to enhance delivery route optimization in our study, taking advantage of its ability to identify the most efficient paths through a swarm intelligence-based search method. This approach was particularly effective in adapting to dynamic changes within the problem environment, reflecting the real-world variability of last-mile deliveries (John 2004).

The pseudocode can be seen below:

```
Initialize pheromone trails
While not termination condition do
     Each ant constructs a solution
     Update best solution (if current solution is better)
     Evaporate pheromone trails
     Deposit pheromone on the trail according to the quality of the best solution
End while
```

Pheromone Update:

$$P_{ij} = \frac{[\tau_{ij}]^{\alpha}.[\eta_{ij}]^{\beta}}{\sum [\tau_{ik}]^{\alpha}.[\eta_{ik}]^{\beta}} \quad (1)$$

$$\tau_{ij} = (1-\rho)\tau_{ij} + \sum \Delta\tau_{ij}^{k} \quad (2)$$

Pij: The probability of moving from node i to node j.

τij: The intensity of the pheromone trail from node i to node j.

ηij: The heuristic information from node i to node j.

ρ: The pheromone evaporation rate

α: Parameter to control the influence of the pheromone trail.

β: Parameter to control the influence of the heuristic information.

$\Delta\tau_{ij}^{k}$: Pheromone deposited by kth ant.

Particle Swarm Optimization (PSO): is a computational strategy designed to refine solutions iteratively based on a specific measure of quality, drawing inspiration from the social behaviors observed in bird flocking or fish schooling. In PSO, each particle, representing a potential solution, modifies its trajectory based on its own experience as well as that of its peers. In our research, we implemented PSO to enhance the efficiency of delivery routes. The strength of PSO lies in its straightforward approach and its swift convergence toward an optimal or near-optimal solution. In our problem scenario, the algorithm demonstrated outstanding efficacy, consistently delivering favorable solutions within practical timeframes (Ai 2009).

The pseudocode can be seen below:

```
Initialize a swarm of particles with random positions and velocities
While not termination condition do
    For each particle in the swarm
        Calculate fitness of the particle
        If the fitness is better than the best fitness of the particle, update the best fitness
        If the fitness is better than the best fitness of the swarm, update the best fitness of the swarm
    End for
    For each particle in the swarm
        Update velocity using the best fitness of the particle and the best fitness of the swarm
        Update position using the new velocity
    End for
End while
```

Velocity Update:

$$v_i(t+1) = Wv_i(t) + c_1 r_1 \left(X_i^{best}(t) - x_i(t)\right) + c_2 r_2 (G_{best}(t) - x_i(t)) \quad (3)$$

Position Update:

$$v_i(t+1) = Wv_i(t) + c_1 r_1 \left(X_i^{best}(t) - x_i(t)\right) + c_2 r_2 (G_{best}(t) - x_i(t)) \quad (4)$$

X_i^{best}: Personal best of the particle number i at iteration number t.

v(t) : Velocity of the particle number i at iteration number t.

x(t) : Position of the particle number i at iteration number t.

Gbes : Global best

r1 , r2 : Random numbers between 0 and 1 open

c_1 : Cognitive coefficient

c_2 : Social coefficient

: Inertia weight

3.3 Deployment of Methods

In this research, we addressed three distinct logistical problems, each involving the implementation of different optimization algorithms to find the most efficient solutions. The algorithms chosen for this study included Genetic Algorithm (GA), Ant Colony Optimization (ACO) and Particle Swarm Optimization (PSO).

Vehicle Routing Problem (VRP): is a complex optimization challenge that involves designing optimal routes for a fleet of vehicles operating from a central depot. The primary goal is to service several dispersed customers and return to the depot, while minimizing the total travel distance or time. This problem integrates various constraints, such as vehicle capacities and customer demands. Each of the three algorithms (GA, ACO and PSO) was adapted to handle these additional constraints, demonstrating their flexibility and effectiveness for solving complex logistical problems.

$$V = |C| \tag{5}$$

$$N = |D| \tag{6}$$

$$\sum D_i \leq \sum C_i \tag{7}$$

$$0 \leq d \leq N \tag{8}$$

V = The number of vehicles must be equal to the number of capacities:

C = set of Capacities

D = set of Demands

D = Depot Index

N = number of Points

VRP with Real-Time Customer Relocation and Smart Locker Adjustments: This variant of the VRP includes the novel challenge of accommodating sudden changes in customer locations relative to "smart lockers". In this scenario, customers might move away from their designated pickup lockers to new locations near different lockers after the delivery routes have been planned. This necessitates real-time adjustments to the routing algorithms to manage the pickup and delivery of parcels. The solution involves identifying which vehicle can most efficiently reroute to the new smart locker, ensuring the parcel is transferred and delivered to where the customer is currently available. This approach highlights the flexibility and responsiveness of the VRP algorithms to dynamically adapt to changing customer positions and availability, mirroring the complexities and demands of modern logistical operations.

$$min \sum_{i,j \in Set_1} D_{ij}.X_{ij} + \sum_{i \in Set_1} D_{iR_t}.X_{iR_t} + \sum_{j \in Set_1} D_{R_t i}.X_{R_t i} \tag{9}$$

$$\sum_{i \in Set_2} d_i + d_U \leq d_{V\,Set_2} \tag{10}$$

$$\sum_{j \in Set_1 \setminus \{c_{11}\}} X_{S_{11} j} = 1 \tag{11}$$

$$\sum_{j \in Set_1 \setminus \{c_{1n}\}} X_{j S_{1n}} = 1 \tag{12}$$

$$S_{Ui} + T_{ij} + Temps\ de\ Service \leq S_{R_t j} \tag{13}$$

$$C_i : Coordonné\ es\ des\ clients.$$

$$D_{\{ij\}} : Distance\ entre\ le\ client\ i\ et\ le\ client\ j.$$

$$T_{\{ij\}} : Temps\ de\ trajet\ entre\ le\ client\ i\ et\ le\ clientj.$$

$$d_i : Demande\ de\ chaque\ client.$$

$$U : ID\ du\ client\ devenu\ indisponible.$$

$$R_T : ID\ du\ client\ devant\ recevoir\ la\ demande\ du\ client\ U.$$

$$N_T : Moment\ de\ la\ notification\ del'indisponibilité.$$

$$Set_1 = \{c_{1i} \mid c_{1i}\ nest\ pas\ livre\ avant\ N_T\}$$

$$Set_2 = \{c_{2i} \mid c_{2i}\ peutaccueillirlademandesupplémentairedeU\}$$

4 Results and Discussion

The system configuration includes a Ryzen 9 7945X CPU, an NVIDIA GeForce RTX 4060 with 8 GB VRAM, and 32 GB of RAM. Visualization of the route was achieved using HTML. The application was developed using PyQt5, and the heuristic was programmed in Python. This heuristic was executed thirty times per instance, with the average outcome presented for analysis.

In this research, we aimed to evaluate the efficiency of various optimization algorithms for solving the Vehicle Routing Problem (VRP). We focused on two main performance indicators: the total tour distance and the algorithm's execution time.

Total Tour Distance: For the VRP, this metric represents the cumulative distance of all the routes taken by each vehicle. We derived these distances from a distance matrix, created using the Bing Maps API to ensure accurate measurements between locations. A lower total tour distance indicates a more efficient solution, reflecting the effectiveness of the algorithm.

Total Tour Distance for VRP:

$$(\text{TTD}) = \sum \sum d(P_i, P_{i+1}) \tag{14}$$

Execution Time: This measures the duration required for the algorithm to generate a solution. It is an important gauge of the algorithm's efficiency, where shorter execution times are preferred. This metric is particularly crucial in real-world scenarios where prompt solutions are vital.

$$(ET) = T_{end} - T_{start} \tag{15}$$

5 Results

In the Vehicle Routing Problem (VRP), four heuristic optimization algorithms were employed: Genetic Algorithm (GA), Ant Colony Optimization (ACO), and Particle Swarm Optimization (PSO). Each vehicle had a consistent capacity across all algorithms. The same demand and depot parameters were used for each algorithm to ensure a fair comparison. Additionally, for VRP with Real-Time Customer Relocation and Smart Locker Adjustments, the unavailable smart locker was number 3, and the reallocation locker was number 24; the notification time was 25 min after the vehicles were dispatched.

The Genetic Algorithm was configured with 100,000 generations, a population size of 100, a mutation rate of 0.1, and a crossover rate of 0.9. The GA took significant time (301.10 s) to solve the VRP and converged at the 80,000th iteration. It found a solution with a total distance of 58.21 km (as shown in Fig. 2). Routes are replanned as depicted in Fig. 3, resulting in a total distance of 59.69 km.

The Ant Colony Optimization algorithm was set with 2000 iterations, 40 ants, and parameters alpha at 0.1, beta at 1, and rho at 0.5. The ACO finished in 226.68 s and converged after 420 iterations, securing a shorter route of 50.23 km (Fig. 4).

Fig. 2 GA route for VRP

Fig. 3 GA after rerouting

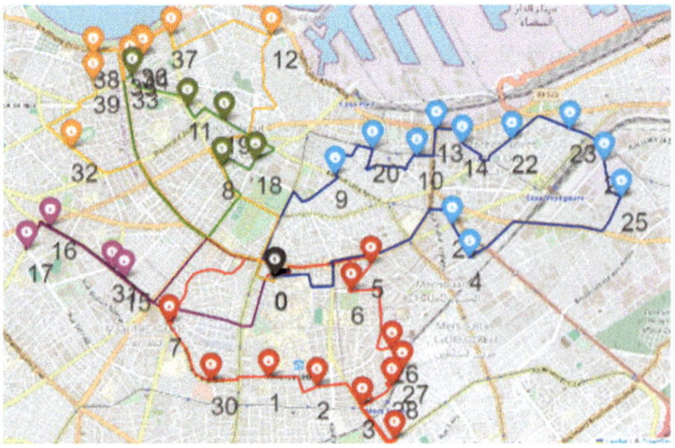

Fig. 4 ACO route for VRP

Fig. 5 ACO after rerouting

As shown in Fig. 5, the routes have been replanned, increasing the total distance to 51.39 km.

Particle Swarm Optimization was configured with 2000 iterations, 40 particles, c1 (personal influence) at 0.1, c2 (social influence) at 0.9, and w (inertia weight) at 0.1. It completed in only 14.55 s, with a total distance of 60.23 km and convergence at 1650 iterations (Fig. 6). The routes have been replanned as illustrated in Fig. 5, leading to a total distance of 63.65 km.

Computational Complexity

Genetic Algorithm (GA): Described as computationally intensive, GA requires managing large populations and numerous generations. The complexity primarily arises from the need to repeatedly evaluate fitness across the population to achieve convergence, a challenge amplified by the urban setting of Casablanca.

Ant Colony Optimization (ACO): ACO's complexity is due to the need to manage multiple ants, each independently constructing routes. The complexity increases with the necessity to update pheromone trails based on solution quality and

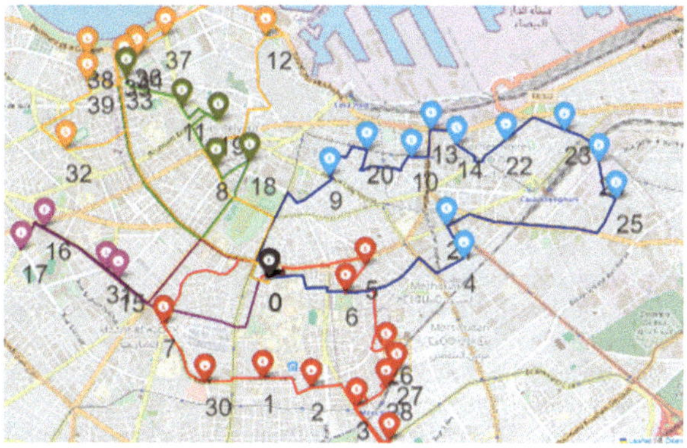

Fig. 6 PSO route for VRP

Fig. 7 PSO after rerouting

adapt to dynamically changing environments typical of urban delivery routes.

Particle Swarm Optimization (PSO): While PSO involves fewer adjustable parameters than GA and ACO, it still demands substantial computational resources to update particle positions based on personal and global bests. This demand escalates with larger problem scales and the additional complexities of real-time route adjustments and smart lockers.

Performance in Different VRP Scenarios:

Standard VRP:

GA: Demonstrated high effectiveness in managing complex route planning with robust solutions, optimizing within a large problem space populated with 40 points in Casablanca.

ACO: Shown to excel in scenarios requiring a good balance between exploration and exploitation, particularly effective in routing with multiple constraints due to its utilization of pheromone trails.

PSO: Noted for its fast convergence and simpler implementation, making it ideal for scenarios with less complex solution landscapes but requiring rapid decision-making.

Dynamic VRP with Real-Time Adjustments:

GA: Adapted well to dynamic changes but was slower to converge, indicating a potential limitation in scenarios requiring quick adaptations to change.

ACO: Its quick adaptation to changes in the routing landscape, driven by real-time updates of pheromone trails, made it ideal for environments where route conditions change frequently.

Table 1 Performance of optimization algorithms VRP and VRP with constraints

Algorithm	VRP Score	VRP with Smart Locker Adjustments Score	VRP with transferring demand directly
GA	58.21 km	59.69 km	61.23 km
ACO	50.23 km	51.39 km	53.68 km
PSO	60.23 km	63.65 km	66.05 km

PSO: Exhibited rapid response capabilities to dynamic changes, which is crucial in urban logistics where delivery conditions can shift unexpectedly (Fig. 7, Table 1).

6 Conclusion

Discussion: Broader Implications and Potential Impact on the Logistics Industry Efficiency and Cost Reduction: The application of advanced algorithms like GA, ACO, and PSO in solving complex Vehicle Routing Problems (VRP) has direct implications for improving the efficiency of logistics operations. By optimizing route planning, these algorithms help reduce fuel consumption, decrease delivery times, and lower overall operational costs. For the logistics industry, which operates on thin margins and faces intense competition, these savings can be significant, impacting the bottom line positively.

Adaptability to Urban Challenges: The integration of smart lockers and real-time rerouting capabilities as demonstrated in the study highlights the potential for modern logistics systems to adapt to urban challenges such as traffic congestion and varied customer availability. This adaptability not only enhances service reliability but also improves customer satisfaction by offering more flexible and convenient delivery options.

Scalability and Future Technologies: The study's findings suggest that these algorithms can be scaled up to handle larger and more complex logistic networks, potentially integrating with emerging technologies such as autonomous delivery vehicles and IoT-enabled infrastructure. This could lead to more autonomous logistics systems where decisions on routing and deliveries are made in real-time, with minimal human intervention.

Environmental Impact: Optimizing routes and incorporating flexible delivery options like smart lockers can significantly reduce the carbon footprint of delivery operations. By minimizing unnecessary travel, reducing congestion, and optimizing vehicle loads, the logistics industry can contribute to broader environmental sustainability goals.

Summary of Findings and Future Research Avenues

Summary of Findings:

The comparative analysis of GA, ACO, and PSO in the context of VRP with smart lockers indicates that while each algorithm has its strengths and limitations, they all contribute effectively to solving complex logistical problems. ACO's capability to adapt quickly to dynamic environments, GA's robustness in exploring large solution spaces, and PSO's fast convergence make them valuable tools in the logistics industry's ongoing efforts to enhance efficiency and responsiveness.

Suggestions for Future Research:

Real-World Testing: Implementing these algorithms in real-world scenarios to test their practical applicability and effectiveness would be a crucial next step. Such studies can help fine-tune the algorithms to meet the specific needs of different geographical and operational contexts.

Advanced Machine Learning Techniques: Exploring machine learning techniques to predict changes in delivery routes and customer availability could further enhance the responsiveness of VRP solutions. Leveraging predictive analytics can preemptively adjust routes even before changes occur, thereby improving efficiency.

Blockchain for Enhanced Security and Traceability: We plan to enhance both security and traceability aspects through blockchain methodology. This initiative has already commenced in our laboratory, aiming to leverage blockchain's inherent strengths in security and data integrity to further fortify our VRP solutions (Nadime et al. 2023).

References

Akhtar, M., Hannan, M.A., Basri, H.: Particle swarm optimization modeling for solid waste collection problem with constraints. In: 2015 IEEE International Conference on Smart Instrumentation, Measurement and Applications, ICSIMA, 7559025, pp: 1–4 (2016)

Akter, S., Nahar, N., ShahadatHossain, M., and Andersson, K.: A new crossover technique to improve genetic algorithm and its application to TSP. In: International Conference on Electrical, Computer and Communication Engineering (ECCE), pp. 1–6 (2019)

Barrie M. Baker, M.A.: Ayechew: A genetic algorithm for the vehicle routing problem (2003)

Chunhua, F., Lijun, Z., Xiaojing, W., Liying, Q.: Solving TSP problem with improved genetic algorithm. AIP Conf. Proc. **1967**, 040057 (2018)

Diah, M., Setyanto, A., Taufiq Luthfi, E.: Systematic literature review of particle swarm optimization implementation for time-dependent vehicle routing problem. JOIN (J. Online Inform.) **7**(1), 38–45 (2022)

Ejjanfi, M-A., Benhra, J.: Flexible routing and optimization for the last mile: an approach based on smart lockers and heuristic techniques, (2024). sciencesconf.org:ola2024:516413

Ejjanfi, M-A., Benhra, J., El Kassimi, E-A, H.: Contribution to forecasting demand and occupancy for smart lockers in the context of Last-Mile Delivery, sciencesconf.org:ola2024:516414 (2024)

Gomes, E ., Iglésias, D., Proença, M.I.D., Lima, A.P., Gaspar, T.M.: PD Applying a genetic algorithm to a m-TSP: case study of a decision support system for optimizing a beverage logistics vehicles routing problem. Electron., **10**(18), pp.2298 (2021)

Huang, S.H., Huang, Y.H., Blazquez, C.A., Chen, C.Y.: Solving the vehicle routing problem with drone for delivery services using an ant colony optimization algorithm. Adv. Eng. Inform. **51**, 101536 (2022)

Jiang, L., Dhiaf, M.M.: A traveling salesman problem with time windows for the last mile delivery in online shopping. Int. J. Prod. Res. **58**(4), 1–12 (2019)

John E. Bell, Patrick R. McMullen.: Ant colony optimization techniques for the vehicle routing problem (2004)

Marinakis, Y., Marinaki, M., Migdalas, A.: A multi-adaptive particle swarm optimization for the vehicle routing problem with time windows. Inf. Sci., **481**, pp. 311–329 (2019)

Moshref-Javadi, M., Hemmati, A., Winkenbach, M.: A comparative analysis of synchronized truck-and-drone delivery models. Comput. Ind. Eng. **162**, 107648 (2021)

Nadime, K. L., Benhra, J., Benabbou, R., Mouatassim, S.: Automating attended home deliveries with smart contracts: a blockchain-based solution for E-commerce logistics. E3S Web of Conferences, 469, 00026 (2023)

Negulescu, S.C., Kifor, C.V., Oprean, C.: Ant colony solving multiple constrains problem: vehicle route allocation. Int. J. Comput., Commun. Control. **3**(4), 366–373 (2008)

The Jin Ai, Voratas Kachitvichyanukul: A particle swarm optimization for the vehicle routing problem with simultaneous pickup and delivery (2009)

Torres-Tapia, W., Montoya-Torres, J., Ruiz-Meza, J., Belmokhtar-Berraf, S.: A matheuristic based on ant colony system for the combined flexible jobshop scheduling and vehicle routing problem. IFAC **55**(10), 1613–1618 (2022)

Veer Tiwari, K., Kumar Sharma, S.: An optimization model for vehicle routing problem in last-mile delivery. Expert Syst. Appl. **222**, 119789 (2023)

Understanding Customer Experiences in the Logistics Sector: Insights from Individual Cargo Companies

İbrahim Çelik, Emrah Aydemir, Samet Güner, and Halil İbrahim Cebeci

Abstract

In recent years, e-commerce platforms offering consumers a wide range of products and various price options have become widely used due to the convenience they provide. The widespread use of e-commerce platforms has positioned logistics companies as significant stakeholders in e-commerce, leading to the emergence of new approaches in logistics services. Both sectors, focusing on customer satisfaction, strive to develop various strategies to enhance service quality, acquire new customers, and encourage repeat purchases. A fast, reliable, and seamless delivery process directly impacts customer satisfaction and loyalty, while delivery issues can lead to customer dissatisfaction and decreased trust in the brand. In this context, consumer experiences shared on various platforms regarding the delivery process emerge as a valuable source of feedback for logistics companies to enhance customer satisfaction. On the other hand, analyzing these reviews through data-driven methods such as text mining and sentiment analysis provides crucial insights for improving the operational processes of logistics companies. In this context, the study aims to analyze customers' views regarding individual cargo companies, focusing on the shortcomings and demands in the delivery process. To this end, user reviews from a widely used online platform containing real user feedback were analyzed using sentiment analysis and machine learning methods. The findings indicate that 82% of user reviews related to cargo companies are negative, with 66% of these reviews focusing on delivery issues such as failed deliveries, delayed deliveries, and not delivered in hand. The study's findings reveal that cargo companies have not been sufficiently successful in their delivery processes and communication with customers.

Keywords

Cargo companies · Logistics issues · Customer experiences · Customer satisfaction · Sentiment analysis · Machine learning

1 Introduction

Recent technological advancements, particularly the widespread use of the internet, have led to significant changes in both social and commercial spheres. One of the most notable developments in the commercial sector is the rapid popularity of electronic commerce (e-commerce) platforms, where individuals can simultaneously view numerous product and price options and easily make purchases. The increased demand for products through e-commerce, coupled with diversified transportation options (such as increased highways and bridges, air transportation facilities, the emergence of new terminals, etc.), has inevitably brought about a transformation in the logistics sector (Hesse 2002). All these developments have, on one hand, increased companies' profitability, while on the other hand, especially raised logistics costs, which are an important stakeholder alongside e-commerce. This situation has necessitated the establishment of new strategies in the logistics sector.

The delivery process, one of the stages of e-commerce, is a crucial factor influencing e-commerce (Bhattacharjya et al. 2016; Vasić et al. 2020). Furthermore, issues related to

This work was supported by Scientific Research Projects Coordinator of Sakarya University under grant number 2024-25-63-43.

İ. Çelik · E. Aydemir (✉) · H. İ. Cebeci
Management Information Systems, Sakarya University, Sakarya, Türkiye
e-mail: emrahaydemir@sakarya.edu.tr

S. Güner
Business, Sakarya University, Sakarya, Türkiye

B. El Bhiri et al. (eds.), *AI and Digital Transformation: Innovations in Supply Chain, Education, and Energy Systems*, Sustainable Artificial Intelligence-Powered Applications, https://doi.org/10.1007/978-3-031-86837-5_5

delivery, such as the consumer's ability to choose the shipping company and the delivery of products undamaged, on time, at a reasonable price, or for free, are closely related to customer satisfaction (Esper et al. 2003; Otim and Grover 2017; Fanani 2023). Therefore, it can be argued that addressing issues related to this process has become crucial for firms aiming to increase customer satisfaction and consequently enhance profitability. Particularly in today's context, the proliferation of platforms where consumers share their positive or negative experiences with the ordering process enables companies to analyze these issues. These shared experiences play a significant role in influencing the preferences of other users (Nadeem et al. 2015), highlighting the necessity for firms to generate solutions to these problems. Based on this premise, the study aims to examine user comments on cargo companies, with a focus on individual cargo transportation, on the Ekşi Sözlük website, one of Turkey's largest internet communities. This examination will employ sentiment analysis methods and utilize machine learning algorithms for classification, thereby aiming to identify both the glitches and demands experienced in the delivery process.

In studies focusing on individual cargo transportation, it has been observed that the aim is often to measure the service quality of cargo companies using various scales, including the SERVQUAL scale, through quantitative methods. In one of these studies, Deniz and Gödekmerdan (2011) aimed to determine customers' thoughts about cargo companies and to identify the problems they encounter while benefiting from these companies' services. Various analyses conducted with data obtained through survey techniques revealed that employees in the cargo company were experienced enough, they did not delay in solving problems, they delivered products in good condition, and they greeted customers correctly. However, customers perceived delays in delivery as a problem. In the study conducted by Atmaca and Turğut (2015), an attempt was made to determine the criteria that customers consider when choosing cargo companies. Seventeen criteria, created based on relevant literature, were first subjected to Chi-square analysis, and then the strength of the relationship between cargo companies and these criteria was examined using Cramer's V coefficient. The analysis revealed a moderate significant relationship between the included companies and the criteria of "Service Price," "Emergency Management and Problem-Solving Skills," and "Company's Economic History, Financial Stability.“ Additionally, the study identified Yurtiçi Cargo, PTT Cargo, and MNG Cargo as the most effective cargo companies, respectively. In another study, Sungur (Sungur 2018) aimed to analyze the impact of cargo companies' service quality on satisfaction. Exploratory Factor Analysis (EFA) and Regression Analysis were applied to the data obtained from participant surveys. The analyses revealed a significant relationship between the reliability of cargo companies and their ability to empathize with satisfaction. Yıldız et al. (2018), on the other hand, focused on the relationship between the service quality customers expect and perceive. Both EFA and Confirmatory Factor Analysis were applied to the survey results, followed by hypothesis testing using Structural Equation Modeling. According to the findings, the perceived service quality of cargo companies was lower than expected. However, factors such as empathy, reliability, responsiveness, and assurance positively and significantly influenced satisfaction. Finally, in Parmaksız's study (2022), non-parametric tests were conducted to determine if there were significant differences in the relationship between cargo companies' service quality, customers' expected and perceived service quality, and certain demographic characteristics. The findings indicated significant differences between the sub-dimensions of cargo companies' service quality and perceived and expected service quality.

2 Methodology

In today's world, the internet contains vast amounts of data for everyone. This dynamic data is constantly increasing. Various methods are employed to make purposeful inferences from the increasing large amounts of unstructured data. Data mining can be defined as the process of discovering meaningful relationships in this large volume of data using computer assistance to make various predictions (Savaş et al. 2012). Text mining, on the other hand, is an interdisciplinary field that encompasses areas such as data mining, machine learning, and statistics. In text mining, data is often obtained from articles, books, emails, blogs, and website content (Han et al. 2012). Text mining can be used for purposes such as text classification, summarization, sentiment analysis, social network analysis, and author identification (Kumar and Ravi 2016). In this context, in the study, data obtained from a website using web scraping techniques has been subjected to analysis after passing through the stages depicted in Fig. 1.

Fig. 1 Data analysis process

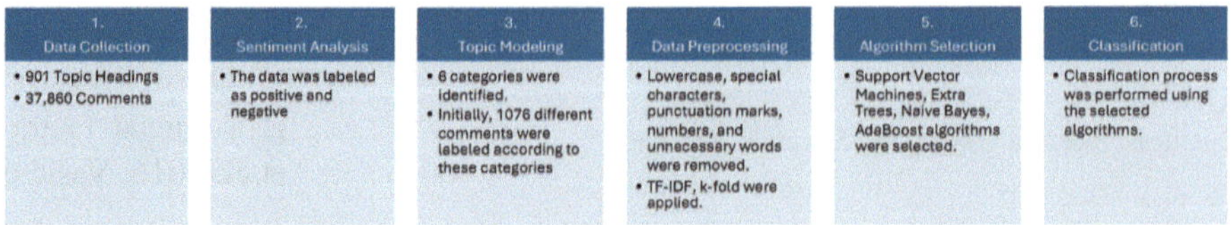

On July 23, 2023, in the data collection phase, firstly, topic titles related to widely used cargo companies in Turkey such as Aras Cargo, Mng Cargo, PTT Cargo, Sürat Cargo, UPS Cargo, and Yurtiçi Cargo were searched on the "Ekşi Sözlük" website. As a result of the research, 901 unique topic titles opened since 1999 were obtained. Subsequently, user comments related to these topic titles were scanned. Using the "Beautiful Soup" library, which is commonly used in web scraping processes and is included in the Python programming language, a total of 37,905 comments were extracted. This library is a toolset that facilitates retrieving information from HTML codes of web pages by navigating between HTML tags to extract the desired information (Crummy. Beautiful Soup 2023). After removing empty and erroneous records, 900 unique topic titles and 37,886 comments to be included in the analysis were obtained. After the comments were extracted, it was checked whether there were duplicate records, and it was found that comments in topic titles containing multiple cargo company names were duplicates. Thus, 45 duplicate comments were deleted, and the comments were gathered under a single topic title. As a result, a total of 37,841 unique comments were obtained.

After the data collection phase was completed, sentiment analysis was conducted to determine whether user comments were positive or negative. Sentiment analysis, also known as opinion mining, classifies any text into expressing a positive, negative, or neutral opinion or thought (Medhat et al. 2014; Hoang et al. 2019). In this study, sentiment analysis was conducted using the BERT (Bidirectional Encoder Representations From Transformers) model, which is a language model that takes into account the context of a word by considering the words that precede and follow it, unlike traditional methods (Devlin and BERT: Pre-training of Deep Bidirectional Transformers for Language Understanding 2019). This method allows researchers to use a pre-trained model in their analysis. Since the research data is in Turkish, the "Bert-base Turkish Sentiment Cased" model, which is based on BERT and has been previously trained and shown high performance, was used in this study (Yıldırım 2022). With this model, the research data was labeled based on whether the sentiment was positive or negative.

This stage was followed by examining the sentiment analysis labeled data and, considering the findings of other studies in the literature (Deniz and Gödekmerdan 2011; Yanar Gürce and Tosun 2017; Burcuoğlu and Erdoğan Yazar 2020; Bayır 2022; Yıldız 2020; Turğut 2016; Topuk 2020; Çullu and Okursoy 2023), it was decided to categorize the comments into the categories in Table 1. This stage constitutes the first step of topic modeling. Topic modeling is the process of labeling texts obtained from a source into predefined categories according to certain methods (Cebeci et al. 2023).

In the second part of the topic modeling process, the obtained comments are labeled according to categories. In this regard, a total of 1,076 different comments were initially labeled according to categories for both the training and test datasets. A comment is encoded as "1" if it belongs to a specific category, and "0" if it does not. A comment can belong to multiple categories. Therefore, as a result of this process, the total number of comments encoded as "1" in all categories was determined to be 1,294 (Table 2).

Table 1 Categories

Cat. no	General category	Subcategory
1	Failed delivery	Comments related to damage, loss, return to seller for various reasons
2	Delayed delivery	Comments about not delivering on time (delayed pick-up/delivery for various reasons)
3	Not delivered in hand	Comments about not delivering to the individual (delivery to someone else, leaving at a different location, customer picking up from the branch)
4	Customer service	Comments about customer service or branch phone calls
5	Corporate issues	Evaluation of corporate image, discourse, advertisement, general brand or system evaluation, comparison, personnel, customer satisfaction
6	Fee-related Iissues	Comments about cargo fees

Table 2 Number of labeled comments

Category	Label count
1st Category	258
2nd Category	235
3rd Category	201
4th Category	200
5th Category	200
6th Category	200
Total	1,294

The study utilized supervised learning algorithms, one of the types of machine learning. Supervised learning involves learning the relationship between a set of input variables and a set of output variables, and using this learned relationship to predict outputs for uncertain sequences (Cunningham et al. 2008). In this method, various preprocessing steps are required to enhance prediction accuracy before analyzing the data. Accordingly, all text expressions were initially converted to lowercase, special characters, punctuation marks, and numbers were removed, and irrelevant words were deleted.

To be able to provide text expressions as input to supervised learning algorithms, they need to be converted into numerical values (Tripathy et al. 2016). For this purpose, the Term Frequency-Inverse Document Frequency (TF-IDF)

Table 3 Example confusion matrix

	Predicted positive	Predicted negative
Actual positive	True Positive (TP)	False Negative (FN)
Actual negative	False Positive (FP)	True Negative (TN)

Table 4 Models and accuracy scores

Models	Accuracy scores					
	1	2	3	4	5	6
Random Forest	0,742	0,745	0,811	0,845	0,793	0,872
Extra Trees	0,746	0,789	0,856	0,878	0,790	0,860
Bagging	0,746	0,723	0,826	0,820	0,760	0,855
Isolation Forest	0,500	0,500	0,500	0,500	0,500	0,500
Logistic Regression	0,740	0,762	0,816	0,853	0,818	0,853
KNN	0,653	0,704	0,764	0,735	0,778	0,780
SVM	0,758	0,753	0,836	0,835	0,830	0,850
Decision Tree	0,725	0,721	0,831	0,810	0,695	0,860
Naive Bayes	0,684	0,721	0,808	0,785	0,845	0,853
Gradient Boosting	0,727	0,774	0,843	0,815	0,745	0,872
AdaBoost	0,715	0,691	0,838	0,823	0,760	0,878
XGBoost	0,725	0,740	0,816	0,800	0,755	0,847
Gaussian NB	0,611	0,621	0,614	0,683	0,710	0,705

method was utilized. TF-IDF is a numerical value indicating the importance of a word in a text sequence (Garreta and Moncecchi 2013). Following this process, the stage of sample selection for separating training and test data was initiated, and the "k-fold" cross-validation technique was preferred among sample selection methods. In this method, the dataset is divided into "k" partitions. These partitions are first used as training and then as test datasets, and predictions are made according to the specified model (Jung and Hu 2015).

In machine learning, one of the performance evaluation metrics used to evaluate models is the confusion matrix (Table 3). The confusion matrix compares predicted values (columns) with actual values (rows) (Beheshti 2022).

Some performance metrics can be calculated using the confusion matrix. The most common ones include accuracy, recall, precision, and the harmonic mean of precision and recall (F1-Score). These metrics are calculated based on the positive and negative values in Table 3 as follows:

$$\text{Accuracy} = \frac{TP + TN}{TP + FP + FN + TN} \tag{1}$$

$$\text{Recall} = \frac{TP}{TP + FN} \tag{2}$$

$$\text{Precision} = \frac{TP}{TP + FP} \tag{3}$$

$$\text{Precision} = 2 * \frac{\text{Precision} * \text{Recall}}{\text{Precision} + \text{Recall}} \tag{4}$$

The data for each category were trained separately with 13 different classification algorithms, and the accuracy scores presented in Table 4 were obtained. According to the results, Support Vector Machines (SVM) achieved higher accuracy scores for comments related to failed deliveries, Extra Trees for delayed deliveries, undelivered items, and customer service comments, Naive Bayes for corporate matters, and AdaBoost for fee-related comments compared to other algorithms.

The Support Vector Machines (SVM) algorithm used in the study draws a line to separate points belonging to different classes on a plane. It aims for the points to be at a maximum distance from this line. The space between the line and the points closest to it is called the "margin". A wider margin indicates better separation, leading to increased model success (Kotsiantis 2007). The Extra Trees algorithm is a variation of the Random Forest algorithm. In this algorithm, there is no bagging stage for decision trees as in the Random Forest algorithm. Instead, the entire training set is used to train all generated decision trees. In this method, extra trees randomly select the best feature and its corresponding value to split the nodes (Geurts et al. 2006). The Naive Bayes classifier is based on Bayes' Theorem. This algorithm makes various probability calculations to determine which class an instance belongs to and assigns it to the class with the highest probability. In real-life applications, it has been successful in many areas such as weather forecasting, credit scoring, and healthcare (Yang 2018). The AdaBoost Algorithm, on the other hand, is considered one of the best boosting algorithms due to its strong theoretical background and successful applications in practice. This algorithm can convert a weak learning algorithm into a strong one with better accuracy (Cao et al. 2013). After selecting these algorithms, which demonstrated high performance in terms of accuracy scores compared to other algorithms, other performance metrics of the algorithms were also checked (Table 5), and the data in the relevant categories were predicted according to these models.

Table 5 Other performance metrics of selected algorithms

Cat	Model	Conf. matrix	Accuracy	Recall	Precision	F1-Score
1	SVM	[176 82] [44 214]	0,758	0,833	0,729	0,771
2	Extra trees	[186 49] [48 187]	0,789	0,795	0,789	0,788
3	Extra trees	[168 33] [26 175]	0,856	0,867	0,837	0,851
4	Extra trees	[185 15] [34 166]	0,878	0,832	0,923	0,873
5	Naive bayes	[175 25] [36 164]	0,845	0,822	0,867	0,839
6	Ada boost	[186 14] [35 165]	0,878	0,827	0,923	0,870

3 Findings

3.1 Distribution of Topic Titles and Comments

On the Ekşi Sözlük website, there are 900 topic titles containing the names of the cargo companies included in the study, and the ratio of the number of titles opened about cargo companies to the total number of topic titles is shown in Fig. 2. According to this, the company with the most topic titles opened is the cargo company coded as "cargo6," with 283 titles opened about it (31%). The company closest to it is the cargo company coded as "cargo1," with 224 topic titles (25%). The company with the fewest topic titles opened is "cargo5," with 30 topic titles (3%). When Fig. 3 is examined, it can be seen that there is generally a similar ranking in the comments sent by users under these titles. The number of comments for the company coded as "cargo6" is 13,402, and its share in all comments is 35%. This is followed by "cargo1" with 7,033 comments and "cargo2" with 6,569 comments. When Figs. 2 and 3 are evaluated together, it can be seen that the ratios of comments made about cargo3 and cargo6 are higher compared to the ratio of opened titles, while this ratio is lower for cargo1 and cargo2. For cargo4 and cargo5, these ratios remain unchanged.

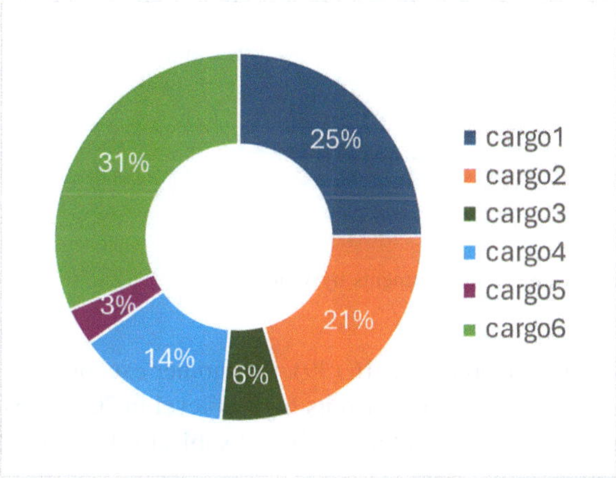

Fig. 2 Rate of opened titles

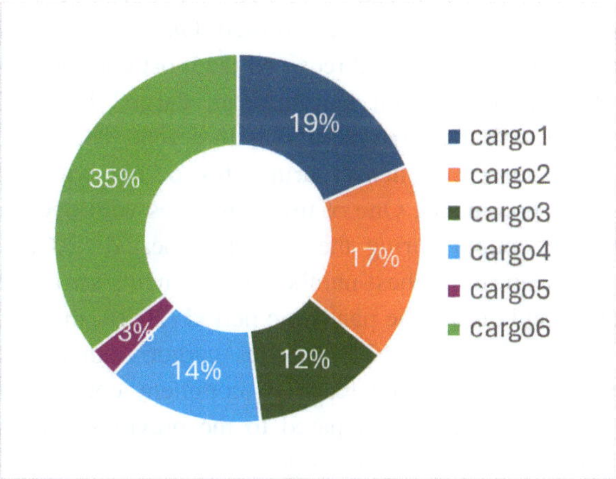

Fig. 3 Rate of comments

The subject titles opened by year about cargo companies are included in Fig. 4. When the aforementioned figure is generally evaluated; it is observed that the number of subject titles entered an increasing trend between 2013 and 2015, peaked in 2020, and decreased from 2021 onwards. From this point on, it can be said that the increase in the number of subject titles started from 2013, when the e-commerce sector began to grow, and reached its peak in 2020 parallel to the increasing e-commerce volume globally due to the Covid-19 pandemic (Özbek 2021). It can be stated that with the end of the pandemic, decreasing order volume, increasing transportation capacity, and development of logistic systems, there is also a decrease in the number of subject titles. During the pandemic period (in 2020), the total number of subject titles opened reached 226, and the most subject titles were opened about cargo2 (67). As an exception to this situation, it can be shown that the number of titles about cargo6 started to increase from 2011, peaked in 2016 (57), and had fewer titles

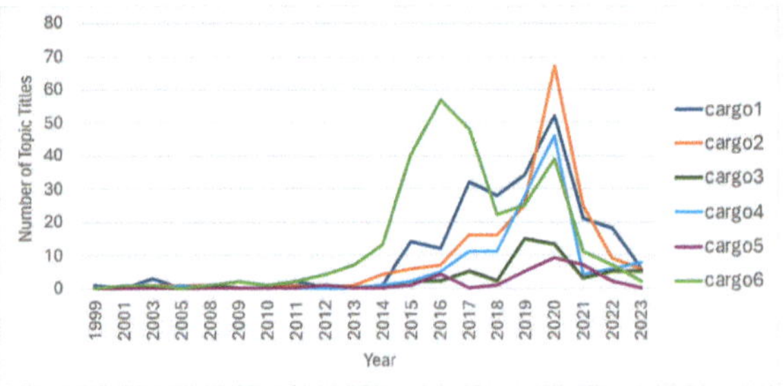

Fig. 4 Number of topic titles by year

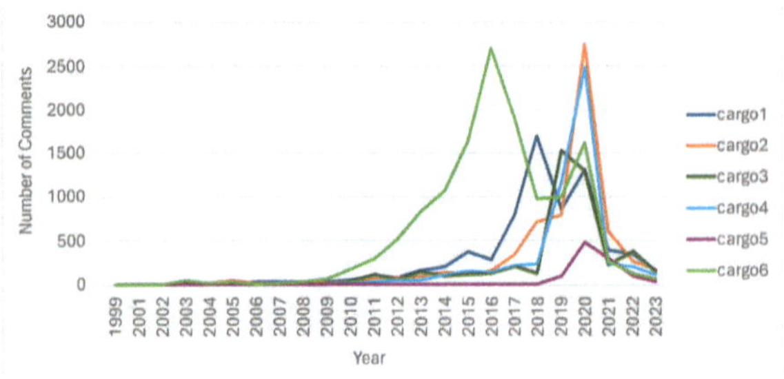

Fig. 5 Number of comments by year

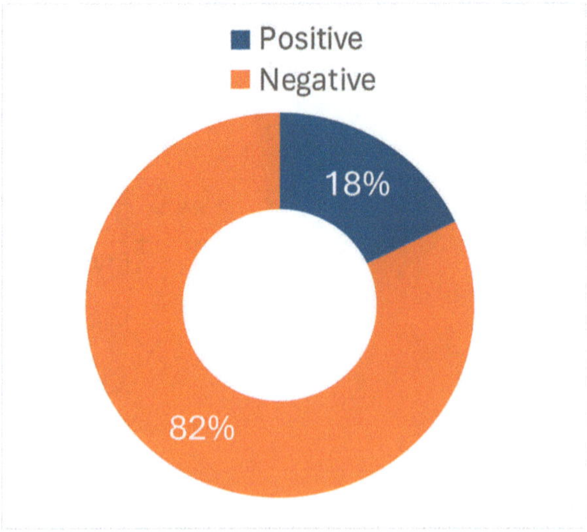

Fig. 6 Positive and negative ratios

in 2020 compared to 2016 (39). As another exception, it can be said that cargo3 reached its highest level in 2019 (15) and showed a decrease in the number of subject titles from 2020 onwards.

A similar trend stemming from the aforementioned reasons is also evident in the comment counts (Fig. 5). In the year with the highest number of comments, 2020, a total of 9,943 comments were posted. On a company basis, during this period, cargo2 received the most comments with 2,746 comments. With this number, cargo2 became the company with the most comments in a year. The company with the least comments during this period was cargo5 with 487 comments. One of the companies with decreasing comment counts during the pandemic period (2020) was cargo6, with the highest number of comments sent in 2016 (2,698), and during the pandemic period (2020), the number of comments was 1,621. Besides cargo6, other companies that showed a decrease in comment counts during the pandemic period compared to the previous period are observed to be cargo1 and cargo3.

3.2 Results of Sentiment Analysis

As a result of sentiment analysis, it was observed that 31,114 comments were coded as negative, accounting for 82% of the total comments, while 6,772 comments were coded as positive, or 18% of the total comments (Fig. 6). When the status of these comments is examined by years, as mentioned in the previous section, it can be said that due to the e-commerce sector's emergence in 2013 as a new sector, the number of negative comments was higher than positive comments (Fig. 7). Upon examining Fig. 7, it is evident that in 2020, which corresponds to the pandemic period and the period with the highest number of opened subject titles and sent comments, both positive and negative comments peaked. During this period, while the number of negative comments was 8,406, the number of positive comments was 1,537. From the aforementioned figure, it is noteworthy that especially after 2021, negative comments decreased faster than positive comments. This can again be attributed to decreasing order volumes compared to the previous period, companies' increasing transportation capacities, and developing logistic systems.

When Fig. 8, which shows the status of user comments as negative or positive in terms of the companies studied, is examined, it is observed that the company with the most negative comments is cargo6 with 10,349 comments, followed by cargo1 with 6,040 comments and cargo2 with 5,699 comments. The company with the least negative comments is cargo5 with 870 comments. The company receiving the most positive comments is again cargo6 with 3,053 comments, followed by cargo1 with 993 comments and cargo3 with 968 comments. The company with the least positive comments in this category is cargo5 with 169 comments.

When the ratio of positive and negative comments of cargo companies to total comments is evaluated (Fig. 9); it is seen that cargo2 has the highest negative comment rate with 86.8%, followed by cargo4 (86.3%) and cargo1 (85.9%). The company with the highest positive comment rate is cargo6 with 22.8%, followed by cargo3 with 21.1%, and cargo5 with 16.3%. The company with the lowest positive comment rate is cargo2 with 13.2%.

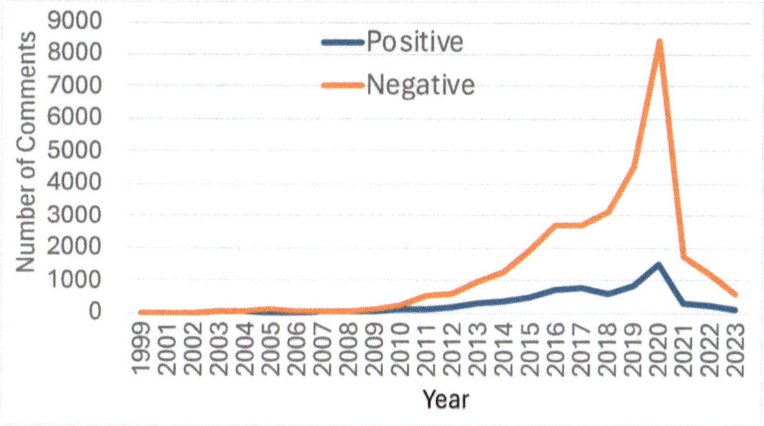

Fig. 7 Change over the years

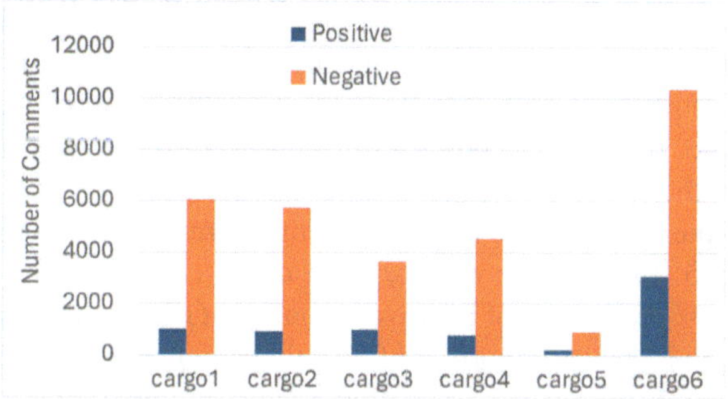

Fig. 8 Positive and negative comments by cargo companies

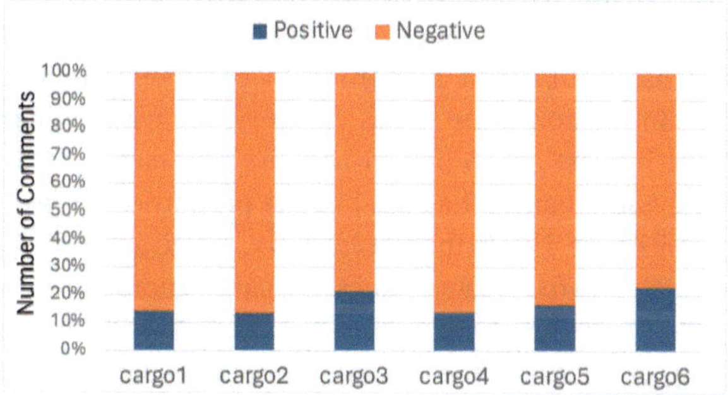

Fig. 9 Positive and negative comment ratios by cargo companies

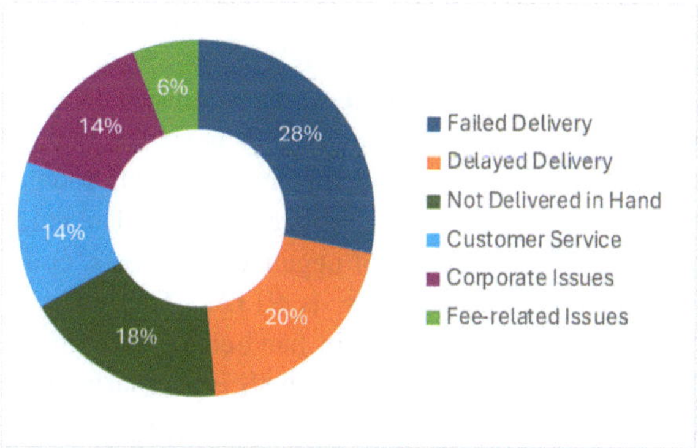

Fig. 10 Comment categories

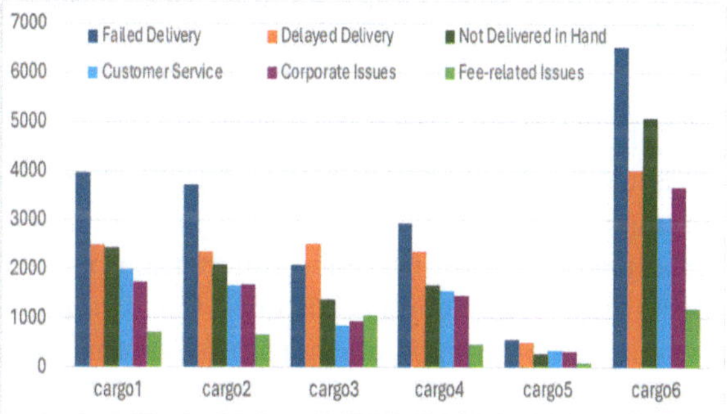

Fig. 11 Classification of comments by cargo

3.3 Results of Classification

The data obtained within the scope of the study were classified using the models with the highest success among the machine learning algorithms applied during the analysis. Accordingly, the category with the most comments related to cargo companies is failed delivery (19,751). This is followed by delayed delivery (14,211), not delivered in hand (12,841), corporate issues (9,805), and customer service (9,452). The category with the least comments sent by users is the fee-related issues category (4,185) (Fig. 10).

The categories of user comments sent to cargo companies are shown in Fig. 11. Accordingly, for all cargo companies except cargo3, the category with the most comments sent by users is failed delivery, and the category with the least comments sent is fee-related issues. For cargo3, users mostly send comments about delayed delivery, while the least comments are sent to the customer service category.

3.4 Results of Sentiment Analysis by Categories

After determining the categories to which the comments are related through the classification process, the proportion of positive and negative comments related to each category was examined (Fig. 12). The category coded as failed delivery contains the highest number of negative comments (17,384). Following this category, delayed delivery with 12,333 comments and not delivered in hand with 11,334 comments are observed. The category with the least negative comments is fee-related issues, with 3,510 comments. When positive comments are evaluated, it is seen that the category with the most positive comments, similar to negative comments, is failed delivery (2,367). Following this category, corporate issues with 2,016 comments and delayed delivery with 1,878 comments are observed, while the least positive comments are sent regarding fee-related issues (675).

The ratio of positive and negative comments to the total number of comments in each category is shown in Fig. 13. According to this, it is found that in almost all categories, more than 80% of the comments are negative. The category with the highest proportion of negative comments, coded as the customer service category, is 91.6%. This category is followed by the not delivered in hand category with 88.3% and the failed delivery category with 88%. Similarly, the category with the highest proportion of positive comments is the corporate issues category, with 20.6%.

Finally, in Fig. 14, the negative and positive comment rates by categories are shown for cargo companies. When the figure is evaluated in general, in negative comments, cargo2 (failed delivery, not delivered in hand, fee-related issues), cargo5 (delayed delivery, customer service), and cargo4 (corporate issues) stand out, while in positive comments, cargo3 (failed delivery, delayed delivery, customer service, fee-related issues) and cargo6 (not delivered in hand, corporate issues) stand out.

According to Fig. 14, the highest proportion of negative comments in the failed delivery category is found in cargo2. In other words, while the proportion of negative comments in the failed delivery category for cargo2 is 90.9%, the proportion of positive comments (also the lowest proportion of positive comments in this category) is 9.1%. The lowest proportion of negative comments (84.8%), and therefore the highest proportion of positive comments (15.2%), in this category belongs to cargo3. In the delayed delivery category, the cargo company with the highest proportion of negative comments (89.9%) and the lowest proportion of positive comments (10.1%) is cargo5. In the same category, the cargo company with the highest proportion of positive comments (18.2%) and therefore the lowest proportion of negative comments (81.8%) is cargo3. In the not delivered in hand category, while the highest proportion of negative comments belongs to cargo5 (95.9%), cargo6 has the highest proportion of positive comments at 14.6%. In the customer service category, cargo5 has the highest proportion of negative comments at 95.7%, while cargo3 has the highest proportion of positive comments at 12.1%. In the corporate issues category, the cargo company with the highest proportion of negative comments is cargo4 at 86%, while cargo6 has the highest proportion of positive comments at 26.1%. In the last category, coded as the 6th category, related to fee-related issues, the cargo company with the highest proportion of negative comments is cargo2 at 92.4%. The highest proportion of positive comments in this category belongs to cargo3 at 24%.

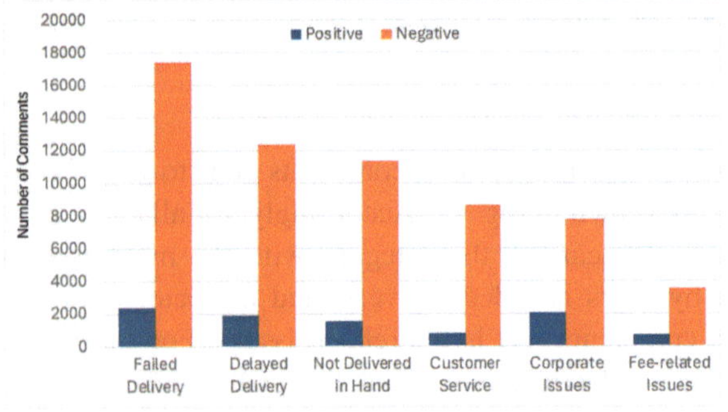

Fig. 12 Sentiment analysis results by category

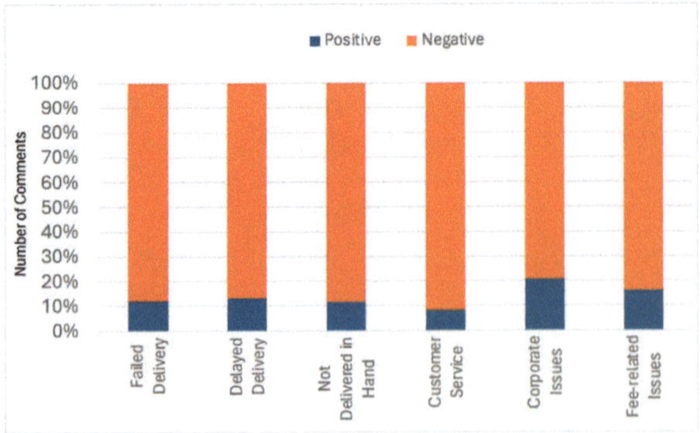

Fig. 13 Sentiment analysis results by category (%)

4 Discussion and Conclusion

The user comments regarding the preferred cargo companies, particularly in the e-commerce sector, were analyzed using sentiment analysis methodology in the study, followed by classification using machine learning algorithms, aiming to identify the issues encountered during the delivery process. According to the research findings, the cargo company with the most topic titles opened and comments sent from 1999 to the date of data collection was cargo6. It was observed that the number of topic titles and comments peaked in 2020. As previously mentioned, this can be

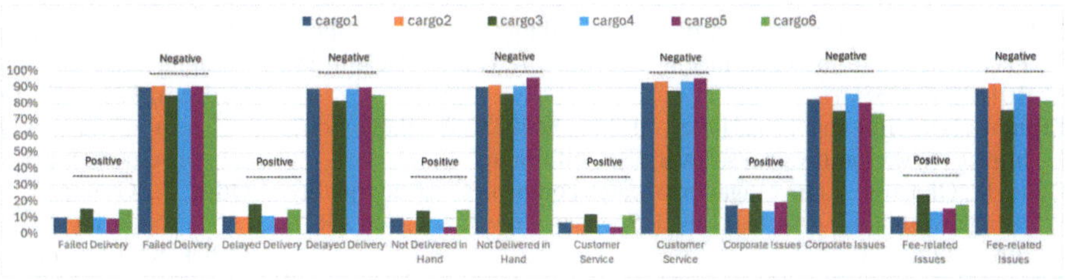

Fig. 14 Sentiment analysis results by cargo companies and categories

attributed to the increased e-commerce volume due to the impact of the Covid-19 pandemic. Although 2020 was the year with the highest number of comments, it is noteworthy that cargo1, cargo3, and cargo6 reached their highest levels of comments before this year, and by 2020, these companies had fewer comments in terms of their own comment counts (Fig. 5). This situation can be interpreted as these companies encountering a similar situation with logistics processes before 2020 and therefore taking some initiatives to solve the problems in advance and being more prepared compared to other companies. Another prominent finding of the study, similar to the study by Atılgan and Yoğurtçu (2021), is that the majority of the comments about cargo companies are negative. From this perspective, it can be said that users send comments related to the problems they encounter more often on the relevant website.

As a result of the study, it was found that the categories to which users send the most comments are respectively failed delivery, delayed delivery, and not delivered in hand categories. While similar results were obtained in the studies of Bayır (2022), Yanar Gürce and Tosun (2017), and Yıldız (Yıldız 2020), Deniz and Gödekmerdan's (2011) study reached the conclusion contrary to the research that cargo companies detect products without damage. It was found in the study that the number of comments related to fees is much lower compared to other categories. This result is similar to Bayır's (2022) study.

Among the most important criteria regarding customers' expectations from cargo companies in the literature are reported to be the fee (Atmaca and Turğut 2015), speed (Yıldız 2020; Turğut 2016), reliability (Yıldız 2020), and emergency management that can be expressed as the ability to deal with problems experienced by customers and find solutions to them (Atmaca and Turğut 2015; Turğut 2016). As a result of the study, it was found that the categories with the highest proportion of negative comments were customer service, not delivered in hand, and failed delivery categories. From this point, it can be said that except for the fee, the expectations of the customers are not sufficiently met.

Logistic activities carried out through e-commerce platforms are a challenging process containing many variables (inter-stakeholder integration, information and tracking systems used, number and quality of personnel, transportation planning and facilities, traffic density, vehicle selection, etc.). In order for this challenging process to succeed, it is critically important to understand customer desires and expectations correctly, clearly identify the problems encountered, and identify the priority areas for improvement. In this context, as a result of the research; it has been observed that users complain about damaged product delivery, loss of products, returning purchased products without delivery, extension of delivery times, products not being delivered to them but left to another person or place, and inability to reach the cargo company or authorities. In this regard, the findings of the study provide important clues for cargo companies that aim to steer user preferences in their favor by adopting a solution-oriented and demand-oriented perspective.

Finally, the study has some limitations such as focusing only on one platform, covering more established and widely used cargo companies, and approaching the topic from the perspective of customers. In this sense, it is considered necessary to conduct similar studies that also address the internal dynamics of companies as well as new generation cargo companies with a flexible structure that offer same-day delivery services and allow delivery options to be changed by users later, which are increasingly becoming widespread today.

References

Atılgan, K.Ö., Yoğurtcu, H.: Sentiment analysis of twitter posts of cargo company customers. Çağ Univ. J. Soc. Sci. **18**(1), 31–39 (2021)

Atmaca, H.E., Turğut, D.: A field research in overall turkey about the determination of criteria for selection of cargo companies. J. Cukurova Univ. Fac. Econ. Adm. Sci. **19**(2), 65–79 (2015)

Bayır, T., A study on E-Complaints regarding new shipping services of E-Commerce businesses: a netnographic analysis on sikayetvar.com. Kast.Onu Univ. J. Fac. Econ. Adm. Sci., **24**(1): p. 57–74 (2022)

Beheshti, N.: Guide to confusion matrices & classification performance metrics. (2022). [cited 2023 09.09.2023]; Available from: https://towardsdatascience.com/guide-to-confusion-matrices-classification-performance-metrics-a0ebfc08408e.

Bhattacharjya, J., et al.: An exploration of logistics-related customer service provision on Twitter. Int. J. Phys. Distrib. Logist. Manag.distrib. Logist. Manag. **46**(6/7), 659–680 (2016)

Burcuoğlu, M., Erdoğan Yazar, E.: Content analysis of customer complaints to cargo companies in a third-party platform. Int. J. Econ. Soc. Res., **16**(1): p. 99–114 (2020)

Cao, Y., et al.: Advance and prospects of adaboost algorithm. Acta Autom. Sin. **39**(6), 745–758 (2013)

Cebeci, H.İ, et al.: Barriers and drivers for biking: What can policymakers learn from social media analytics? J. Transp. Health **28**, 101542 (2023)

Crummy. Beautiful Soup. 2023 [cited 2023 06.09.2023]; Available from: https://www.crummy.com/software/BeautifulSoup/.

Çullu, B., Okursoy, A.: Investigation of cargo companies' service quality using text mining. Anadolu Univ. J. Soc. Sci. Univ. J. Soc. Sci. **23**(2), 399–422 (2023)

Cunningham, P., Cord, M., Delany, S.J.: Supervised learning, in machine learning techniques for multimedia: case studies on organization and retrieval, Cord, M., Cunningham, P., Editors. Springer Berlin Heidelberg: Berlin, Heidelberg. p. 21–49 (2008)

Deniz, A., Gödekmerdan, L.: A study on customers' attitudes and feelings towards cargo firms' services. J. Grad. Sch. Soc. Sci. **15**(2), 379–396 (2011)

Devlin, J., et al.: BERT: Pre-training of deep bidirectional transformers for language understanding. Minneap., Minn.: Assoc. Comput. Linguist. (2019)

Esper, T.L., et al.: The last mile: an examination of effects of online retail delivery strategies on consumers. J. Bus. Logist.logist. **24**(2), 177–203 (2003)

Fanani, Y.D.: Shopee pay free shipping promotion moderation on customer satisfaction. J. Soc. Sci. Bus. Stud. **1**(1), 27–41 (2023)

Garreta, R., Moncecchi, G.: Learning scikit-learn: machine learning in python. Packt Publishing Ltd (2013)

Geurts, P., Ernst, D., Wehenkel, L.: Extremely randomized trees. Mach. Learn. **63**(1), 3–42 (2006)

Han, J., Kamber, M., Pei, J.: Data mining : concepts and techniques. 3rd ed. Morgan Kaufmann series in data management systems. Massachusetts: Morgan Kaufmann Publishers (2012)

Hesse, M.: Shipping news: the implications of electronic commerce for logistics and freight transport. Resour. Conserv. Recycl.. Conserv. Recycl. **36**(3), 211–240 (2002)

Hoang, M., Bihorac, O.A., Rouces, J., Analysis, A.-B., using BERT.: Turku. Linköping University Electronic Press, Finland (2019)

Jung, Y., Hu, J.: A K-fold averaging cross-validation procedure. J. Nonparametric Stat. **27**(2), 167–179 (2015)

Kotsiantis, S.B.: Supervised machine learning: a review of classification techniques. Inform **31**, 249–268 (2007)

Kumar, B.S., Ravi, V.: A survey of the applications of text mining in financial domain. Knowl.-Based Syst..-Based Syst. **114**, 128–147 (2016)

Medhat, W., Hassan, A., Korashy, H.: Sentiment analysis algorithms and applications: A survey. Ain Shams Eng. J. **5**(4), 1093–1113 (2014)

Nadeem, W., et al.: Engaging consumers online through websites and social media: A gender study of Italian Generation Y clothing consumers. Int. J. Inf. Manage. **35**(4), 432–442 (2015)

Otim, S., Grover, V.: An empirical study on Web-based services and customer loyalty. Eur. J. Inf. Syst. **15**(6), 527–541 (2017)

Özbek, S.: E-Commerce in the world and turkey: a descriptive evaluation. J. Soc. Sci. **54**, 539–548 (2021)

Parmaksız, Ö.F.: Measuring the service quality during the activities of cargo companies with the servqual model. Unpublished master's thesis, Harran University (2022)

Savaş, S., Topaloğlu, N., Yılmaz, M.: Data mining and application examples in Turkey. Istanb. Commer. Univ. J. Sci. **11**(21), 1–23 (2012)

Sungur, M.: The effect of service quality in customer satisfaction in logistics companies: an application in cargo companies. İstanbul Aydın University: Unpublished master's thesis (2018)

Topuk, N.: Evaluation of Firms Operating in the Cargo Sectors in Turkey Using EDAS and MOORA. Marmara University: Unpublished master's thesis (2020)

Tripathy, A., Agrawal, A., Rath, S.K.: Classification of sentiment reviews using n-gram machine learning approach. Expert Syst. Appl. **57**, 117–126 (2016)

Turğut, D.: Evaluation of firms operating in the cargo sector in Turkey by using a multi-criteria decision-making method. Gazi University: Unpublished master's thesis (2016)

Vasić, N., et al.: Satisfaction is a function of users of logistics services in e-commerce. Technol. Anal. & Strat. Manag. **33**(7), 813–828 (2020)

Yanar Gürce, M., Tosun, P.: Consumer complaints about shipment services: a content analysis. J. Bus. Res., **3**(9): p. 177–196 (2017)

Yang, F.J.: An implementation of naive bayes classifier. In: 2018 International Conference on Computational Science and Computational Intelligence (CSCI) (2018)

Yıldırım, S.: Bert-base Turkish Sentiment Model. (2022). [cited 2023 08.09.2023]; Available from: https://huggingface.co/savasy/bert-base-turkish-sentiment-cased

Yıldız, B.: Content analysis of service quality of cargo companies. Karamanoglu Mehmetbey Univ. J. Soc. Econ. Res. **22**(38), 42–57 (2020)

Yıldız, B., Şemsettin, Ç., Aslan, H.: The effect of service quality of cargo companies on customer satisfaction. Int. J. Soc. Hum.ities Sci. Res. (JSHSR) **5**(21), 829–843 (2018)

Towards a Pedagogical Revolution: Intelligent Learning Systems in the Service of Higher Education

Imane Oudmine and Sara Oudmine

Abstract

This paper explores the transformative impact of Intelligent Learning Systems (ILS) in higher education through several case studies. The Pittsburgh Science of Learning Center demonstrates the implementation of advanced educational tools to promote student-centered learning. The value of Virtual Environments (VEs) and Virtual Reality (VR) technologies in providing cost-effective and immersive training experiences is highlighted, showcasing their potential in specialized fields such as aviation and medicine. Arizona State University's extensive use of adaptive learning technologies underscores the benefits of personalized learning paths and immediate feedback, which enhance student engagement and success. Additionally, Carnegie Mellon University's research in Intelligent Tutoring Systems (ITS) reveals the effectiveness of sophisticated tutoring models in various instructional domains. By addressing current educational challenges and leveraging innovative technologies, these institutions exemplify a commitment to modernizing educational practices and improving learning outcomes.

Keywords

Intelligent learning systems · Virtual environments · Adaptive learning · Intelligent tutoring systems · Higher education

1 Introduction

Intelligent learning systems are coming to reality today not only because of the advancement of state-of-the-art technologies but also because of the hard work of many people in the field. Such systems have been proved to be successful in different domains and areas of instruction. In this paper, we present a number of case studies that demonstrate successful implementations of such learning systems. The case studies include background information, a description of the systems, the techniques employed, and the underlying AI techniques. The implementations are evaluated based on their capabilities, limitations, and lessons learned. Case studies are described throughout the paper, and in the end, a conclusion section is included.

What are intelligent learning systems? Intelligent learning systems are inventive systems that make use of innovative techniques from the artificial intelligence domain to develop practical learning methods and tools. Many vendors of such tools address their systems as intelligent tutoring systems, expert systems, neural network systems, etc. These systems utilize AI techniques to enhance the instructional learning methods and tools. Intelligent learning systems incorporate intelligence both at the software level as well as at the instructional level to provide sufficient support to the learners. These systems adapt to the learners' profiles and react to their changing behavioral patterns over time during the course of learning. Furthermore, these systems also incorporate some degree of user modeling.

I. Oudmine (✉)
Faculty of Legal, Economic, and Social Sciences, Ait Melloul, Morocco
e-mail: i.oudmine@uiz.ac.ma

S. Oudmine
Faculty of Legal, Economic, and Social Sciences, Agadir, Morocco

B. El Bhiri et al. (eds.), *AI and Digital Transformation: Innovations in Supply Chain, Education, and Energy Systems*, Sustainable Artificial Intelligence-Powered Applications, https://doi.org/10.1007/978-3-031-86837-5_6

2 Foundations of Intelligent Learning Systems

Intelligent Learning Systems are systems that combine machine learning, user modeling, and learner control in order to deliver instruction that is adapted to the needs of the individual. The central design principle of an intelligent learning system is to form a mental model of the learner and to use this model to control the instructional process in a way that is maximally beneficial to the learner. Any such system must perform three essential functions: it must be able to model the learner, to model the subject matter to be learned and to plan instructional activities that bridge the gap between the learner's current state of knowledge and a desired state. From a technological point of view, Intelligent Learning Systems are advanced applications that require sophisticated computing environments and AI technologies (Haydari and Yılmaz 2020).

The complexity of the learning process and the diversity of the learners call for a flexible, learner-centered educational system, which is able to take into consideration the unique characteristics of each learner and provide him or her with the most suitable educational experience. Recent developments in the areas of machine learning and the emergence of the knowledge society have great potential for advancing higher education (Zhao et al. 2021). It is argued that learning has to change, in order to prepare students for living and working in the knowledge society. The paper then presents the foundations of intelligent learning systems by stressing their key components and theoretical underpinnings.

2.1 Definition and Components

One of the major challenges in the field is developing effective methods for constructing, updating, and utilizing the learner model, which would allow for effective personalization and customization. This challenge is even more formidable when we consider collaborative learning or the effect of groups of learners. Since teaching and learning are also highly complex processes, involving a large number of interacting factors, ILS development requires the development and deployment of complex models that embody advanced practical and theoretical understanding of both teaching and learning (Bygstad et al., 2022). As a result, the construction and implementation of ILS are inherently complex, and the most advanced ILS are currently developed using what is termed a "discipline-based approach" that involves collaboration between computer and educational experts, and content experts from the relevant domain of study (Coman et al. 2020).

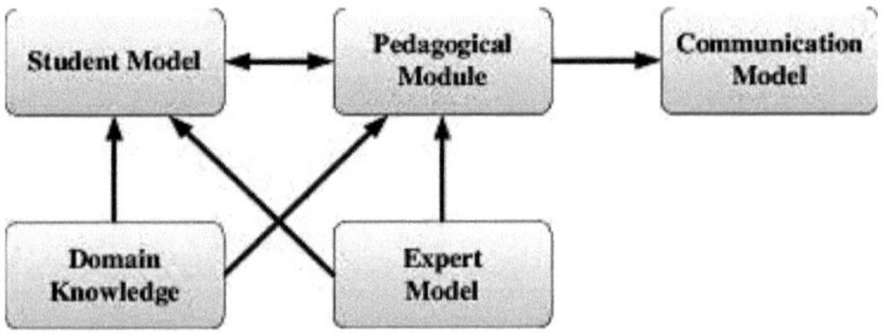

Fig. 1 Main components of intelligent learning systems

Intelligent Learning Systems (ILS) are systems that provide specialized support to learning and teaching processes, and operate at the frontier of both computer science and education. Specifically, ILS are "knowledge-based systems that support learning, instruction, or both." They consist of two main components. The knowledge component is responsible for capturing and representing the expertise of humans capable of effective teaching or tutoring in some domain, as well as knowledge related to how learners learn in that domain. The second component is the learner model, representing the knowledge state of each learner being supported by the system. In addition, ILS may have further components, such as the tutoring component, the user interface, and other more specialized components depending on the particular system and its capabilities (Fig. 1).

2.2 Theoretical Frameworks

Vygotsky's work has also come to the fore in a new context through technology via the concept of the digital native. Advances in computing, artificial intelligence, and neuroscience are, we think, converging to reshape technology's role in education.

Our fundamental pedagogical paradigm comes from Vygotsky's social constructivism. Vygotsky argued that learning is linked to social interaction and development. He proposed that students learn best in environments that promote collaboration and conversation among peers with varying levels of ability. From this perspective, the presence of a more knowledgeable individual is crucial, as knowledge is co-constructed (Zhai et al. 2021). As we have learned more through brain research about how people learn, we have added cognitive insights to our philosophical underpinnings by merging the work of Vygotsky and Piaget with an understanding of how the brain functions. This approach suggests that knowledge is constructed as the result of problem-solving activities and that higher mental processes, including thinking, understanding, and reasoning, develop best in the context of supportive interaction with both expert instructors and peers through collaboration and conversation.

3 Applications in Higher Education

Computer-based assessment can be far more reliable than human marking. The system can mark students' work instantaneously well, at least for many types of task. This speed of marking can be a significant advantage, allowing more rapid feedback to students and perhaps even allowing students to be re-assessed at a later stage of their work, moving some way towards continuous assessment. Moreover, because ILT systems can store data about student interactions, they can record every answer entered by a student. This provides grounds for a different style of assessment, analysing a student's competence on the basis of his or her pattern of responses to a series of questions on a particular topic. Such 'adaptive assessment' can be quite sophisticated, asking further questions in areas where a student performs poorly on the first few, for example, and can in principle at least, assess the student's knowledge more reliably than a single ad hoc task (Liu & Liu, 2023).

One important application of ILT in HE (Higher Education) is, then, the automation of assessment, which can be implemented in several ways. Basic automated assessment can be made by the computer checking answers entered by students against a model answer. This model may be prepared by the course tutor and can incorporate rules for partial matching, accepting synonyms, and so on. It is relatively simple to generate this kind of model on the basis of explicit knowledge. Another kind of model can be induced by machine learning techniques. In effect, the computer 'learns' what the correct answer should look like on the basis of a number of examples provided by the tutor. It can then use this learned model to assess new answers. Assessments of this kind become more reliable as the number of examples considered by the system increases, but there is a trade-off between reliability and the number of model-based examples needed to achieve it.

3.1 Personalized Learning

One way in which intelligent learning systems can make online education more personal is through rapid and sophisticated assessment of student work. In a small class, an instructor can provide individualized feedback to every student on writing assignments, programs, or other complex products. In a large lecture, multiple choice questions are often used as a compromise; they can be graded automatically and provide at least some feedback to students. In online education, questions of many different types can be scored automatically, allowing for more diverse and sophisticated forms of assessment. Furthermore, even complex student products, such as essays, can be evaluated automatically with reasonable accuracy. This can allow for rapid student feedback even on open-ended assignments (Alam 2022). One significant benefit of this approach is that students receive immediate feedback, which can help them understand their strengths and areas for improvement in real-time. Additionally, the use of intelligent learning systems in assessment can provide insights into student learning patterns and individual needs, allowing instructors to tailor their teaching methods to better meet the needs of each student. This personalized approach to assessment and feedback contributes to a more effective and engaging online learning experience (Chen et al. 2020).

Several massive-scale online higher education platforms, both for-profit and nonprofit, have emerged. In some courses, enrollments exceed hundreds of thousands of students. An apparent contradiction of such mass education is that it is delivered with a "one-size-fits-all" pedagogy wherein a single sequence of learning activities is offered to the entire student cohort. In contrast, small-scale education, whether in a traditional college or an online seminar, allows for highly personalized instruction. Intelligent Learning Systems can resolve this contradiction. Robust, scalable systems have been developed to assist students and instructors in online education. The assistance can be extremely responsive and can be based on sophisticated models of student knowledge.

3.2 Adaptive Assessment

In its most common form, adaptive assessment selects items based on the learner's responses to previous items. The observed response, be it correct or incorrect, to the item presented indicates the learner's underlying ability to the assessor. Selection of the next item takes this relationship into account. If, for example, a relatively easy item is answered incorrectly, it may be an indication that a learner's ability is at a level higher than that represented by the item, labeled, for example, as of medium difficulty. The hypothesis is that the learner is performing below his or her ability level and, according to this hypothesis, a more difficult item, labeled, for example, as being of high difficulty, should be presented. The actual response of the learner to the first item, whether correct or incorrect, informs the system about the learner's ability and guides the system in selecting the next appropriate item.

Adaptive testing or assessment, in its current form, is a computer-based method that ensures that questions (items) selected for a learner are matched to the ability of that learner. The rationale underlying adaptive testing is that the assessment process will be made more efficient if irrelevant questions are not presented to a learner. More importantly, confidence in the precision of an assessment also increases given the relationship between reliability and the number of items presented to a learner. In an ideal case, adaptive testing is so efficient that only one item is needed to establish the learner's ability within a specified margin of error. This one item is referred to as a "stopping rule," and it is the Holy Grail of adaptive testing (Minn 2022).

3.3 Virtual Laboratories

The advent of virtual laboratories marks a major transformation in higher education, introducing new opportunities for interactive and immersive learning experiences. By leveraging advanced learning technologies, educators can provide students with valuable practical training and enhance their understanding of complex concepts. The integration of intelligent tutoring systems and virtual laboratories has the potential to transform the way students learn and engage with course materials. As the field continues to evolve, further research and innovation will be crucial in harnessing the full potential of these advanced learning technologies in higher education (Radhamani et al. 2021).

In higher education, laboratory work is crucial in the natural sciences, engineering, and other disciplines. It provides students with practical experience that complements theoretical concepts taught through lectures and textbooks. Similarly, in computer science, programming assignments are essential to reinforce class discussions on algorithms and data structures. Despite the importance of these practical activities, intelligent learning systems have had limited involvement in higher education until recently. However, there has been a surge of interest and activity in the development of virtual laboratories. These virtual labs aim to replicate hands-on experiences and are designed to enhance student learning in a wide range of subject areas (Marks & Thomas 2022).

The deployment of intelligent tutoring systems and other advanced learning technologies in higher education is still in its early stages. However, it holds immense potential to revolutionize teaching methods and reshape the entire educational process. One area of exciting development is virtual laboratories, which showcase the power and limitations of intelligent learning systems in higher education. These computer-based approximations of hands-on physical or practical work are being created for various subject areas using different computer technologies. They are often integrated with other courseware components, such as intelligent tutoring systems.

4 Case Studies of Intelligent Learning Systems in Higher Education

4.1 Pittsburgh Science of Learning Center: Enhancing Learning with AI

The Pittsburgh Science of Learning Center (PSLC) demonstrates the transformative potential of Intelligent Learning Systems (ILS) in higher education. By leveraging adaptive technologies, the center focuses on personalizing learning experiences for students, addressing individual needs more effectively than traditional methods.

One notable outcome is the reported 30% reduction in the time needed for students to achieve proficiency in mathematics courses when using AI-driven cognitive tutors. These systems dynamically adjust the complexity of exercises based on student performance, ensuring optimal learning paths tailored to each learner's progress. This innovative approach fosters deeper engagement and understanding among students, particularly in challenging subjects like STEM disciplines.

4.2 Virtual Environments in Advanced Training

Virtual Environments (VEs) have become indispensable tools for practical, immersive education, particularly in fields requiring high-stakes training such as medicine and aviation. These environments allow learners to simulate real-world scenarios without the associated risks or high costs.

In surgical training, VR platforms have been shown to decrease error rates by 40% compared to traditional methods, equipping trainees with the skills needed to perform procedures confidently. Similarly, aviation students who practiced with VR-based flight simulators demonstrated a 35% improvement in handling emergency situations, highlighting the effectiveness of these tools in preparing students for real-world challenges. These findings illustrate how VEs not only enhance skill acquisition but also improve overall learner confidence and readiness.

4.3 Arizona State University: Leading Innovation in Adaptive Learning

Arizona State University (ASU) has pioneered the integration of adaptive learning technologies across its curriculum, setting a benchmark for innovation in higher education. Through platforms like SmartBook and Knewton, ASU has transformed the educational experience for its diverse student body.

Students enrolled in courses utilizing these tools report a 20% increase in retention rates and an improvement in academic performance, particularly in foundational subjects such as biology and mathematics. By providing real-time feedback and personalized learning pathways, these technologies address knowledge gaps efficiently, enabling students to progress at their own pace. Moreover, the inclusion of multimedia elements and gamified content has significantly increased student engagement, making learning more interactive and enjoyable.

4.4 Intelligent Tutoring Systems: The Carnegie Mellon Model

Carnegie Mellon University has long been at the forefront of Intelligent Tutoring Systems (ITS) research. These systems, designed to provide targeted and adaptive support to learners, have proven particularly effective in enhancing comprehension and problem-solving abilities.

For example, students using ITS for introductory programming courses completed 30% more assignments and reported higher confidence in coding tasks compared to peers in traditional classes. By tailoring instruction to individual needs and offering immediate, constructive feedback, ITS fosters a supportive learning environment that promotes sustained academic growth. These successes emphasize the importance of integrating technology into the learning process to address diverse learner profiles effectively.

4.5 Opportunities and Challenges

The integration of ILS presents both significant opportunities and notable challenges. On the one hand, institutions that adopt these technologies report improved educational outcomes, such as reduced dropout rates and enhanced student engagement. For instance, adaptive learning platforms have been shown to reduce attrition rates by up to 25% in certain programs, underscoring their role in retaining students.

On the other hand, barriers such as high implementation costs, faculty training needs, and resistance to technological change persist. Additionally, concerns about data privacy and the ethical use of AI in education require careful consideration. Addressing these challenges will require collaborative efforts between educators, policymakers, and technology developers to ensure that ILS can be implemented equitably and sustainably.

5 Discussion

The integration of Intelligent Learning Systems (ILS) in higher education represents a transformative shift in how learning and teaching are approached. Building on the findings presented in the case studies, this discussion delves deeper into the implications, challenges, and future directions of ILS implementation.

Revisiting the Impact of Adaptive Learning Systems

As highlighted in the case studies, adaptive learning systems have proven effective in personalizing education, particularly in institutions like Arizona State University (ASU). These systems not only enhance student retention and academic performance but also address knowledge gaps in real time. The success observed at ASU serves as a benchmark, demonstrating that tailoring learning experiences to individual needs can result in measurable improvements in educational outcomes. This aligns with broader findings in educational research, where personalized approaches have consistently outperformed one-size-fits-all models.

However, the broader adoption of these systems faces challenges, such as high initial development costs and the need for faculty training. Institutions must balance the significant upfront investment with the long-term benefits, particularly in terms of reduced dropout rates and enhanced student success.

Role of Virtual Environments in Skill Development

The effectiveness of Virtual Environments (VEs) in fields such as medicine and aviation underscores their potential to bridge the gap between theoretical knowledge and practical application. As evidenced by the 40% reduction in surgical errors and 35% improvement in handling emergencies among aviation students, VEs provide a safe and cost-effective alternative to traditional hands-on training.

These findings reaffirm the value of immersive learning technologies in equipping students with critical skills in a controlled, risk-free setting. However, challenges such as the high cost of VR hardware and the need for consistent updates to ensure technological relevance must be addressed. Future research should focus on scaling these technologies for broader access, particularly in resource-constrained institutions.

Intelligent Tutoring Systems: A Model for Scalable Personalized Learning

Carnegie Mellon University's success with Intelligent Tutoring Systems (ITS) offers valuable insights into the scalability of personalized learning. By enabling students to complete more assignments and boosting their confidence in problem-solving, ITS exemplifies how AI can complement traditional teaching methods.

Table 1 Comparative Analysis of ICT Integration in Higher Education Institutions

Institution	System	Types of activities	Technologies used	Advantages	Limitations
Pittsburgh Science of Learning Center (PSLC)	Intelligent Learning Systems (ILS)	Educational games, role-play simulations, virtual laboratories	ICT, adaptive learning technologies	Dynamic and interactive learning, improved learning outcomes	High initial investment, training required for faculty
University of Sydney	Various	Innovative teaching methods	ICT, constructivist pedagogies	Gradual voluntary adoption, enthusiastic faculty	Initial slow uptake, dependent on faculty interest
University of Edinburgh	Various	Formal support programs for faculty	ICT, training programs	Well-equipped faculty, effective integration of technologies	Time-consuming, resource-intensive
MIT	Digital learning initiatives	On-campus and online education	Comprehensive digital learning tools	Enhanced education quality, broad reach	High budget required, ongoing maintenance
Carnegie Mellon University	Open Learning Initiative (OLI)	Adaptive learning technologies	Intelligent Learning Systems (ILS)	Improved individual learning outcomes, personalized learning paths	Development effort required, resource quality variance

Despite these successes, ITS implementation requires significant resources and expertise. Collaboration between educational institutions and technology providers is crucial to develop user-friendly ITS that are accessible to a broader audience. Additionally, there is a need for robust frameworks to evaluate their long-term impact on learning outcomes and employability.

Insights from Comparative Institutional Practices

A comparative analysis of ICT integration across institutions provides further evidence of the varied methodologies and outcomes associated with adopting Intelligent Learning Systems. The following table summarizes these practices (Table 1).

This table illustrates the diverse approaches institutions have taken to incorporate ICT in their teaching practices, highlighting both the benefits and challenges of these efforts.

Strategic Frameworks and Challenges

In addition to comparing institutional efforts, the following strategic framework outlines best practices for implementing Intelligent Learning Systems effectively (Table 2).

Table 2 Strategic framework for ICT integration in higher education

Aspect	Description
Objective	Leverage ICT to transform teaching and learning at the university
Academic Development Unit (ADU)	Equip faculty with Intelligent Learning Systems (ILS) to support student-centered, constructivist pedagogies
Educational innovations	- ILS: Provides adaptive learning technologies tailored to individual student needs
Implementation strategy	- Informal Training: Initiate with informal training by experts for interested faculty, promoting voluntary adoption - Formal Support: Provide structured support programs for faculty development, inspired by successful models from other universities
Institutional commitment	- Enthusiastic support from leadership, backed by budget allocations

This framework underscores the need for institutional commitment, innovative teaching methods, and structured faculty training to maximize the potential of ILS.

Challenges and Ethical Considerations

While the advantages of ILS are well-documented, their implementation raises ethical concerns, particularly regarding data privacy and algorithmic bias. Adaptive systems rely heavily on learner data, necessitating stringent measures to protect privacy and prevent misuse. Moreover, algorithmic decisions must be transparent and free from biases that could disadvantage certain student groups.

Institutions must prioritize ethical guidelines and robust governance frameworks to address these challenges. Ensuring that all stakeholders—students, educators, and developers—are involved in the design and implementation process can help mitigate potential risks.

Future Directions

The evidence presented underscores the transformative potential of ILS in higher education. Moving forward, three key areas warrant attention:

1. **Interdisciplinary Collaboration**: Institutions should foster partnerships between educators, technologists, and policymakers to develop scalable and ethical ILS solutions.
2. **Research and Evaluation**: Longitudinal studies are needed to assess the impact of ILS on cognitive retention, student satisfaction, and employability.
3. **Accessibility and Equity**: Efforts must be made to democratize access to these technologies, ensuring that resource-constrained institutions and underrepresented student populations benefit equally.

6 Conclusion

The adoption of Intelligent Learning Systems (ILS) in higher education is reshaping traditional pedagogical frameworks, offering innovative solutions to longstanding challenges in teaching and learning. Case studies from institutions such as the Pittsburgh Science of Learning Center and Arizona State University demonstrate the tangible benefits of ILS, including personalized learning experiences, improved student engagement, and enhanced academic outcomes. Similarly, the integration of Virtual Environments and Intelligent Tutoring Systems highlights the potential of technology to bridge the gap between theory and practice, equipping learners with critical skills in a controlled, cost-effective manner.

However, the successful implementation of these systems requires addressing key challenges, including resource allocation, faculty training, and ethical considerations surrounding data privacy and algorithmic fairness. Institutions must adopt strategic frameworks that prioritize interdisciplinary collaboration, robust evaluation mechanisms, and equitable access to technology.

The evidence presented in this paper underscores the transformative potential of ILS to modernize higher education, fostering adaptive, interactive, and student-centered learning environments. As institutions continue to innovate, future research should focus on the long-term impacts of these technologies on cognitive retention, employability, and equity in education. By leveraging these insights, higher education can position itself as a leader in preparing students for the demands of a rapidly evolving knowledge economy.

References

Akyuz, Y. (2020). Effects of intelligent tutoring systems (ITS) on personalized learning (PL). Creative Education.

Al Rekhawi, H.A.: Android applications development intelligent tutoring system (2020)

Alam, A.: Harnessing the power of AI to create intelligent tutoring systems for enhanced classroom experience and improved learning outcomes. In Intell. Commun. Technol. Virtual Mob. Netw. (pp. 571–591). Singapore: Springer Nature Singapore (2023)

Appel, T., et al.: VR headsets in anatomical study (2020)

Azadegan, A., Dooley, K.: A typology of supply network resilience strategies: Complex collaborations in a complex world. J. Supply Chain. Manag. (2021)

Cao, J., Yang, T., Lai, I. K. W., Wu, J.: RETRACTED: Student acceptance of intelligent tutoring systems during COVID-19: The effect of political influence. Int. J. Electr. Eng. & Educ., 60(1_suppl), 2495–2509 (2023)

Deeva, G., Bogdanova, D., Serral, E., Snoeck, M., De Weerdt, J.: A review of automated feedback systems for learners: Classification framework, challenges and opportunities. Comput. Educ.. Educ. **162**, 104094 (2021)

Ewing, L.: The University of Edinburgh's Institute for Academic Development (2021)

Haydari, A., Yılmaz, Y.: Deep reinforcement learning for intelligent transportation systems: a survey (2020)

Hayter, C.S., Link, A.N.: Governance mechanisms enabling inter-organizational adaptation: Lessons from grand challenge R&D programs. Sci. Public Policy (2020)

Holmes, W., Porayska-Pomsta, K.: The University of Sydney's adoption of innovative teaching practices (2022).

Kochmar, E., Vu, D.D., Belfer, R., Gupta, V., Serban, I.V., Pineau, J.: Automated data-driven generation of personalized pedagogical interventions in intelligent tutoring systems. Int. J. Artif. Intell. Educ.Artif. Intell. Educ. **32**(2), 323–349 (2022)

Kurni, M., Mohammed, M.S., Srinivasa, K.G.: Intelligent tutoring systems. In: A beginner's guide to introduce artificial intelligence in teaching and learning, pp. 29–44. Springer International Publishing, Cham (2023)

Lacka, M., et al.: The use of VR technology in flight simulators (2021)

Li, X., Jackson, T.: University of Southern California's use of VR simulations in medical training (2024)

Marks, A., Thomas, P.: Medical training through VEs (2022)

McGovern, J., et al.: United States Navy's Underway Replenishment training program (2020)

Meyer, M.W., Norman, D.: Changing design education for the 21st century. She Ji: J. des., Econ., Innov. **6**(1), 13–49 (2020)

Potra, S., et al.: Carnegie Mellon University's Open Learning Initiative (2021)

Reich, J. (2020).: Failure to disrupt: Why technology alone can't transform education

Saifuddin, N., Yin, L.: The educational potential of VEs

Scavarelli, A., et al.: Experiential constructivism in VEs (2021)

Siripongdee, K., et al.: MIT's digital learning initiatives (2020)

Southworth, J., Migliaccio, K., Glover, J., Reed, D., McCarty, C., Brendemuhl, J., & Thomas, A.: (2023)

Troussas, C., Krouska, A., Alepis, E., Virvou, M.: Intelligent and adaptive tutoring through a social network for higher education. New Rev. Hypermedia Multimed. **26**(3–4), 138–167 (2020)

Whitcomb, J., Singh, R.: University of Melbourne's early intervention program (2021)

White, G.: Adaptive learning technology relationship with student learning outcomes. Comput. Educ.: Artif. Intell. **4**, 100127 (2020)

Zhang, L., Basham, J.D., Carter, R.A., Jr., Zhang, J.: Exploring factors associated with the implementation of student-centered instructional practices in US classrooms. Teach. Teach. Educ. **99**, 103273 (2021)

Zhao, X., Zhong, S., Guo, X.: Digital competence and its impact on higher education and knowledge society development (2021)

Preparing for Industry 5.0: Inclusive Curricular Approaches and Digital Empowerment

Pallavi Dhamak, Padmanabha Aital, Anand Daftardar, and Bhaveshkumar Pasi

Abstract

This paper explores the necessary change in education to meet the requirements of Industry 5.0, focusing on the need of an inclusive curricular pedagogy. The purpose mentioned promotes the acceptance of diversity and the elimination of digital disparities, with the goal of providing learners with the necessary skills for success of Industry 5.0. The exploration encompasses important issues, such as ensuring equitable access to digital resources, aligning technology with teaching methods, adapting to different cultures and languages, empowering teachers, and implementing comprehensive assessments within inclusive educational frameworks. In addition, a comprehensive examination of three key pedagogical methods in Higher Education Institutions (HEIs)—E-learning, Project-Based Learning and Blended Learning—offers valuable insights into the worldwide transformation of teaching and learning practices in Industry 5.0 era. The conclusion underlines the importance of promoting inclusive curricular pedagogy and emphasises the profound effect of these pedagogies have on preparing learners for Industry 5.0.

P. Dhamak (✉) · P. Aital
Department of Technology Management, Mukesh Patel School of Technology Management and Engineering, Shri Vile Parle Kelavani Mandal Narsee Monjee Institute of Management Studies, Vile Parle, Mumbai, India
e-mail: pallavivivekdongare@gmail.com
URL: http://www.springer.com/gp/computer-science/lncs

A. Daftardar
Department of Civil Engineering, Mukesh Patel School of Technology Management and Engineering, Shri Vile Parle Kelavani Mandal Narsee Monjee Institute of Management Studies, Vile Parle, Mumbai, India

B. Pasi
Department of Mechanical Engineering, Vishwaniketan's Institute of Management Entrepreneurship and Engineering Technology (ViMEET), Navi Mumbai, India

Keywords

Industry 5.0 · Digital Equity · Project-Based Learning

1 Introduction

The recent development of novel technologies, often linked to the Fifth Industrial Revolution or Industry 5.0, has significantly influenced many facets of society, including job opportunities and education (Martínez-Gutiérrez et al. 2024). The current period is distinguished by the amalgamation of digital, physical, and biological systems, resulting in un- paralleled changes in industries worldwide (Mohamed Hashim et al. 2024). With the increasing prevalence of technologies like artificial intelligence, machine learning, robots, and the Internet of Things, the skills demanded by the workforce are also changing. Given these alterations, it is essential to readjust our educational institutions in order to synchronise with the requirements of the contemporary labour market (Bispo and Amaral 2024). This entails restructuring educational programmes to include the competencies and information necessary for the age of Industry 5.0. The conventional method of teaching may no longer be enough in equipping pupils with the necessary skills to navigate the complex world of the digital era (Calvetti et al. 2024). The primary objective of this recalibration is to advance an inclusive approach to curriculum and teaching methods. Inclusive education endeavours to provide equitable access to educational opportunities for all students, irrespective of their socio-economic status, cognitive aptitude, or personal situations. Within the framework of Industry 5.0, inclusive curricular pedagogy acknowledges the significance of adapting to various learning styles, experiences, and requirements (Karapanagiotis and Reimers 2024). It highlights the need of including under-represented groups, such as women, minorities, and those

B. El Bhiri et al. (eds.), *AI and Digital Transformation: Innovations in Supply Chain, Education, and Energy Systems*, Sustainable Artificial Intelligence-Powered Applications, https://doi.org/10.1007/978-3-031-86837-5_7

with disabilities, who may encounter obstacles in obtaining educational and job opportunities in the technology industry (Erp et al. 2024). Nevertheless, implementing a comprehensive educational approach that includes all students in the current age of digital technology poses several obstacles. The difficulties include the following (Ramírez Molina et al. 2024).

- Digital Divide: The digital divide refers to the discrepancies in accessing technology and digital resources, which may worsen existing educational disparities. Students hailing from underprivileged circumstances may face a dearth of resources and facilities necessary for engaging in digital education.
- Skills Gap: The skills gap arises when the fast rate of technological progress surpasses the capacity of educational institutions to modify their curriculum accordingly. There is an increasing disparity between the skills that are being taught in schools and the abilities that companies in new sectors want.
- Diversity and Inclusion: Despite ongoing initiatives to foster diversity and inclusion in STEM (science, technology, engineering, and mathematics) disciplines, women, minorities, and those with disabilities remain insufficiently represented. It is essential to tackle systemic obstacles and prejudices in order to provide an inclusive learning environment.
- Teacher Training: Educators may need more instruction and professional growth opportunities to proficiently incorporate new technology into their teaching methodologies. Developing teachers' digital literacy abilities is crucial for promoting innovation and creativity in the classroom.
- Equitable Access: Ensuring equal and impartial access to high-quality education for every student is crucial for fostering upward social mobility and economic well-being. To tackle this issue, it is necessary to confront and rectify systemic disparities while allocating resources to programmes that uplift marginalised groups.

When facing these difficulties, it is crucial to acknowledge the complex interconnection between technology, education, and diversity. To optimise educational interventions for varied learners, it is crucial to understand and tackle the many elements that impact learning results (Rodrigues et al. 2024). This encompasses the implementation of educational methods that prioritise the needs and interests of the learners, using advanced technology that can adapt to individual learning styles, and fostering places that encourage cooperation and shared learning experiences. The ultimate objective of promoting inclusive curricular pedagogy is to enhance the provision of services to all individuals participating in the education system (Bonello et al. 2024). By cultivating a culture that values diversity and encourages creative thinking, we can provide students with the necessary skills to succeed in the rapidly evolving and interconnected realm of Industry 5.0 (Adela 2024). Ultimately, the goal of advancing inclusive curricular pedagogy is to provide better services to all stakeholders involved in the education system. By fostering a culture of inclusivity and innovation, we can prepare students to thrive in the dynamic and interconnected world of Industry 5.0. The objectives of the current study strongly emphasise the progress of inclusive curricular pedagogy, particularly in the context of Industry (Knoth et al. 2024).

2 Literature Survey

An inclusive curriculum approach acknowledges the diversity of learners' origins, learning styles, experiences, and skills. By recognising this variability, educators may customise their teaching approaches to fit various learning requirements, guaranteeing equal opportunities for all students to achieve success (Imjai et al. 2024). This method promotes a nurturing and entirety learning atmosphere in which students feel appreciated and encouraged to actively engage in their education. Additionally, it fosters empathy, comprehension, and reverence for varied view-points, hence cultivating people who are versatile and successful in multicultural and diverse cultures (Padovano et al. 2024). The term "digital divide" pertains to inequalities in technological access and digital resources, which may worsen pre-existing discrepancies in education. In the age of Industry 5.0, when technology assumes a pivotal role in education and work, it is imperative to tackle the digital gap in order to provide equitable opportunities for all learners (Piccarozzi et al. 2024). Initiatives aimed at narrowing the digital gap may include facilitating the availability of computers, internet connection, and digital educational materials in disadvantaged areas. Moreover, digital literacy programmes have the potential to provide learners with the necessary abilities to proficiently navigate and use technology, irrespective of their socioeconomic status (Pacher and Glinik 2024). Through the removal of obstacles associated with the digital divide, educators may provide an environment where all learners have equal and fair access to educational opportunities and resources. The ultimate objective of adopting an inclusive curricular methodology that aligns with the concepts of Industry 5.0 is to provide learners with the essential skills needed to succeed in a fast-changing environment. The qualities included are critical thinking, flexibility, creativity, teamwork, and a global viewpoint (Poschauko et al. 2024).

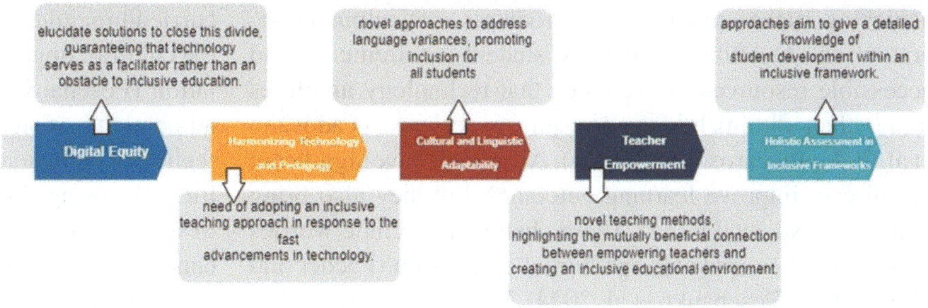

Fig. 1 Key concerns

In the rapidly evolving landscape of Industry 5.0, traditional methods of memorization and passive learning are inadequate. Instead, learners must nurture critical thinking abilities to analyse, assess, and resolve intricate problems. Furthermore, adaptability is paramount as technological advancements continuously reshape industries and job roles (Lagorio and Cimini 2024). Learners must be ready to adjust to new technologies and changing circumstances throughout their lives. In an interconnected world where cross-cultural communication and collaboration are increasingly important, a global perspective is indispensable. Educators can foster this perspective by incorporating global issues, viewpoints, and experiences into the curriculum (Pasi et al. 2021a). By nurturing these skills, inclusive curriculum pedagogy equips learners to navigate the challenges and opportunities of the digital age, empowering them to become lifelong learners and active contributors in shaping the future (Pasi et al. 2020a). Following are the key concerns of the in the rapidly evolving landscape of Industry 5.0 as shown in Fig. 1.

2.1 Digital Equity

As technology becomes an integral part of education, there exists nevertheless considerable anxiety about the digital divide. It is crucial to tackle the in- equalities in device availability, internet connection, and knowledge of digital technology. Disparities in device access pertains to the inequitable allocation of electronic devices, such as laptops, tablets, or cell phones, among students (Pasi et al. 2020b). Within several areas, some students may have limited access to these gadgets, either as a result of budgetary limitations or insufficient institutional backing. The lack of access to devices poses obstacles to engaging in digital learning activities and obtaining educational information (Pasi et al. 2020c). To tackle this problem, some options may include providing devices to students who need them, establishing programmes that lend out devices, or collaborating with organisations to get devices at reduced prices or via donations (Pasi et al. 2021b). Digital literacy encompasses the proficiency in using digital tools and technology to efficiently discover, assess, generate, and convey knowledge. A significant number of students, especially those from underprivileged families, may have a deficiency in fundamental computer literacy abilities, which might impede their academic achievements and limit their future prospects (Martínez-Gutiérrez et al. 2024; Mohamed Hashim et al. 2024). To bridge the digital literacy disparities, it is necessary to provide instruction and materials to kids, educators, and parents on the navigation of digital platforms, utilisation of productivity software, discernment of online information, and safeguarding of personal privacy and security on the internet (Bispo and Amaral 2024; Calvetti, et al. 2024). Incorporating digital literacy into the curriculum and providing continuous professional development for instructors are crucial elements of this endeavour. In order to tackle the issue of digital equality in education, a comprehensive strategy is needed that includes ensuring access to devices and internet connection, as well as fostering digital literacy skills among students, instructors, and family's curriculum (Karapanagiotis and Reimers 2024; Erp, et al. 2024). Through the implementation of precise measures aimed at closing these disparities, we can guarantee that technology serves to improve rather than impede the objective of providing inclusive education to all youngsters (Ramírez Molina et al. 2024; Rodrigues et al. 2024).

2.2 Harmonizing Technology and Pedagogy

The incorporation of emergent technology necessitates a careful equilibrium between novelty and inclusiveness. Emerging technologies like Artificial Intelligence (AI) and Virtual Reality (VR) have exceptional prospects to transform learning experiences in today's educational environment. Artificial Intelligence (AI) has the ability to tailor learning paths for individual students, provide immediate feedback, and streamline administrative duties (Bonello et al. 2024; Adela 2024). On the other hand, Virtual Reality (VR) can generate immersive simulations and virtual excursions to other locations. Educators have the ability to incorporate these technologies into their teaching methods in order to include students in interactive and hands-on learning activities, which may accommodate various learning styles and priorities (Knoth et al. 2024; Imjai et al. 2024). Nevertheless, achieving successful integration requires meticulous deliberation of educational objectives, student requirements, and accessible resources to guarantee that technology amplifies rather than diminishes the learning experience (Padovano et al. 2024; Piccarozzi et al. 2024). AI and VR have significant promise to improve learning outcomes, but they also bring ethical and societal issues. These challenges include worries about prejudice, privacy, and digital citizenship (Pacher and Glinik 2024; Poschauko et al. 2024). Educators must be diligent in eliminating prejudice in AI algorithms and ensuring that AI-driven learning systems are impartial, transparent, and responsible. This may include rigorously assessing the data used for training AI models and aggressively addressing biases in both content and evaluation. When using VR, educators must address concerns around representation and diversity in virtual settings (Lagorio and Cimini 2024; Pasi et al. 2021a). It is crucial to ensure that all students feel included and adequately represented in their learning experience. Furthermore, the promotion of responsible digital citizenship involves providing students with education on digital ethics, privacy rights, and the appropriate utilisation of technology. This education empowers students to traverse the world of technology securely and ethically.

2.3 Cultural and Linguistic Adaptability

The Industry 5.0 era is distinguished by the presence of classrooms that are diverse in terms of language and cultural background (Pasi et al. 2020a, 2020b). In the age of the Industry 5.0, classrooms are becoming more diversified, mirroring the interconnectivity of global communities. The presence of cultural and linguistic diversity enhances the learning environment by providing students with a wide range of viewpoints, experiences, and methods of acquiring knowledge. Educators may embrace this variety by integrating multicultural viewpoints into the curriculum, emphasising the contributions made by other cultures and groups (Pasi et al. 2020c). This facilitates the development of cultural awareness and comprehension among learners, cultivating empathy and reverence for a wide range of identities. Establishing a culturally responsive classroom atmosphere entails recognising and validating students' cultural origins and experiences, integrating teaching re- sources that are culturally appropriate, and offering occasions for students to share their cultural heritage with their classmates (Pasi et al. 2021b). Curriculum flexibility is crucial for successfully engaging learners who come from varied cultural and language backgrounds. This involves creating adaptable educational materials and procedures that may be customised to accommodate the requirements and preferences of various student demographics (Pacher and Glinik 2024). Educators should give priority to incorporating inclusion and representation into the curriculum, guaranteeing that the materials accurately depict the cultural, linguistic, and ethnic variety of the student population (Poschauko et al. 2024). This not only confirms the authenticity of students' identities but also strengthens their feeling of inclusion and active participation in academic activities. By using differentiated teaching, educators may cater to a wide range of learning styles, preferences, and skills, enabling all students to effectively engage with and excel in the curriculum, regardless of their background or language competence (Lagorio and Cimini 2024).

2.4 Teacher Empowerment

Teacher preparation has to change along with pedagogy. The Industry 5.0 is bringing about a shift in educational methods due to the fast improvements in technology and changes in the workforce (Pasi et al. 2021a). Educators must possess the requisite skills and expertise to adeptly traverse this ever-changing terrain. Teacher training programmes should provide priority to professional development oppor-tunities that specifically target the integration of new technologies, including Artificial Intelligence (AI), augmented reality (AR), and data analytics, into the process of teaching and learning (Pasi et al. 2020a). In addition, educators need instruction in digital literacy, critical thinking, problem-solving, and collaboration skills in order to adequately prepare students for the requirements of the Fifth Industrial revolution economy (Pasi et al. 2020b). This entails comprehending how to use technology to optimise education, tailor learning experiences, and provide individualised learning paths for students. As educational methods adapt to technological progress and shifting educational models, educators must investigate inventive ways to instruction and knowledge acquisition (Pasi et al. 2020c). Inquiry-based learning, project-based learning, flipped classrooms, and competency-based education are pedagogical approaches that encourage student engagement, critical thinking, and creativity in the era of Industry 5.0. By integrating these innovative teaching methods into their instruction, educators can establish dynamic and interactive learning settings that cultivate curiosity, experimentation, and collaboration among students (Pasi et al. 2021b).

3 Research Findings

In the ever-changing field of higher education, it is crucial to embrace new teaching methods in order to address the changing requirements of students. This section presents a thorough worldwide examination of three important teaching methods adopted in higher education institutions (HEIs) in the era of Industry 5.0. These are E-learning (Web-based learning), Project-based learning (PBL) and Blended

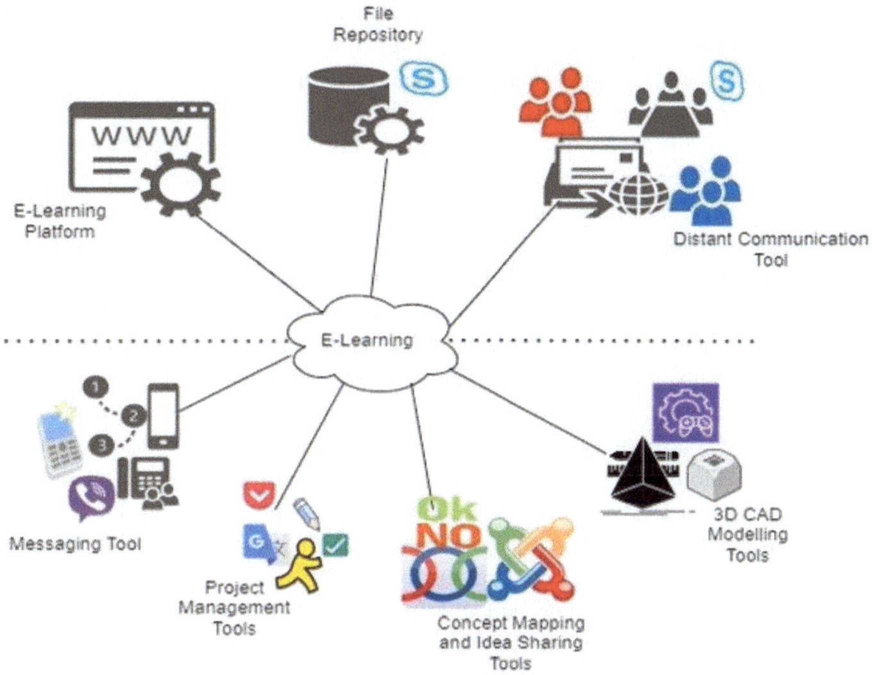

Fig. 2 E-learning

Learning (BL). Every method provides a distinct viewpoint on how teaching and learning are transformed to improve student involvement, analytical thinking and overall educational achievements.

3.1 E-learning (Web-Based Learning) Pedagogy

E-learning, or web-based learning, utilises digital technology to transmit instructional information and enable learning experiences over the internet. This pedagogical technique allows learners to remotely access instructional materials, resources, and activities, irrespective of their physical location. It surpasses the limitations of conventional classrooms, providing students with the freedom and convenience to study at their own speed and schedule. E-learning encompasses diverse modalities, such as synchronous virtual classrooms where students and instructors engage in real-time interaction via video conferencing or chat platforms, as well as asynchronous self-paced modules where learners access pre-recorded lectures, readings, and assignments at their convenience. The widespread use of e-learning has revolutionised the field of education, enhancing accessibility and inclusivity in learning possibilities. The democratisation of education has occurred via the dismantling of obstacles to access, enabling people from all backgrounds and geographic areas to engage in learning and enhance their skills. E-learning has also promoted lifetime learning and professional growth, enabling people to get fresh information and abilities to adjust to evolving social and economic needs. In addition, e-learning has created opportunities for collaborative learning and information sharing, promoting the formation of global communities of learners and educators who interact and share ideas internationally as shown in Fig. 2. In general, e-learning has had a significant influence on education by providing exceptional adaptability, availability, and prospects for continuous learning in the era of digital technology. To put it concisely, e-learning, also known as web-based learning, is an innovative teaching method that utilises digital technology to make education more accessible and increase learning possibilities worldwide. The flexibility, accessibility, and capacity for cooperation of online platforms have fundamentally transformed the methods of teaching and learning in higher education institutions.

3.2 Project-Based Learning Pedagogy

Project-based learning (PBL) is an educational method that involves students actively participating in immersive learning experiences, where the emphasis is on student-centered learning rather than teacher-led teaching. Project-Based Learning (PBL) involves students participating in genuine, real-life assignments that need them to resolve intricate issues, do research, and employ theoretical understanding to practical situations. Project-Based Learning (PBL) prioritises dynamic learning, investigation, and cooperation, providing students the chance to delve into subjects extensively and develop their abilities in critical analysis and problem-solving. Projects in problem-based learning (PBL) often include the integration of ideas from other subject areas and are designed to mirror real-world issues and circumstances. This interdisciplinary approach fosters students' ability to establish links across many fields and effectively use their acquired knowledge in practical and significant ways. Project-Based Learning (PBL) fosters student independence and accountability for their learning, enabling them to explore subjects they find intriguing, establish objectives, and choose the methods and completion of their projects. Through active participation in practical, project-oriented tasks, students develop crucial 21st-century competencies like effective communication, cooperative teamwork, innovative thinking, and adaptability. They acquire the skills to collaborate efficiently in groups, articulate their thoughts persuasively, and adjust to dynamic situations. Project-Based Learning (PBL) enhances students' comprehension of academic content by offering them chances to use abstract principles in practical situations. By using this experiential learning strategy, students are able to better retain and apply information due to their increased perception of its relevance and practicality. In general, problem-based learning (PBL) fundamentally changes the process of acquiring knowledge, enabling students to take an active role in their education and equipping them for achievement in a swiftly evolving society. Project-Based Learning (PBL) enhances critical thinking abilities by prompting students to scrutinise issues, assess evidence, and devise innovative resolutions. Collaboration is an essential element of Project-Based Learning (PBL), whereby students engage in cooperative teamwork to generate ideas, allocate responsibilities, and accomplish common objectives. The inherent authenticity of PBL projects fosters a more profound comprehension of the subject matter by establishing meaningful connections between theoretical ideas and practical real-world applications. Project-Based Learning (PBL) equips students with the necessary skills to navigate the demands of a swiftly evolving labour market, fostering crucial abilities such as adaptation, resilience, and problem-solving. Illustrations of prosperous PBL deployments across a wide range of fields demonstrate the adaptability and efficacy of this instructional method in promoting student involvement, drive, and accomplishment. Project-based learning (PBL) is an educational method that fosters active, inquiry-based learning and equips students with the necessary skills for success in the modern day. Through active participation in practical assignments, students develop essential cognitive abilities such as analytical thinking, teamwork, and the ability to resolve complex issues, all while enhancing their comprehension of the academic content. The efficacy of the Problem-Based Learning (PBL) method in promoting student learning and equipping them for the demands of a constantly changing workforce is underscored by the successful application of PBL in several fields.

3.3 Blended Learning Pedagogy

Blended Learning integrates conventional in-person teaching with online elements, resulting in a versatile and individualised learning encounter. Blended Learning is an educational method that integrates conventional in-person teaching with online elements, resulting in a hybrid learning experience. Blended learning is a combination of face-to-face classroom activities with online learning activities. These online activities include accessing digital materials, completing assignments, engaging in conversations, and getting feedback. The integration of in-person and virtual components enables a versatile and adjustable educational encounter that accommodates various learning styles, preferences, and requirements. Blended Learning is the seamless integration of technology into the classroom, using digital tools and resources to improve teaching and learning experiences. Technology-driven elements of blended learning may include learning management systems (LMS), multimedia presentations, interactive simulations, virtual laboratories, online conversations, and collaborative tools. By integrating technology into teaching, educators may create dynamic and captivating learning experiences that promote active engagement, cooperation, and customised learning paths for students. Blended Learning offers a well-rounded method of teaching that accommodates various learning styles and interests. Some students excel in direct contact with teachers and classmates, while others enjoy the freedom and independence provided by online learning settings. Blended Learning enables educators to use the advantages of both conventional classroom teaching and online learning, offering students the chance to engage in practical tasks, collaborate with peers, get immediate feedback, and study at their own speed. Blended Learning (BL) is an educational approach that integrates conventional in-person teaching with online elements, resulting in a versatile and customised learning encounter. Through the seamless integration of technology in the classroom, a balanced approach for various learning styles, and the demonstration of successful blended learning implementations, educators can harness the advantages of blended learning to establish dynamic and flexible educational environments that cater to the needs of diverse learners.

4 Conclusions

This research paper emphasises the crucial significance of promoting inclusive curricular pedagogy within the age of Industry 5.0. The stated objective and worries underscore the need for educators and politicians to together design an educational system that is prepared for the future. This system must not only adapt to fast technology changes but also prioritise inclusion, guaranteeing that all learners have equitable access to high-quality education. The paper provides a comprehensive examination of the significant influence of different teaching methods in the era of Industry 5.0 such as E-learning, Project-Based Learning and Blended Learning in the education process in Higher Education In-stitutions (HEIs) globally. Through a comprehensive analysis of the advantages, difficulties, and international differences of different teaching methods, educators, policymakers, and stakeholders may make well-informed choices to provide stimulating and efficient learning settings to understand Industry 5.0 concepts. Through the use of inclusive curricular methodology, educators may design educational experiences that accommodate the varied needs and backgrounds of students, therefore promoting a feeling of inclusion and empowerment. In addition, Higher Education Institutions (HEIs) may effectively equip learners to successfully adapt to future problems and seize opportunities by using cutting- edge teaching methods such as E-learning, Project-Based Learning and Blended Learning. Essentially, this paper promotes a comprehensive approach to education to adopt Industry 5.0 that not only welcomes technological progress but also prioritises diversity, equality, and excellence. Through collaborative efforts, educators and policymakers may provide a solid groundwork for an educational system that prepares learners to succeed in the Industry 5.0 environment by promoting inclusive curricular methodology. To summarise, it is crucial to prioritise the development of inclusive curricular pedagogy in the age of Industry 5.0. In light of the ongoing transformation of education and the workforce by technology, it is imperative for educators and policymakers to collectively address the stated objectives and concerns. By doing this, they may provide the foundation for a future-oriented educational system that not only keeps up with technological progress but also firmly embraces diversity. This paper has emphasised the significance of equipping learners with the necessary skills to effectively manage the difficulties and enthusiastically embrace the possibilities that lie ahead in the future. To do this, educators and policymakers must give priority to inclusion while designing curricula and selecting teaching methods. This entails ensuring that educational opportunities are readily available to all learners, irrespective of their background, ability, or circumstances. E-learning provides learners with flexibility and accessibility, enabling them to access educational resources and engage in activities at their preferred speed and convenience. Project-Based Learning promotes the development of critical thinking skills, encourages cooperation, and allows for the practical application of information in real-world situations. Blended Learning is a teaching approach that integrates in-person education with online elements, offering a versatile and individualised learning encounter. By capitalising on the advantages of these teaching methods and tackling their difficulties, educators, policymakers, and stakeholders may establish stimulating and efficient learning settings that equip learners for triumph in the Industry 5.0. By working together and fostering creativity, they have the ability to mould an educational system that is prepared for the future and provides learners with the necessary abilities, information, and mentality to succeed in a rapidly evolving environment.

Acknowledgements The authors want to thank the editors for their valuable advices on this paper. We thank the reviewers for their careful and insightful work on this paper.

Disclosure of Interests. The authors declare no conflict of interest, financial or otherwise.

References

Adela, V. et al.: Bookkeeping practices and SME performance: The intervening role of owners' accounting skills. Heliyon **10**, (2024)

Bonello, A., Francalanza, E., Refalo, P.: The realities of achieving a smart, sustainable, and Inclusive shopfloor in the age of Industry 5.0. in Procedia computer science vol. 232, pp. 2406–2415 (Elsevier B.V., 2024)

Bispo, L.G.M., Amaral, F.G.: The impact of Industry 4.0 on occupational health and safety: A systematic literature review. J. Safety. Res., (2024) https://doi.org/10.1016/j.jsr.2024.04.009

Calvetti, D. et al.: Experiencing education 5.0 for civil engineering. In: Procedia computer science vol. 232, pp. 2416–2425. Elsevier B.V., (2024)

Imjai, N., Aujirapongpan, S., Yaacob, Z.: Impact of logical thinking skills and digital literacy on Thailand's generation Z accounting students' internship effectiveness: Role of self-learning capability. Int. J. Edu. Res. Open. **6**, (2024)

Karapanagiotis, P., Reimers, P.: Why does the schooling gap close while the wage gap persists across country income comparisons? J. Econ. Dyn. Control. **159**, (2024)

Knoth, N., Tolzin, A., Janson, A., Leimeister, J.M.: AI literacy and its im- plications for prompt engineering strategies. Comp Educ: Artific Intelli. **6**, 100225 (2024)

Lagorio, A., Cimini, C.: Towards 5.0 skills acquisition for students in industrial engineering: the role of learning factories. In: Procedia Computer Science vol. 232, pp. 317–326. Elsevier B.V., (2024)

Martínez-Gutiérrez, A., Díez-González, J., Perez, H., Araújo, M.: Towards in- dustry 5.0 through metaverse. Robot. Comput. Integr. Manuf. **89**, (2024)

Mohamed Hashim, M.A., Tlemsani, I., Mason-Jones, R., Matthews, R., Ndrecaj, V.: Higher education via the lens of industry 5.0: Strategy and perspective. Soc Sci Huma Open. **9**, (2024)

Padovano, A., Cardamone, M., Woschank, M., Pacher, C.: Exploring human-centricity in industry 5.0: Empirical insights from a social media discourse. In: Procedia computer science vol. 232 1859–1868. Elsevier B.V., (2024)

Piccarozzi, M., Silvestri, C., Fici, L., Silvestri, L.: Metaverse: a possible sustainability enabler in the transition from Industry 4.0 to 5.0. In: Procedia computer science vol. 232, pp. 1839–1848. Elsevier B.V., (2024)

Pacher, C., Glinik, M.: Fostering entrepreneurial mindsets in deep tech disciplines: Exemplary development of a toolkit. In: Procedia computer science vol. 232, pp. 1309–1318. Elsevier B.V., (2024)

Poschauko, V.C., Kreuzer, E., Hirz, M., Pacher, C.: Engineering education goes lifelong learning: modularized technical vocational education and training program for the automotive sector. In: Procedia Computer Science vol. 232, pp. 1799–1808. Elsevier B.V., (2024)

Pasi, B.N., Mahajan, S.K., Rane, S.B.: A method for performing forging operation: a perspective of industry 4.0. Recent. Pat. Mech. Eng. **14**, 423–435 (2021a)

Pasi, B.N., Mahajan, S.K., Rane, S.B.: Redesigning of smart manufacturing system based on iot: perspective of disruptive innovations of industry 4.0 paraDIGM. www.tjprc.org SCOPUS Indexed Journal. editor@tjprc.org www.tjprc.org (2020)

Pasi, B.N., Mahajan, S.K., Rane, S.B.: Smart supply chain management: A perspective of industry 4.0. Int. J. Adv. Sci. Technol. **29**, 3016–3030 (2020b)

Pasi, B.N., Mahajan, S.K., Rane, S.B.: The current sustainability scenario of Industry 4.0 enabling technologies in Indian manufacturing industries. Int. J. Prod. Perform. Manag. **70**, 1017–1048 (2020c)

Pasi, B.N., Mahajan, S.K., Rane, S.B.: Development of innovation ecosystem framework for successful adoption of industry 4.0 enabling technologies in Indian manufacturing industries. J. Sci. Technol. Policy Manag., (2021b) https://doi.org/10.1108/JSTPM-10-2020-0148

Ramírez Molina, R.I., Santamaria Ruiz, M.J., Ramírez Molina, R.J., Lay Raby, N.D., Severino-González, P.: Innovation systems in industry 5.0: Theoretical and methodological bases. In: Procedia computer science vol. 231, pp 595–600. Elsevier B.V., (2024)

Rodrigues, V., Breda, Z., Rodrigues, C.: The implications of industry 4.0 for the tourism sector: A systematic literature review. Heliyon, e31590 (2024) https://doi.org/10.1016/j.heliyon.2024.e31590

van Erp, T. et al.: Industry 5.0: A new strategy framework for sustainability management and beyond. J Clean Prod., 142271 (2024) https://doi.org/10.1016/j.jclepro.2024.142271

Production Leveling Implementation Methodologies: Comparison and Review

Meryem Boutbagha and Laila El Abbadi

Abstract

This work aims to delve into the implementation of production leveling by comparing the used methodologies in three different contexts (batch production, Job shop production, and mixed model assembly line in high-volume production). A review is conducted to synthesize the available literature in scientific publications. The study also included three important Toyota Production Systems (TPS) books to discover how its pioneers explained the concept of production leveling and its implementation. The analysis showed us that despite the differences in the three studied contexts, the implementation of production leveling contains three common phases: selection of the products to be leveled, grouping of the chosen products into families, the elaboration of the leveling pattern (defining the quantity and the sequence of the mix of products).

Keywords

Production leveling · Heijunka · Implementation · Toyota production system · Goal-chasing algorithm

1 Introduction

Production leveling, also known as production smoothing or heijunka, is the solution Toyota adopted to adapt production to variable demand (Monden 2012). Variation in demand complicates the development of appropriate production plans due to the uncertainty it causes (Nyaga et al. 2007), hence the unfortunate consequences on service performance and customer satisfaction. Indeed, production leveling, as the name implies, aims to "lower the peaks and raise the valleys in production as much as possible so that the flow surface is smooth" (Ohno 1988), that is, to minimize fluctuation (quantity variance) in the production line (Monden 2012).

Production leveling is a mixed production system (Coleman and Vaghefi 1994) based on producing mixed products in small quantities, as a result of dividing equally the customer orders collected during an interval of time (Boutbagha and Abbadi 2022; Liker 2004), instead of producing according to the order of customer demand, which implies that the volume and the type of the products are both leveled (Boutbagha and Abbadi 2022, 2024).

Notably, a successful implementation of production leveling depends on minimization of the setup processes (Boutbagha and Abbadi 2022, 2024) guaranteed by the use of techniques such as SMED (Coleman and Vaghefi 1994; Boutbagha and Abbadi 2022), Single Minute Exchange of Die.

In literature, production leveling has not received much attention, although being the Toyota Production System (TPS) most overarching principle (Coleman and Vaghefi 1994; Boutbagha and Abbadi 2022). It is advantageous for businesses wherein a multitude of models has to be produced, such as the automotive industry (Fortuny-Santos and Ruiz de Arbulo 2024). In this respect, the present work aims to compare methodologies of implementing production leveling (heijunka) in different contexts. The review takes into consideration previously published scientific works (journal articles, conference papers) in addition to three of the most influencing books in TPS history, namely: « Toyota Production System: Beyond Large-Scale Production » of Taiichi Ohno, « The Toyota Way » of J.F. Liker, "Toyota Production System: An Integrated Approach to Just-In-Time" of Yasuhiro Monden. This is of great importance because it allows us to understand how the concept of heijunka, and its implementation obviously, evolved from its infancy to nowadays.

M. Boutbagha (✉) · L. El Abbadi
Laboratory of Engineering Sciences, National School of Applied Sciences, Ibn Tofail University, Kenitra, Morocco
e-mail: meryem.boutbagha@uit.ac.ma

B. El Bhiri et al. (eds.), *AI and Digital Transformation: Innovations in Supply Chain, Education, and Energy Systems*, Sustainable Artificial Intelligence-Powered Applications, https://doi.org/10.1007/978-3-031-86837-5_8

The remainder of this paper is structured into five sections. After the introduction comes the "Methods" section, in which we explained how the study has been conducted. The third section "Results" discusses the content of the three pre-cited TPS books and the available literature. The fourth section "Discussion" contains the answer to the research question based on the results. The last section of the review is the "Conclusion" wherein the major conclusions are given, and the perspectives of future research are proposed.

2 Methods

As mentioned above, this work is a review of the methodologies of production leveling implementation in various contexts and thus answers the following research question: Does production leveling have a one-size-fits-all methodology?

The relevant literature included in this study consists of journal articles and conference papers published between 2012 and 2023. The review was limited to articles written in English and was done using the following keywords: Heijunka, production smoothing, production leveling, mixed model production, and implementation. After searching different databases and basing on analyzing the abstracts, we only kept the relevant papers that included details about production leveling implementation methods. That way, the data set of articles used in the present work is 14 (except the three books).

3 Results

The book of Taiichi Ohno (Ohno 1988) is known to be the first that talked about TPS as the writer is the one that set up its foundations. Taiichi Ohno used the terms load smoothing and production leveling rather than Heijunka to describe the adopted solution to obtain "zero fluctuation at the final assembly line, or the last process" (Ohno 1988). In addition, Ohno mentioned that production leveling is a keystone to set up the Kanban system (Ohno 1988). However, Ohno has not given any detail about how to implement production leveling. This may be since the book is "more a narration than a real model or pattern"(Chiarini et al. 2018). Contrarily, J.F. Liker used the term "Heijunka" and described it as the basis of TPS and as its fourth principle, for the reason that it is essential to eliminate waste (Muda), unevenness (Mura) and overburden (Muri) (Boutbagha and Abbadi 2022; Liker 2004; Elrhanimi et al. 2020). The production leveling, according to J.F. Liker, consists of taking the actual customer demand, define the pattern of leveling (the volume and the mix of products), then build an everyday levelled schedule (Liker 2004).

Neither Ohno nor Liker have given details about the methodology of implementing production leveling but Monden has, in his book entitled "Toyota Production System: An Integrated Approach to Just-In-Time" (Monden 2012). He used the term production smoothing and defined it as the way to "diminish as much as possible the quantity variance in a production line" (Monden 2012).

The process of smoothing the production, according to Monden, consists of firstly smoothing the production quantity, then defining the pattern/the sequence of the mixed models so that the consumption speed of each part on the production line be constant (Monden 2012). To this purpose, Toyota uses the Goal-Chasing Algorithms I and II then later the Goals-Coordinating Method (which is nothing but an evolution of the Goal-Chasing Method (Monden 2012)), that Monden explained exhaustively in his above-mentioned book.

Based on an objective function (1) that uses different parameters related to quantities of parts necessary to assemble the total production quantity Q, the Goal Chasing Algorithms aim to select products, one by one, in order to generate the sequence of the mixed models by solving the objective function Q times (Mane et al. 2002). In other words, the algorithms "find a product sequence one position at a time by solving the objective function" (Mane et al. 2002). The main limitation to those algorithms is the extremely high number of parts used in production, especially in automotive industry, which makes them difficult to apply (Monden 2012).

$$D_{Ki*} = \min\{D_{ki}\}/i\epsilon_{S_{k-1},i} \tag{1}$$

$$Where\ D_{ki} = \sqrt{\sum_{j=1}^{\beta}\left(\left(K.\frac{N}{Q}\right) - X_{j,k-1} - b_{ij}\right)^2} \tag{2}$$

The function (1) aims to minimize the distance (difference) that separates the total necessary quantity of a part a_j used to produce the products of a determined sequence from 1st to Kth designated by $X_{j,k}$ and the average necessary quantity of a part a_j for producing k units of products, which is [$(K.N_j)/Q$] (Monden 2012). Q being the total production quantity of all products, as above stated.

Thus, for each position k of the sequence (k= {1, 2,……,Q}, D_{ki} is computed for each product i. The product i* whose D_{ki*} is the minimum among all the D_{ki} is the one that will be put in the position k of sequence (Monden 2012).

Monden is considered the one who initiated the Production Smoothing Problems (PSP) that arose in the context of JIT manufacturing, and his model was later adapted by other

Table 1 Production contexts according to the combination of quantity and variety of products (Groover 2020)

		Quantity		
		Low	Medium	High
Variety	Hard	Job shop (airlines industry)	Batch production	Flow line production-mixed model production line
	Soft	–	Cellular manufacturing	Quantity production
				Flow line production-single model production line

Table 2 Details of papers, per category, included in the study

Categories	References
Job shop production	(Bohnen et al. 2011; Birkmann and Deuse 2007; Bohnen et al. 2013)
Batch production	(Araujo and Queiroz 2010a; Araujo and Queiroz 2010b)
High volume production-mixed model assembly line	(Żywicki et al. 2017; Rewers et al. 2017; Rewers et al. 2018; Rewers et al. 2019a; Rewers 2019; Rewers and Diakun 2021; Rewers et al. 2021)

authors such as Miltenberg (Miltenburg 1989) and Kubiak (Kubiak 1993). Monden's Model aims to reduce the variability of the consumption rates of the sub-assemblies at the final stage (Monden 2012), which Kubiak called Out-put Rate Variation ORV (Kubiak 1993), whereas Miltenberg's focuses on the product level (Miltenburg 1989) (Product Rate Variation PRV) as it is cited above.

To sum up, Monden worked on smoothing the usage of sub-assemblies (usage goal) by minimizing the deviation of actual and specified ideal consumption amounts (Monden 2012), caused by the inability of the production resources to process multiple products at the same time (Yavuz and Akçali 2007).

We will devote the second part of the discussion to the analysis of the available literature that studied the methodologies of implementing production leveling in different contexts, depending on the combination of both the quantity of production and the variety of products. Table 1 gives more details about the possible existing combinations. However, the focus of this study will be on the following production contexts: Job shop production, batch production, and high-volume production. It should be pointed out that the leveling of production was initiated at first in assembly lines (Yavuz and Akçali 2007). Table 2 shows details about the articles included in this study.

3.1 Production Leveling in Job Shop Production: Family-Oriented Leveling Pattern

The authors in Bohnen et al. (2011); Birkmann and Deuse 2007; Bohnen et al. 2013) were interested in studying the production leveling in a low volume and high mix production context, which is also known as "job shop production": The products are highly variated, the equipment should be multi use and the work force with high and flexible skills (Groover 2020; Bohnen et al. 2011). In addition to those requirements, the setting up of production leveling needs the level of inventory to be supervised and the changeover time to be quick (Bohnen et al. 2011).

The methodology of implementing production leveling as it is proposed by the authors contains four main steps and is based on principles of Group Technology (GT), that makes possible the classification of products to form groups or families considered as a basis for leveling (Birkmann and Deuse 2007). It also aims to make it possible to shorten the large number of product types (Bohnen et al. 2011). The methodology is summarized in Table 3.

In the same context, the job shop production context, the authors in Kulsum and Pratiwi (2019) implemented production leveling in a company and compared its benefits to another method called the "Non delay method". To that purpose, they used a "sequencing rule" formula to calculate the position of the corresponding product type in the sequencing pattern.

3.2 Production Leveling in Batch Production

We talk about batch production when the product variety is high and the annual average quantity is medium (Groover 2020). Each group/batch of a product type is produced at one time, then the setup/changeover is made to produce a batch of another product type, et cetera (Groover 2020).

Table 3 Implementing production leveling in job shop production

Steps	Sub-steps
Analysis and constitution of the leveling model	• Definition of the volume to be levelled • Determination of the "leveling period" for production scheduling (one day, three days period, one shift…) • Determination of specific characteristics of leveling (whether the leveling concerns only the volume or both the volume and the mix)
Formation of product families for leveling	• Selection of grouping criteria: The most relevant are the manufactured-oriented ones such as the process time, the setup time, the operation sequences…
	• Application of different clustering algorithms: The first step is the preparation of raw data to ensure the constitution of a matrix that contains the attributes of each product type depending on the grouping/clustering criteria ✔ Then a measure, called "Proximity measure", is defined to quantify the similarities of the objects, described as: $d(X_j, X_l) = \sqrt{\sum_{i=1}^{p} (x_{ji} - x_{li})^2}$ X_j and X_l are two product types while x_{ji} and x_{li} are the values of the attribute i for the products j and l ✔ To choose the most appropriate algorithm, a validation function is chosen and is defined as the geometric mean of the desirability functions w_i (validation criteria) that correspond to the criteria chosen by the authors: such as the Homogeneity of the formed families
Creation of the levelling pattern	• The manufacturing interval is defined by Every Family Every Interval (EFEI). The EFEI depends on the demand forecasts and the available capacity for changeover • Two products families are defined: "Runners", which are the most concerned by the leveling pattern and "Strangers" • The minimal total setup time is the decisive factor in defining the sequence in the leveling pattern which is kept constant for the "leveling period"
Realization and continuous improvement	• The continuous improvement in the leveling pattern is made gradually by comparing the current and the target condition

In their works (Araujo and Queiroz 2010a, 2010b), the authors focused on studying the application of leveling production in a batch production system as it is an appropriate system of a wide range of manufacturing processes, including the automotive industry. They also state that the used methodology to implement production leveling is suitable for all types of manufacturing systems that have similarities with the batch production system, in their processing operations (Araujo and Queiroz 2010a). Indeed, the authors' methodology combines the traditional framework for operations planning (with changes at tactical and operational levels) and the basic concepts of Lean Manufacturing (Araujo and Queiroz 2010b). A case study is conducted in a company, which was named Company A, and a comparison of the situation before (previous state) and after (future state) implementation is done using performance indicators (Araujo and Queiroz 2010a). The process of production leveling as it is described by Araujo and Queiroz 2010a; Araujo and Queiroz 2010b) is presented in Fig. 1.

3.3 Production Leveling in High-Volume Production

The authors in Żywicki et al. (2017); Rewers et al. 2017; Rewers et al. 2018; Rewers et al. 2019a; Rewers 2019; Rewers and Diakun 2021; Rewers et al. 2021) stated that the majority of the works interested in production leveling proposed general-purpose methodologies making them hard to apply in real life conditions (Rewers et al. 2019b). For that reason, there was a necessity for taking into account those conditions when developing new methodologies (Rewers et al. 2019b). Consequently, the authors own methodology consists of six steps (Rewers et al. 2018) presented in the table (see Table 4).

The proposed methodology was developed according to their own experience and not on any pre-established studies (Rewers et al. 2017). Remarkably, each of their works discusses one of the below-mentioned steps.

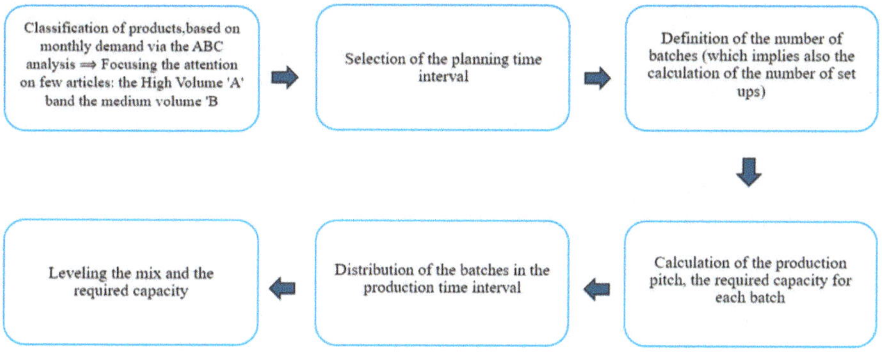

Fig. 1 The leveling production process in the batch production context

Table 4 Leveling production in high-volume production

Steps	Sub-steps
Selection of the products to be leveled	The classification takes an interest in products under regular production and brings out three categories: "fast moving" (key products), "medium moving" and "slow moving" products. It was based on three main criteria which are: rotation of products (the speed at which the products are sold), sales of products and the minimum, economic production batch (Rewers et al. 2019b)
Grouping the selected products into families	The authors had recourse to a data analytics program called ANAPRO to form the product families, which are defined as groups/clusters of products sharing the same technological features (Żywicki et al. 2017). It is also possible to use a matrix of technological similarities of products (Rewers et al. 2019a)
Analysis of the frequency and number of customers orders	The analysis should be performed in order to define the mean demand of the previously selected products (Rewers et al. 2019a)
Calculation of stock levels	It is about determining the stock levels at the supermarket (Żywicki et al. 2017)
Definition of the value of EPEI and the batch size	This fifth step consists of defining the production batch size (the quantity) and production frequency (time) regarding the selected products for the leveling (Rewers et al. 2021)
Constitution of the levelled plan by defining the sequence and the quantity of the products to be produced	Priority rules (Shortest task time, nearest task due date, First In First Out…etc.) are defined to constitute the production plan, and are assessed to select which of them meets the requirements. The assessment is done via weighted criteria, on a simulation computer program developed by the authors To constitute a levelled production plan, three factors should be taken into account: the production intervals, the sequence of the mix of products, and the lot size (the quantity of each model to be produced) (Rewers 2019)

4 Discussion

Toyota called production leveling the concept of levelling out the production schedule and considered it as the foundation to its production system. Previous works such as (Korytkowski et al. 2013; Gupta et al. 2021; Aoki and Katayama 2018) stated that the implementation of this concept is advantageous, whether it is in manufacturing or service sectors.

The research question presented in this article is: "Does production leveling have one size-fits-all methodology?". The study has clearly shown that, despite the different contexts in which the production leveling has been applied, the proposed methodologies have got common main phases:

- Before elaborating the levelled production plan, one should first choose the products to be levelled. This selection is important in the way that it helps focusing on a special range of products, those that are regularly sold or most required by the customers. The selection could be done basing on an ABC analysis, Glenday Sieve technique (and obviously other classification methods) and using different criteria such as: sales of products.
- The second common phase is grouping the chosen products into families that have technological similarities. That is to say, the products that have similarities in the manufacturing process are clustered in groups to reduce the changeover time.
- The elaboration of the levelling pattern: It is about defining the quantity of products to be levelled and establishing the sequence of the mix of products, i.e., the order in which the different products will be manufactured, then defining the time interval during which the same sequence will be repeated.

With regard to the books included in this study, it is only Monden's book that discussed exhaustively how to implement heijunka. He especially talked about the third phase "the elaboration of the levelling pattern" and explained in detail how the sequence of the mix model is defined basing on the famous algorithm developed by Toyota, "The Chasing Algorithm" and using the following parameters:

- Q: Total production quantity of all products A_i.
- N_j: Total necessary quantity of a part a_i to be consumed for producing all products A_i.
- X_{jk}: Total necessary quantity of the part a_j to be utilized for producing the products of determined sequence from first to Kth

  K $(N/Q_j) \cdot =$ Average necessary quantity of the part a_j for producing K units of products.

5 Conclusion

From the analysis that has been carried out, we tried to answer the following research question: Does production leveling have one size-fits-all methodology?

Summing up the results, we can clearly make two statements:

- Although production leveling has been applied first to limited production diversity environments, it has evolved to cover other environments as it is explained in the sections below.
- Even though every context wherein production leveling has been implemented has got its own specifications, the proposed methodologies have got commonalities: We should first choose the products that will be leveled and then group them into families depending on their similarities. The elaboration of the leveled production plan is the last common phase, in which we define the quantity and the sequence of the mix of products. Finding the right sequence of products (sequencing problems) constitutes the most important (but not sufficient) part of the process of implementing production leveling. The only concern is about the applicability of the proposed methodologies in real production contexts.

The limited set of articles included in the present review emphasizes the existing gaps in the literature and indicates a need for further research for a thorough exploration and understanding of the topic.

In our future research, we intend to concentrate on studying production leveling from a practical point of view to find out how it is applied and in what form. Contrary to our previous works, which are based on literature reviewing, the next stage will be addressed to practicians of production leveling in a manufacturing context. More research is also needed to explore the applicability of different scheduling techniques (Petri net, fuzzy logic, neighborhood search methods… etc.) in defining the sequence of the mix of products.

References

Aoki, R., Katayama, H.: Heijunka operation management of agri-products manufacturing by yield improvement and cropping policy. In: Xu, J., Gen, M., Hajiyev, A., Cooke, F.L. (eds.), Proceedings of the eleventh international conference on management science and engineering management. Springer International Publishing, Cham, pp. 1407–1416 (2018)

de Araujo, L.F., de Queiroz, A.A.: A conceptual model for production leveling (Heijunka) implementation in batch production systems. In: Vallespir, B., Alix, T. (eds.), Advances in production management systems. New challenges, new approaches, pp. 81–88. Springer, Berlin, Heidelberg, (2010a)

de Araujo, L.F., de Queiroz, A.A.: Production Leveling (Heijunka) implementation in a batch production system: A case study. In: Vallespir, B., Alix, T. (Eds.), Advances in production management systems. New challenges, new approaches, pp. 105–112. Springer, Berlin, Heidelberg, (2010b)

Birkmann, S., Deuse, J.: Using a group technology approach to level a low volume and high mix production, vol. 6, (2007)

Bohnen, F., Maschek, T., Deuse, J.: Leveling of low volume and high mix production based on a Group Technology approach. CIRP J. Manuf. Sci. Technol. **4**, 247–251 (2011). https://doi.org/10.1016/j.cirpj.2011.06.003

Bohnen, F., Buhl, M., Deuse, J.: Systematic procedure for leveling of low volume and high mix production. CIRP J. Manuf. Sci. Technol. **6**, 53–58 (2013). https://doi.org/10.1016/j.cirpj.2012.10.003

Boutbagha, M., Abbadi, L.E.: Heijunka-Levelling customer orders: A systematic literature review. Int. J. Prod. Manag. Eng. (2024). https://doi.org/10.4995/ijpme.2024.19279

Boutbagha, M., El Abbadi, L.: Production leveling or Heijunka: A bibliometric study. In: 2022 IEEE 3rd International Conference on Electronics, Control, Optimization and Computer Science (ICECOCS), pp. 1–5 (2022)

Chiarini, A., Baccarani, C., Mascherpa, V.: Lean production, Toyota Production System and Kaizen philosophy: A conceptual analysis from the perspective of Zen Buddhism. The TQM Journal **30**, 425–438 (2018). https://doi.org/10.1108/TQM-12-2017-0178

Coleman, B.J., Vaghefi, M.R.: Heijunka (?): A key to the Toyota production system—ProQuest. (1994). https://www.proquest.com/openview/a513208ef093a1e2314d74213a0bf4c2/1?pq-origsite=gscholar&cbl=36911. Accessed 9 Jun 2022

Elrhanimi, S., Abbadi, E.L.L.: Assessment model of lean effect (AMLE). The TQM Journal **33**, 1020–1048 (2020). https://doi.org/10.1108/TQM-02-2019-0039

Fortuny-Santos, J., Ruiz de Arbulo, P.: Heijunka and Jidoka in Spanish Research. A systematic literature review. In: Bautista-Valhondo, J., Mateo-Doll, M., Lusa, A., Pastor-Moreno, R., (eds.), Proceedings of the 17th International Conference on Industrial Engineering and Industrial Management (ICIEIM)—XXVII Congreso de Ingeniería de Organización (CIO2023). Springer Nature Switzerland, Cham, pp. 421–426 (2024)

Groover, M.P.: Fundamentals of modern manufacturing: materials, processes, and systems. John Wiley & Sons (2020)

Gupta, P., Kumar, S.: Productivity improvements in an Indian automotive OEM using Heijunka, A lean manufacturing approach: a case study. In: Sachdeva, A., Kumar, P., Yadav, O.P., et al. (eds.) Operations Management and Systems Engineering, pp. 161–173. Springer, Singapore (2021)

Korytkowski, P., Wisniewski, T., Rymaszewski, S.: Multivariate simulation analysis of production leveling (heijunka)—a case study. IFAC Proc. **46**, 1554–1559 (2013). https://doi.org/10.3182/20130619-3-RU-3018.00285

Kubiak, W.: Minimizing variation of production rates in just-in-time systems: A survey. Eur. J. Oper. Res. **66**, 259–271 (1993). https://doi.org/10.1016/0377-2217(93)90215-9

Kulsum, M.Y., Pratiwi, A.S.: Comparison of job shop production scheduling by using the non-delay method and the Heijunka method at PT XYZ. IOP Conf Ser: Mater Sci Eng **673**, 012089 (2019). https://doi.org/10.1088/1757-899X/673/1/012089

Liker, J.K.: The Toyota way: 14 Management principles from the World's Greatest Manufacturer, Reissue édition. McGraw-Hill Professional, New York, NY (2004)

Mane, A., Nahavandi, S., Zhang, J.: Sequencing production on an assembly line using goal chasing and user defined algorithm. In: Proceedings of the winter simulation conference, vol. 2, pp. 1269–1273 (2002)

Miltenburg, J.: Level schedules for mixed-model assembly lines in just-in-time production systems. Manage. Sci. **35**, 192–207 (1989). https://doi.org/10.1287/mnsc.35.2.192

Monden, Y.: Toyota production system: An integrated approach to just-in-time, 4th Edn. Routledge & CRC Press, (2012). https://www.routledge.com/Toyota-Production-System-An-Integrated-Approach-to-Just-In-Time-4th-Edition/Monden/p/book/9781439820971. Accessed 12 Apr 2023

Nyaga, G.N., Closs, D.J., Rodrigues, A., Calantone, R.J.: The impact of demand uncertainty and configuration capacity on customer service performance in a configure-to-order environment. J. Bus. Logist. **28**, 83–104 (2007). https://doi.org/10.1002/j.2158-1592.2007.tb00059.x

Ohno, T.: Toyota production system: Beyond large-scale production

Rewers, R.: Planning the inflow of products for production levelling. Mach. Technol. Mater. **13**, 439–442 (2019)

Rewers, P., Diakun, J.: A heijunka study for the production of standard parts included in a customized finished product. PLoS ONE **16**, e0260515 (2021). https://doi.org/10.1371/journal.pone.0260515

Rewers, P., Hamrol, A., Żywicki, K., et al.: Production leveling as an effective method for production flow control—experience of polish enterprises. Procedia Engineering **182**, 619–626 (2017). https://doi.org/10.1016/j.proeng.2017.03.167

Rewers, P., Trojanowska, J., Diakun, J., et al.: A Study of priority rules for a levelled production plan. In: Hamrol, A., Ciszak, O., Legutko, S., Jurczyk, M. (eds.) Advances in manufacturing, pp. 111–120. Springer International Publishing, Cham (2018)

Rewers, P., Osiński, F., Żywicki, K.: Classification of products in production levelling. In: Machado, J., Soares, F., Veiga, G. (eds.) Innovation, engineering and entrepreneurship, pp. 411–417. Springer International Publishing, Cham (2019b)

Rewers, P., Czaja, M., Janczura, K., Diakun, J.: Determination of the production frequency and batch size for the manufacturing process. In: Tonkonogyi, V., Ivanov, V., Trojanowska, J., et al. (eds.) Advanced manufacturing processes II, pp. 72–82. Springer International Publishing, Cham (2021)

Rewers, P., Bozek, M., Kulus, W.: Increasing the efficiency of the production process by production levelling. Manag. Prod. Eng. Rev. **10**, 93–100 (2019). https://doi.org/10.24425/mper.2019.129572

Yavuz, M., Akçali, E.: Production smoothing in just-in-time manufacturing systems: a review of the models and solution approaches. Int. J. Prod. Res. **45**, 3579–3597 (2007). https://doi.org/10.1080/00207540701223410

Żywicki, K., Rewers, P., Bożek, M.: Data analysis in production levelling methodology, pp. 460–468 (2017)

Enhancing Candidate Evaluation in Recruitment Interviews Based ResNext-101 for Facial Expression Recognition

Kaoutar Khayyiba, Aouatif Amine, and Bouchra Nassih

Abstract

Facial expression recognition (FER) has a significant impact on evaluating candidates in recruitment interviews, as it provides valuable information about their emotional states and non-verbal cues. In this paper, a new approach was proposed that relies on ResNeXt-101 to evaluate candidates in recruitment interviews. We perform an experimental evaluation of two facial expression databases, CK+ and FER + . Our suggested approach, which shows a considerable enhancement in the recognition rate compared to other used methods, gets 99.49 and 85.08% in terms of accuracy on CK+ and FER+ respectively.

Keywords

Facial expression recognition · ResNeXt-101 · Transfer learning · Recruitment

1 Introduction

The field of facial expression recognition (FER) has become increasingly attractive because of its diverse range of practical uses in healthcare (Silva et al. 2021), driving (Cui et al. 2020), education (Aly et al. 2023), and human-computer interaction (Chowdary et al. 2023). One important area where this technology is being applied is in candidate evaluation during job interviews, where recruiters rely on verbal cues and facial expressions to assess a candidate's personality, as these are difficult to fake (Degroot and Gooty 2009). Researchers in industrial-organizational psychology and computer science have proposed that artificial intelligence could outperform human abilities in personality recognition or prediction during job screenings. The rationale behind this is that AI techniques applied to audiovisual datasets can provide greater reliability and predictive accuracy compared to human raters (Celiktutan and Gunes 2015; Langer et al. 2017).

Significant progress in deep learning (DL) and machine learning (ML) has greatly enhanced the potential of FER systems. Among these techniques, convolutional neural networks (CNNs) are well-known for their great performance in image processing and first impression inference from camera images. Recently, there has been significant research demonstrating the effectiveness and promise of deep learning in FER applications.

Hossain et al. (2019) employed a pre-trained CNN model along with two deep sparse auto-encoders for feature extraction from facial and speech inputs. They then utilized a support vector machine to assign emotion labels in a secure edge and cloud computing environment.

Li et al. (2021) suggested using ResNet-50 architecture as the basic network. They employed a convolutional neural network to extract features and used batch normalization (BN) along with the ReLU activation function to enhance the model's convergence ability.

Zhang et al. (2022) developed a transformer-based multimodal information fusion architecture that uses dynamic multimodal functions to fully exploit emotional knowledge from different viewpoints and static viewpoints.

Wang et al. (2020) developed a diversified attention network with a hierarchical pyramid structure. This network improves the contextual information of features. Hierarchical

K. Khayyiba (✉) · A. Amine
Engineering Sciences Laboratory, National School of Applied Sciences, Ibn Tofail University, Kenitra, Morocco
e-mail: kaoutar.khayyiba@uit.ac.ma

A. Amine
e-mail: aouatif.amine@uit.ac.ma

B. Nassih
Faculty of Economics and Management, Ibn Tofail University, Kenitra, Morocco
e-mail: Bouchra.nassih1@uit.ac.ma

B. El Bhiri et al. (eds.), *AI and Digital Transformation: Innovations in Supply Chain, Education, and Energy Systems*, Sustainable Artificial Intelligence-Powered Applications, https://doi.org/10.1007/978-3-031-86837-5_9

multi-scale local features with attention mechanisms are integrated into this network to improve the efficiency of face recognition.

Ma et al. (2021) introduced a method that uses two parallel ResNet-18 networks to extract features: one processes the original image, while the other handles an image transformed by local binary patterns. They employed a visual transformer to integrate the features extracted by both networks and capture their interrelationships. Despite enhancing accuracy, this approach also results in a significant increase in computational complexity compared to alternative techniques.

Kim et al. (2022) proposed a hybrid method utilizing a Vision Transformer (ViT) as its core architecture. They incorporated a squeeze module to simplify the computational complexity by reducing feature dimensions. By effectively merging global and local features, this approach improved facial expression recognition performance.

Despite the success of models such as ResNet (He et al. 2016), Inception (Szegedy et al. 2016) and VGGNet, there is still a need for architectures that can more efficiently capture complex facial variations. In this context, our approach presents the ResNeXt-101 model, a novel architecture in FER that incorporates cardinality to improve performance. We have fine-tuned the model's conv4, conv5 and FC layers, achieving high accuracies on CK+ and FER+ datasets, making it an ideal candidate for integration with existing recruitment software.

The paper is organized as follows: In Sect. 2, the proposed approach is described, including system architecture, databases, and data pre-processing. Section 3 details the model architecture of ResNeXt. The experimental results and performance analysis are outlined in Sect. 4. Finally, Sect. 5 provides the conclusion of the paper.

2 Proposed Approach

In this study, we propose a method for facial expression recognition, illustrated in Fig. 1. We started with a dataset containing a variety of facial expression images, grouped into seven emotions in the CK+ dataset and eight emotions in the FER+ dataset. Before training our model, we resized the images and applied normalization and data augmentation techniques to introduce diversity. We then used a pre-trained ResNeXt 101 model that was previously trained on an extensive dataset. This process involved updating the weights of the Conv4, Conv5 and Fully Connected (FC) layers to improve performance on our dataset. The resulting emotion recognition model was then used to classify the expressions.

To assess the performance of our emotion recognition model, an image from the test dataset is fed into the system. The model analyses the image. Provides the identified emotion, showing how well the proposed approach can recognize expressions. The details of the different steps of our approach are described.

2.1 Databases

In this work, we have used two well-known publicly available datasets, FER+ and CK+. Figures 2 and 3 illustrate examples of facial expressions extracted from these datasets.

FER+ Database. The FER+ dataset (Barsoum et al. 2016), introduced by Barsoum et al. builds on the FER2013 dataset (Goodfellow et al. 2013), which initially collected 35,887 images through Google's image search API. FER+ enriches these images by providing reliable emotion labels. The dataset contains 28,709 images for training, 3,589 for validation and

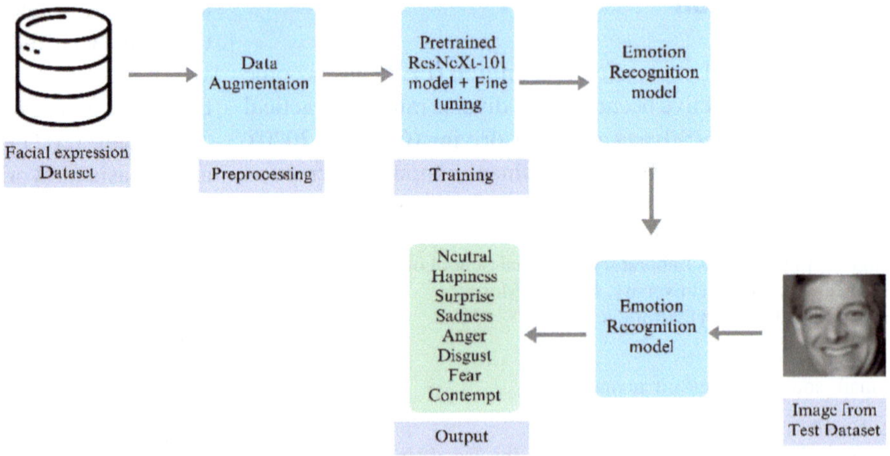

Fig. 1 Architecture of the proposed method

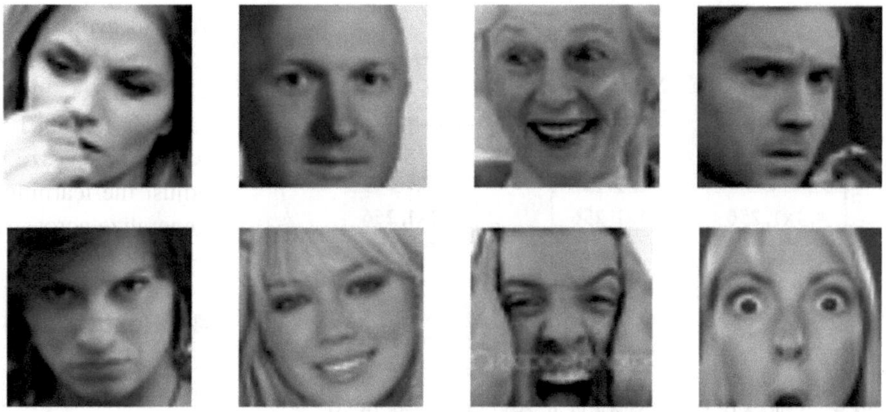

Fig. 2 Examples of training samples of the FER+ dataset

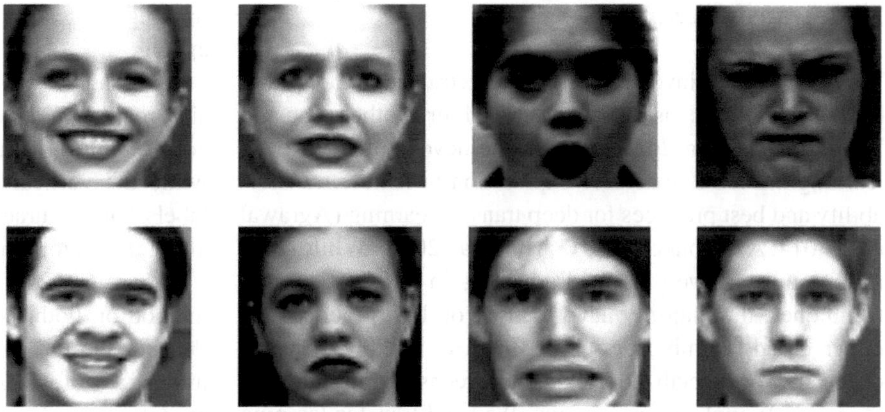

Fig. 3 Examples of training samples of the CK+ dataset

3,589 for testing. Every image is assigned one of the eight emotions: neutral, surprise, happiness, sadness, fear, disgust, anger, and contempt.

CK+ Database. The second database used in this study is the Extended Cohn-Kanade Dataset (CK+), a popular choice for facial expression recognition research due to its controlled laboratory environment (Lucey et al. 2010). As described by Lucey et al. (2010), CK+ consists of 593 sequences from 123 subjects. With emotion labels categorized into seven universal emotions. The sequences are captured in grayscale images. All the sequence images include detailed annotations such as facial landmarks, and FACS codes.

2.2 Data Pre-Processing

Effective preprocessing plays a crucial role in enhancing the performance of FER (Canedo and Neves 2019). In our research, we implemented a variety of data enhancement techniques, for FER+: Grayscale to RGB conversion, resizing images to 224×224 pixels to match the input size required by the ResNeXt-101 model, Random Horizontal Flip, Random Rotation, Colour Jittering, Random Resized Crop, Normalization. Similarly, for CK+, images were resized to 224×224 pixels, random horizontal flip, normalization and random rotation. The CK+ dataset was then split into training and test sets using 80–20%.

3 ResNeXt Model

3.1 ResNeXt-101 Architecture

To conduct our studies, we used the ResNeXt-101 model developed by Xie et al (2017). The ResNeXt architecture, which won second place in the ILSVRC 2016 competition, is based on increased cardinality rather than depth or width, and introduces the term 'cardinality', which distinguishes it from traditional ResNet (He et al. 2016) models in the mainstream. Cardinality refers to the size of the set of transformations in a building block, ResNeXt uses the same layer repetition strategy as VGG and ResNets, but introduces the split-transform-merge approach in a straightforward and scalable manner as shown in Fig. 4 (Xie et al. 2017). ResNeXt offers several variants designed to explore different configurations and achieve better performance in different tasks.

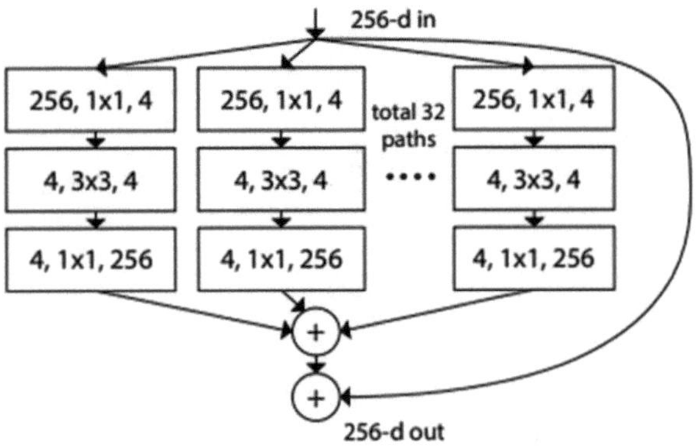

Fig. 4 A block of ResNeXt Xie et al. 2017 (*Source*, p. 1493)

3.2 Transfer Learning

Transfer learning involves utilizing a pretrained model previously used for one task that is now used for another (Pan and Yang 2009). From 2012 to 2016, the development of deep learning has led to extensive exploration of feature transferability and best practices for deep transfer learning (Agrawal et al. 2014; Azizpour et al. 2015; Huh et al. 2016). While some recent studies have presented advances in transfer learning techniques, the majority have focused on best practices for either broad (Kornblith et al. 2019; Li et al. 2002; Plested et al. 2107) or specialized (Heker and Greenspan 2004; Raghu et al. 2019) application domains. We used transfer learning to adapt a pre-trained ResNeXt-101 model for the task of facial expression recognition. We updated conv4 and conv5 to extract features that are particularly relevant to our task and to prevent overfitting, dropout layers were added after conv4 and conv5 with a dropout rate set to 0.5. In addition, the fully connected (FC) layer was adjusted to match the number of classes corresponding to the specific emotions represented in our datasets.

3.3 Parameter Tuning

Parameter tuning was a key part of our method to recognize facial expressions using the ResNeXt-101 model. We explored different hyperparameters to optimize performance, for the FER+ dataset we used Stochastic Gradient Descent (SGD) as our optimizer, setting the learning rate to 0.01 and incorporating momentum. This accelerated convergence and smoothed the optimization process for the FER+ dataset, along with weight decay to prevent overfitting. We implemented ReduceLROnPlateau, which reduces the learning rate when a metric validation loss stops improving. This helps to fine-tune the model, we analyzed batch sizes to find the ideal balance between training speed and performance, a mini-batch size of 128 was chosen based on experimentation to manage memory constraints. And to avoid overfitting and identify the ideal number of training epochs, we implemented early stopping. On the CK+ dataset, we used the Adam optimizer for a learning rate set to 0.001. The model was trained for 35 epochs with a batch size of 32, and a step scheduler was used to adjust the learning rate, improving the model's capacity to generalize across various expressions.

4 Results and Discussions

In this section, we provide a comprehensive examination of the performance metrics obtained from the CK+ and FER+ datasets. Figure 5 illustrates the accuracy and loss for both training and testing of our method in the CK+ dataset over 35 epochs.

We evaluate how well our method works using the confusion matrix on different datasets, where the horizontal rows correspond to the actual emotional labels of the data tested, while the vertical columns indicate the model's predicted labels. The accuracy for those classes is shown by the values along the major diagonal.

Figure 6 shows the confusion matrices for our proposed method on both the CK+ dataset (a) and the FER+ dataset (b). For the CK+, the results show accuracy with a recognition rate of 100% for most expressions. In addition, our model excelled in identifying 'surprise' with a recognition rate of 97.78%, illustrating how well our model performs in recognizing expressions.

The confusion matrix for the FER+ dataset demonstrates varied results across the different emotional categories. The expressions 'Happiness' and 'Neutral' were easily identified with recognition rates of 93.50% and 91.41% respectively. While 'surprise' had an accuracy rate of 84.77%, 'anger' was recognized at a rate of 79%. However, emotions such as 'sadness' (64% recognition rate), 'disgust' and 'fear' had lower recognition rates due to an unbalanced distribution in the dataset.

Finally, a detailed comparison of our method with other existing techniques on the CK+ and FER+ datasets is provided in Tables 1 and 2.

As illustrated in Table 1, our method surpasses all existing approaches (Mollahosseini et al. 2016; Fei et al. 2020; Podder et al. 2021; Cai et al. 2021; Zhang et al. 2017), achieving a high accuracy rate of 99.49% in the CK+ dataset, which can be attributed to the effective transfer learning approach and fine-tuning of the ResNeXt-101 model.

The FER+ dataset serves as another challenging benchmark for expression recognition. Table 2 provides a comparison of our method with other existing approaches on the FER+ dataset. Furthermore, our method shows significant success on the FER+ dataset, outperforming the approaches used by Li et al. (2018) with an accuracy of 85.08%.

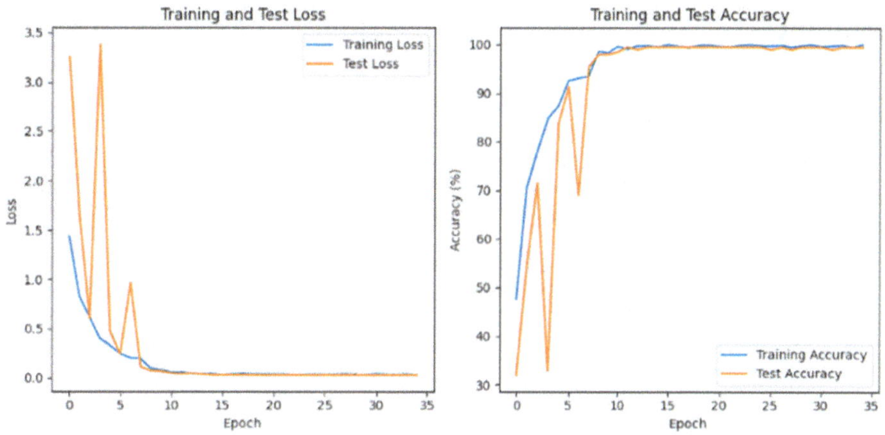

Fig. 5 Training and testing accuracy and loss of our proposed method on the CK+ dataset

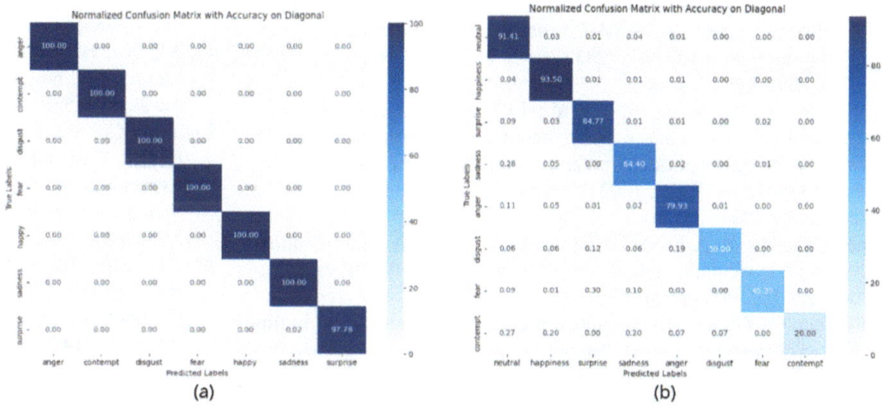

Fig. 6 Confusion matrices of our proposed method on **a** CK+ dataset and **b** FER+ dataset

Table 1 Comparison with other methods on the CK+ dataset

Methods	Accuracy (%)
DNN (Mollahosseini et al. 2016)	93.2
AlexNet + FC6 + LDA (FEI et al. 2020)	94.70
Lightweight CNN (Podder et al. 2021)	96.12
IF-GAN (Cao et al. 2021)	97.52
PHRNN-MSCNN (Zhang et al. 2017)	98.50
Our approach	**99.49**

Table 2 Comparison with other methods on the FER+ dataset

Methods	Accuracy (%)
ResNet18 (Li et al. 2018)	83.1
ResNet18 + FC (Li et al. 2018)	83.3
TFE-JL (Li et al. 2018)	84.3
Our approach	**85.08**

5 Conclusions

In this paper, we presented an effective approach to recognizing facial expressions. We used transfer learning to train a ResNeXt 101 model on FER+ and CK+ datasets to evaluate its accuracy. By tuning its Conv4, Conv5 and Fully Connected (FC) layers, we were able to improve its accuracy. Our method showed accuracy rates of 99.49% on CK+ and 85.08% on FER+. These results highlight the capability of ResNeXt-101-based systems to recognize facial expressions accurately. They suggest that this technology could enhance the recruitment process by providing objective insights into the emotional states and non-verbal communication abilities of candidates.

Building on these promising results, our next steps involve investigating the capabilities of alternative ResNeXt models, such as ResNeXt-152, to see if they offer better accuracy in this area. In addition, we will explore the potential of combining ResNeXt with different deep learning architectures, possibly using ensemble learning techniques, to achieve even higher recognition rates.

References

Agrawal, P., Girshick, R., Malik, J.: Analyzing the performance of multilayer neural networks for object recognition. In: Computer Vision–ECCV 2014: 13th European conference, Zurich, Switzerland, September 6–12, 2014, Proceedings, Part VII 13, pp. 329–344. Springer International Publishing, (2014). https://doi.org/10.1007/978-3-319-10584-0_22

Aly, M., Ghallab, A., Fathi, I.S.: Enhancing facial expression recognition system in online learning context using efficient deep learning model. IEEE Access, **11**, 121419–121433 (2023). https://doi.org/10.1109/ACCESS.2023.3325407

Azizpour, H., Razavian, A.S., Sullivan, J. et al.: Factors of transferability for a generic convnet representation. IEEE Trans. Pattern Anal. Mach. Intell., **38**(9), 1790–1802 (2015). https://doi.org/10.1109/TPAMI.2015.2500224

Barsoum, E., Zhang, C., Ferrer, C.C., et al.: Training deep networks for facial expression recognition with crowd-sourced label distribution. In: Proceedings of the 18th ACM international conference on multimodal interaction, pp. 279–283 (2016). https://doi.org/10.1145/2993148.2993165

Cai, J., Meng, Z., Khan, A.S. et al.: Identity-free facial expression recognition using conditional generative adversarial network. In: 2021 IEEE International Conference on Image Processing (ICIP), pp. 1344–1348. IEEE, (2021). https://doi.org/10.1109/ICIP42928.2021.9506593

Canedo, D., Neves, A.J.: Facial expression recognition using computer vision: A systematic review. Appl. Sci., **9**(21), p. 4678 (2019). https://doi.org/10.3390/app9214678

Celiktutan, O., Gunes, H.: Automatic prediction of impressions in time and across varying context: Personality, attractiveness and likeability. IEEE Trans. Affect. Comput., **8**(1), 29–42 (2015). https://doi.org/10.1109/TAFFC.2015.2513401

Chowdary, M.K., Nguyen, T.N., Hemanth, D.J.: Deep learning-based facial emotion recognition for human–computer interaction applications. Neural Comput. Appl., **35**(32), 23311–23328 (2023). https://doi.org/10.1007/s00521-021-06012-8

Cui, Y., Ma, Y., Li, W. et al.: Multi-EmoNet: a novel multi-task neural network for driver emotion recognition. IFAC-PapersOnLine, **53**(5), 650–655 (2020). https://doi.org/10.1016/j.ifacol.2021.04.155

DeGroot, T., Gooty, J.: Can nonverbal cues be used to make meaningful personality attributions in employment interviews? J. Bus. Psychol., **24**, 179–192 (2009). https://doi.org/10.1007/s10869-009-9098-0

Fei, Z., Yang, E., Li, D.D.U., et al.: Deep convolution network based emotion analysis towards mental health care. Neurocomputing, **388**, 212–227 (2020). https://doi.org/10.1016/j.neucom.2020.01.034

Goodfellow, I.J., Erhan, D., Carrier, P.L., et al.: Challenges in representation learning: A report on three machine learning contests. In: Neural information processing: 20th International Conference, ICONIP 2013, Daegu, Korea, November 3–7, 2013. Proceedings, Part III 20, pp. 117–124. Springer Berlin Heidelberg, (2013). https://doi.org/10.1007/978-3-642-42051-1_16

He, K., Zhang, X., Ren, S. et al.: Deep residual learning for image recognition. In: Proceedings of the IEEE conference on computer vision and pattern recognition, pp. 770–778 (2016). https://doi.org/10.48550/arXiv.1512.03385

Heker, M., Greenspan, H.: Joint liver lesion segmentation and classification via transfer learning. (2020). arXiv preprint arXiv:2004.12352. https://doi.org/10.48550/arXiv.2004.12352

Hossain, M.S., Muhammad, G.: Emotion recognition using secure edge and cloud computing. Information Sciences, **504**, 589–601 (2019). https://doi.org/10.1016/j.ins.2019.07.040

Huh, M., Agrawal, P., Efros, A.A.: What makes ImageNet good for transfer learning? (2016). arXiv preprint arXiv:1608.08614. https://doi.org/10.48550/arXiv.1608.08614

Kim, S., Nam, J., Ko, B.C.: Facial expression recognition based on squeeze vision transformer. Sensors, **22**(10), 3729 (2022). https://doi.org/10.3390/s22103729

Kornblith, S., Shlens, J., Le, Q.V.: Do better imagenet models transfer better? In: Proceedings of the IEEE/CVF conference on computer vision and pattern recognition, pp. 2661–2671 (2019). https://doi.org/10.48550/arXiv.1805.08974

Langer, M., König, C.J., Krause, K.: Examining digital interviews for personnel selection: Applicant reactions and interviewer ratings. Int. J. Sel. Assess., **25**(4), 371–382 (2017). https://doi.org/10.1111/ijsa.12191

Li, B., Lima, D.: Facial expression recognition via ResNet-50. Int. J. Cogn. Comput. Eng., **2**, 57–64 (2021). https://doi.org/10.1016/j.ijcce.2021.02.002

Li, H., Chaudhari, P., Yang, H. et al.: Rethinking the hyperparameters for fine-tuning. (2020). arXiv preprint arXiv:2002.11770. https://doi.org/10.48550/arXiv.2002.11770

Li, M., Xu, H., Huang, X., et al.: Facial expression recognition with identity and emotion joint learning. IEEE Trans. Affect. Comput., **12**(2), 544–550. https://doi.org/10.1109/TAFFC.2018.2880201

Lucey, P., Cohn, J.F., Kanade, T. et al.: The extended cohn-kanade dataset (ck+): A complete dataset for action unit and emotion-specified expression. In: 2010 IEEE computer society conference on computer vision and pattern recognition-workshops, pp. 94–101. IEEE, (2010). https://doi.org/10.1109/CVPRW.2010.5543262

Ma, F., Sun, B., Li, S.: Facial expression recognition with visual transformers and attentional selective fusion. IEEE Trans. Affect. Comput., **14**(2), 1236–1248 (2021). https://doi.org/10.1109/TAFFC.2021.3122146

Mollahosseini, A., Chan, D. and Mahoor, M.H.: Going deeper in facial expression recognition using deep neural networks. In: 2016 IEEE Winter conference on applications of computer vision (WACV), pp. 1–10. IEEE, (2016). https://doi.org/10.1109/WACV.2016.7477450

Pan, S.J., Yang, Q.: A survey on transfer learning. IEEE Trans. Knowl. Data Eng., **22**(10), 1345–1359 (2009). https://doi.org/10.1109/TKDE.2009.191

Plested, J., Shen, X., Gedeon, T.: Non-binary deep transfer learning for image classification. (2021). arXiv preprint arXiv:2107.08585. https://doi.org/10.48550/arXiv.2107.08585

Podder, T., Bhattacharya, D., Majumdar, A.: Dew computing-inspired mental health monitoring system framework powered by a lightweight CNN. In: Disruptive technologies for big data and cloud applications: Proceedings of ICBDCC 2021, pp. 309–319. Springer Nature Singapore, Singapore (2022). https://doi.org/10.1007/978-981-19-2177-3_31

Raghu, M., Zhang, C., Kleinberg, J. et al.: Transfusion: Understanding transfer learning for medical imaging. Adv. Neural Inf. Process. Syst., **32**, (2019). https://doi.org/10.48550/arXiv.1902.07208

Silva, J.L., Oliveira, I., Topolniak, Z. et al.: A CNN approach implemented to emotional facial expression recognition for the prevention of autistic meltdowns. In: 2021 2nd Sustainable Cities Latin America Conference (SCLA), pp. 1–6. IEEE, (2021). https://doi.org/10.1109/SCLA53004.2021.9540183

Szegedy, C., Vanhoucke, V., Ioffe, S. et al.: Rethinking the inception architecture for computer vision. In: Proceedings of the IEEE conference on computer vision and pattern recognition. pp. 2818–2826 (2016). https://doi.org/10.48550/arXiv.1512.00567

Wang, Q., Wu, T., Zheng, H. et al.: Hierarchical pyramid diverse attention networks for face recognition. In: Proceedings of the IEEE/CVF conference on computer vision and pattern recognition, pp. 8326–8335 (2020). https://doi.org/10.1109/CVPR42600.2020.00835

Woo, S., Park, J., Lee, J.Y et al.: Cbam: Convolutional block attention module. In: Proceedings of the European conference on computer vision (ECCV), pp. 3–19 (2018). https://doi.org/10.48550/arXiv.1807.06521

Xie, S., Girshick, R., Dollár, P., et al.: Aggregated residual transformations for deep neural networks. In: Proceedings of the IEEE conference on computer vision and pattern recognition, pp. 1492–1500 (2017). https://doi.org/10.48550/arXiv.1611.05431

Zhang, K., Huang, Y., Du, Y. et al.: Facial expression recognition based on deep evolutional spatial-temporal networks. IEEE Trans. Image Process., **26**(9), 4193–4203 (2017). https://doi.org/10.1109/TIP.2017.2689999

Zhang, W., Qiu, F., Wang, S. et al.: Transformer-based multimodal information fusion for facial expression analysis. In: Proceedings of the IEEE/CVF conference on computer vision and pattern recognition, pp. 2428–2437 (2022). https://doi.org/10.48550/arXiv.2203.12367

Bayesian Deep Learning for Multi-disease Detection in Retinal Imaging with Uncertainty Quantification

Qaddour Mariem, Yassine Benjelloune Touimi and Khalid Minaoui

Abstract

This research highlights the importance of timely and precise identification of retinal diseases to prevent vision impairment and improve clinical outcomes. We apply Deep Learning (DL) techniques to classify and predict seven common eye conditions. Retinal images are critical for diagnosis, offering essential insights for ophthalmologists due to the retina's responsiveness to changes in microvascular structures. DL has advanced medical imaging significantly, especially in tasks like classification, segmentation, detection, and prediction of retinal diseases. However, the complexity of DL models can result in decision-making errors, necessitating the estimation of model uncertainty, which traditional DL models cannot provide. Therefore, Bayesian DL methods are being increasingly used. In this study, we develop a simple Convolutional Neural Network (CNN) model for retinal disease classification and apply Bayesian CNN methods using Variational Inference (VI) and Monte Carlo dropout (MC-Dropout) to obtain the posterior predictive distribution. Using the Dataset 'Derbi_Hackathon_Retinal _Fundus_Imag_Dataset,' our results show that the proposed models outperform state-of-the-art models, with test accuracies of 84% for CNN, 85% for BCNN-VI, and 84% for MC-Dropout. We also compute the predictive distribution to quantify model uncertainty, demonstrating the potential benefits of Bayesian DL methods in enhancing the accuracy and reliability of disease diagnosis and treatment in medical image analysis.

Keywords

Bayesian deep learning • Convolutional neural network • Uncertainty quantification • Retinal diseases

1 Introduction

The advent of Deep Learning (DL) has revolutionized many fields (LeCun et al. 2015). However, DL models often struggle with uncertainty in predictions, which is critical in areas like medicine. To address this, Bayesian Deep Learning (BDL) integrates Bayesian probability with DL to offer uncertainty-aware predictions (Ghahramani 2015). Unlike conventional DL, BDL provides posterior distributions for model parameters, essential for precise uncertainty assessment in medical applications (Abdullah et al. 2022). BDL employs Bayes' theorem to generate posterior distributions, quantifying uncertainties in both model and data (Hüllermeier and Waegeman 2021). Methods like Variational Inference (VI) and Monte Carlo Dropout (MC-Dropout) offer scalable alternatives to Markov Chain Monte Carlo (MCMC) for large datasets (Abdullah et al. 2022).

This research uses a Bayesian architecture with VI to classify retinal diseases using the Dataset from Kaggle. The focus is on leveraging VI for accurate classification and effective uncertainty quantification. Key objectives include

1. Use CNN architecture for classifying retinal fundus images.
2. Implement Bayesian Convolutional Neural Networks (BCNN) using Retina-Net with VI and MC-Dropout.
3. Enhance classification by optimizing hyperparameters and refining training.
4. Create a BCNN architecture that classifies retinal images and estimates prediction uncertainty.

Q. Mariem (✉) · Y. B. Touimi · K. Minaoui
LRIT Laboratory, Faculty of Sciences, Mohammed V University, Rabat, Morocco
e-mail: Mariem.qaddour@um5r.ac.ma

Y. B. Touimi
e-mail: yassine.touimi@um5s.net.ma

K. Minaoui
e-mail: khalid.minaoui@um5r.ac.ma

B. El Bhiri et al. (eds.), *AI and Digital Transformation: Innovations in Supply Chain, Education, and Energy Systems*, Sustainable Artificial Intelligence-Powered Applications, https://doi.org/10.1007/978-3-031-86837-5_10

2 Related Work

Recent years have seen a surge in research exploring the use of deep learning, Bayesian deep learning, and uncertainty-aware models in diagnosing medical conditions using publicly available datasets. This section reviews some of the latest studies in this field.

2.1 Deep Learning Applied to Retinal Diseases

Majumder and Kehtarnavaz [2021] introduced a multitask deep learning approach for diagnosing the five stages of diabetic retinopathy (DR). Their model includes both classification and regression components, featuring a modified densely connected deep neural network. Evaluations on the APTOS and EyePACS Kaggle datasets showed weighted Kappa values of 0.88 and 0.90, respectively.

Alahmadi [2022] developed a deep learning framework for DR image classification, integrating content recalibration and an adaptive mechanism for identifying informative image regions. The model's superior performance was demonstrated on the APTOS 2019 dataset.

Qaddour et al. [2023] introduced Retina-Net, a new CNN architecture, and compared it with pretrained models. EfficientNet-B3 showed high precision and recall, with validation and test precision rates of 80.892% and 79.360%, respectively. In contrast, VGG-19 had lower rates of 29.446% and 28.200%. Retina-Net achieved validation and test rates of 39.650% and 35.470%, performing competitively with EfficientNet-B3 and significantly outperforming VGG-19 in ocular disease classification.

2.2 Bayesian Deep Learning Applied to Diabetic Retinopathy

Band et al. (2022) benchmarked Bayesian deep learning (BDL) methods for detecting diabetic retinopathy (DR) using EyePACS and APTOS datasets. They evaluated Bayesian techniques like Mean-Field Variational Inference (MFVI), Structured Mean-Field Variational Inference (SFVI), Monte Carlo Dropout (MC-Dropout), Maximum a Posteriori (MAP), and deep ensembles. The study used extensive computational resources, including 20-GPU days and 100-TPU days, and evaluated 400 hyper-parameter configurations across six random seeds.

Ahsan et al. (2021) developed a hybrid model combining a Bayesian CNN with MC-Dropout and an Active Learning (AL) module for DR classification. This model addressed prediction uncertainty and leveraged unlabeled data. Their evaluation showed the hybrid model's superiority over state-of-the-art methods, highlighting its effectiveness in scenarios with limited labeled data.

3 Proposed Method

Preprocessing enhances retinal image contrast to highlight features for diagnosing retinal diseases. We utilize CNN and Bayesian CNN models for detection and classification. The CNN identifies patterns from a large dataset, while the Bayesian CNN accounts for prediction uncertainty by estimating the posterior distribution of its parameters. Model uncertainty quantification by the Bayesian CNN is crucial in medical diagnosis for gauging prediction confidence through the predictive distribution (Fig. 1).

3.1 Dataset Used

In our study, we used the "Derbi.Hackathon.Retinal.Fundus. Image Dataset" from Kaggle, containing 3785 multi-labeled color images of the retinal fundus. The dataset includes 7 categories: Opacity, diabetic_Retinopathy, Glaucoma, Macular_edema, Macular_Degeneration, Retinal_vascular_Occlusion, and Normal retinal images. We split the dataset into 80% for training, 10% for testing, and 10% for validation. Each image was centered on either the optic disk or the macula. Expert clinicians annotated the images, classifying them as "normal" or "abnormal," with the latter including various retinal anomalies (Fig. 2).

3.2 CNN Model

This method uses Convolutional Neural Networks (CNNs) to detect and classify retinal diseases, categorized into 7 disease classes. CNNs automatically extract features crucial for disease identification and classification (Ismael et al. 2021; Sungheetha and Sharma 2021). The CNN architecture comprises convolutional, pooling, and fully connected layers. Convolutional layers extract features like edges, curves, and blood vessels. Figure 3 shows the CNN architecture for identifying retinal disease patterns. Cross-entropy loss optimizes model parameters during training. The CNN model includes five convolutional layers, four Maxpool layers, four batch normalization layers, two dropout layers, and two fully connected layers. Convolutional layers use ReLU activation; the final dense layer uses Softmax activation for class probabilities. Maxpool layers reduce dimensions, batch normalization enhances training efficiency, and dropout prevents overfitting. Fully connected layers combine features for classification. Table 1 details layer parameters, and Fig. 4 shows the archi-

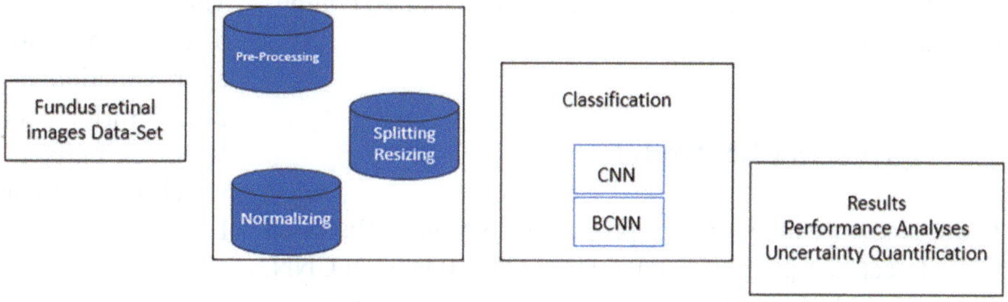

Fig. 1 Block diagram of the proposed method

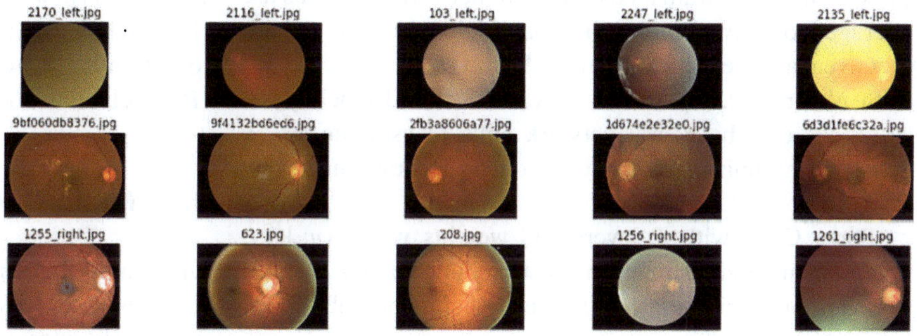

Fig. 2 Different images of retinal fundus in my dataset

LAYER	Parameters
CONV-1	64 filter 11*11, padding = 5, stride=4
MAX-POOLING	2*2
BATCH-NORMALIZATION	
CONV-2	192filter 5*5, padding = 2, stride=2
MAX-POOLING	2*2
BATCH-NORMALIZATION	
CONV-3	256 filter 3*3, padding = 1, stride=4
MAX-POOLING	2*2
BATCH-NORMALIZATION	
DROPOUT	0.1
FCL	256
DROPOUT	0.2

Fig. 3 Details of the proposed CNN architecture

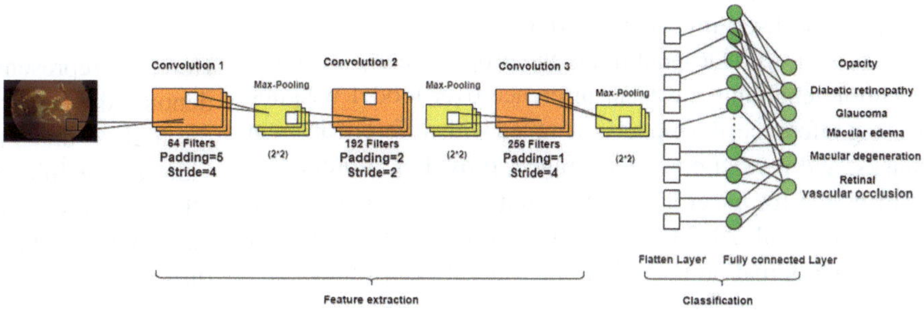

Fig. 4 Details of the proposed CNN architecture

tecture. This design aims for high accuracy with a simple model, suitable for real-world scenarios with limited computational resources.

3.3 Bayesian CNN Model

In this study, we transform the proposed CNN into a Bayesian CNN to capture prediction uncertainty. In a Bayesian CNN, weights and biases are random variables, not deterministic. This allows for variability and uncertainty estimation. We define a prior distribution for weights and biases and obtain a posterior distribution using Bayes' rule, generating multiple weight sets to train an ensemble of networks. Outputs are averaged to derive a probability distribution for each weight.

Given a supervised learning dataset $\mathcal{D} = \{\mathbf{x}_n, y_n\}_{n=1}^N$ with N samples, input features $\mathbf{x}_n \in \mathbb{R}^m$, and labels $y_n \in \{1, 2, \ldots, C\}$, the neural network aims to estimate $y_n = f(\mathbf{x}_n)$ by minimizing prediction error and generalizing to new data.

For a CNN with L layers and weights $\mathbf{w} = \{w_i\}_{i=1}^L$, Bayesian methods estimate the posterior uncertainty $P(\mathbf{w}|\mathcal{D})$ with a prior distribution $P(\mathbf{w})$. The prediction probability for a test input $\mathbf{x}^*$ is:

$$\begin{aligned} P(y|\mathbf{x}^*, \mathcal{D}) &= \mathbb{E}_{P(\mathbf{w}|\mathcal{D})}[P(y|\mathbf{x}^*, \mathbf{w})] \\ &= \int_{\mathbf{w}} P(y|\mathbf{x}^*, \mathbf{w}) P(\mathbf{w}|\mathcal{D}) d\mathbf{w} \end{aligned} \tag{1}$$

Computing this integral exactly is intractable, so approximation methods like MCMC (Chib 2001; Papamarkou et al. 2022; Salimans et al. 2013), MC-Dropout (Gal and Ghahramani 2016; Myojin et al. 2020; Sadr et al. 2022), and Variational Inference (VI) (Graves 2011; Paisley et al. 2012) are used. In our model, we use both VI and MC-Dropout methods.

3.3.1 MC Dropout

MC Dropout is an extension of standard dropout (Srivastava et al. 2014) used in deep learning to estimate model prediction uncertainty. In standard dropout, nodes are randomly dropped during training with probability p to enhance feature robustness and prevent overfitting. At inference, all nodes are retained for deterministic predictions.

Bayesian MC Dropout treats the dropout rate p as a distribution, enabling Bayesian inference over model weights. During inference, the model is sampled with different dropout masks, and predictions are averaged, providing a predictive distribution for uncertainty estimation (Gal and Ghahramani 2016; Gal et al. 2017). This method ensures comprehensive neuron participation in predictions without overfitting.

3.3.2 Variational Inference

Variational Inference (VI) approximates the posterior distribution $p(\mathbf{w}|\mathcal{D})$ of a Bayesian model by fitting an approximation $q_\theta(\mathbf{w})$ to it (Blundell et al. 2015; Graves 2011; Shridhar et al. 2019). The goal is to minimize the Kullback-Leibler (KL) divergence $\mathrm{KL}(q_\theta(\mathbf{w}) \| p(\mathbf{w}|\mathcal{D}))$, which measures information loss between distributions. The Evidence Lower Bound (ELBO) $\mathcal{L}(\theta)$ is minimized:

$$\mathcal{L}(\theta) = -\mathbb{E}_{q_\theta(\mathbf{w})}[\log p(\mathcal{D}|\mathbf{w})] + \mathrm{KL}(q_\theta(\mathbf{w}) \| p(\mathbf{w})) \tag{2}$$

Here, $\mathbb{E}_{q_\theta(\mathbf{w})}$ denotes the expectation over the variational distribution, capturing how well it explains observed data, and $\mathrm{KL}(q_\theta(\mathbf{w}) \| p(\mathbf{w}))$ quantifies the proximity to the true posterior $p(\mathbf{w}|\mathcal{D})$. Optimization typically involves gradient descent to determine θ, refining the weight distribution parameters (Blundell et al. 2015; Gal and Ghahramani 2016).

3.4 Model Uncertainties

Uncertainty quantification is crucial in predictive modeling to assess the certainty of predictions. Bayesian modeling commonly addresses Aleatoric and Epistemic uncertainties (Hüllermeier and Waegeman 2021). While powerful, Deep Learning (DL) methods often struggle with expressing uncertainty. Bayesian deep learning resolves this issue by enabling networks to quantify uncertainty in predictions (Abdar et al. 2021).

We propose a Bayesian CNN for retinal disease classification, utilizing Variational Inference (VI) and MC-Dropout to estimate uncertainty. The posterior distribution of network weights quantifies uncertainty. The mean predictive probability is computed using Eq. 1 by marginalizing the approximate posterior distribution of weights $q(\mathbf{w})$ with MC integration over T samples (Gal and Ghahramani 2016):

$$\begin{aligned} p(y = c|\mathbf{x}^*, \mathcal{D}) &\approx \frac{1}{T} \sum_{t=1}^{T} p(y = c|\mathbf{x}^*, \mathbf{w}_t) \\ &\approx \frac{1}{T} \sum_{t=1}^{T} p_c^{(t)} = p_c \end{aligned} \tag{3}$$

Here, $\mathbf{w}_t$ represents the vector of estimated weights, and c denotes the true class. The final classification is based on the highest mean prediction probability. Our approach provides a principled framework to incorporate uncertainty into DL models, which is critical in medical applications such as classifying retinal fundus images. Its efficacy is demonstrated through experimental results in Sect. 4.

3.5 Evaluating Metrics

After implementing and training the models, we assess performance using several evaluation metrics. These include accuracy, recall, precision, and F1-score. A normalized confusion matrix offers a comprehensive view of classification performance. The Receiver Operating Characteristic (ROC) curve and the area under the ROC (AUC) analyze classification task performance. For the Bayesian CNN model, entropy measures prediction uncertainty, aiming to reduce uncertainty and minimize entropy. These metrics provide a robust framework for performance assessment. The metrics used are as follows:

$$\text{Accuracy} = \frac{TP + TN}{TP + TN + FP + FN} \tag{4}$$

$$\text{Precision} = \frac{TP}{TP + FP} \tag{5}$$

$$\text{Recall} = \frac{TP}{TP + FN} \tag{6}$$

$$F1\ \text{score} = 2 \times \frac{\text{Precision} \times \text{Recall}}{\text{Precision} + \text{Recall}} \tag{7}$$

where TP, TN, FP, and FN represent True Positive, True Negative, False Positive, and False Negative respectively, in the classification of predicting the presence or absence of retinal diseases in medical images. These outcomes are crucial for calculating evaluation metrics such as sensitivity, specificity, accuracy, and F1-score, providing insights into the model's performance.

4 Experimental Results and Discussion

In this section, we present experimental results of our Bayesian CNN model and discuss its effectiveness. Our analysis evaluates the model's classification capability and uncertainty quantification, highlighting areas for improvement and proposing future research directions. Through comprehensive experiments, we assess performance and provide detailed analysis of findings, comparing results with existing literature.

4.1 Data Preprocessing

Data processing was conducted in Python using TensorFlow and TensorFlow Probability on Kaggle Notebook. The dataset of images was imported using OpenCV transformation functions and split into training (80%), testing (10%), and validation (10%) sets. Directories were managed using "os.makedirs" to handle datasets. All images were resized to 224×224 pixels and normalized to a range of 0 to 1 to achieve uniform distribution and eliminate feature bias.

4.2 Training Process for CNN Model

The CNN model underwent 130 epochs of training with the Adam optimizer, using learning rates ranging from 0.0001 to 1e-5. Categorical cross-entropy was used to gauge model performance. Hyperparameters and input image sizes are detailed in Table 1.

4.3 Training Process for Bayesian CNN Model

The Bayesian CNN (BCNN) utilized the same architecture with MC-Dropout applied at a probability of 0.2 across all layers. Training lasted for 40 epochs. For the VI model, TensorFlow Probability implemented Convolutional 2DFlipout with default prior and posterior values. It employed OneHotCategorical layer due to one-hot encoding, trained for 39 epochs with Adam optimizer and learning rate of 1e-3, using negative log-likelihood (nll) loss. Table 1 summarizes hyperparameters and input image sizes.

4.4 Models Performances

This section evaluates our models for retinal disease classification using a retinal image dataset. We analyze prediction accuracy and model uncertainty, comparing with state-of-the-art studies.

4.4.1 Classification Performance

Tables present performance of models for diagnosing seven ophthalmic diseases (opacity, diabetic retinopathy, glaucoma, macular edema, macular degeneration). In Table 2, CNN shows precision 84%, recall 77%, F1-score 80%; BCNN-MC Dropout achieves precision 97%, recall 89.2%, F1-score 87.2%. BCNN-VI: precision 85%, recall 77%, F1-score 79%. In Table 3, models' accuracy on training/testing data and training loss: CNN, BCNN-VI test accuracy 84.51%, 84.50%; BCNN-MC Dropout 83.3%. Training accuracy: CNN, BCNN-VI 85%; BCNN-MC Dropout 84%. Training loss: BCNN-MC Dropout 0.26, CNN 0.35, BCNN-VI 0.40. In summary, BCNN-MC Dropout excels in precision, CNN in balanced precision, recall, F1-score. BCNN-VI competes well, especially in recall, F1-score. Results underline architecture and training choice importance. The ROC curves in Fig. 6b–c and AUC values indicate that the classification model has varying performance across different disease classes. The "macular edema" class shows the best results, while the "glaucoma," "macular degeneration," and "opacity"

Table 1 Summary of hyperparameters configuration and input image size

Hyperparameter	CNN	BCNN-MC	BCNN-VI
Epochs	130	40	39
Optimization	Adam	Adam	Adam
Learning rate	1×10^{-5}	1×10^{-5}	1×10^{-3}
Input size	$224 \times 224 \times 3$	$224 \times 224 \times 3$	$224 \times 224 \times 3$
Loss function	Categorical cross-entropy	Categorical cross-entropy	Negative log-likelihood
Activation function	Softmax and ReLU	Softmax and ReLU	ReLU

Table 2 Performance des modèles

Modèle	Précision (%)	Rappel (%)	F1-score (%)
CNN	84	77	80
BCNN-MC Dropout	97	89.2	87.2
BCNN-VI	85	77	79

Table 3 Performance des modèles (suite)

Modèle	Train Acc (%)	Train loss	Test Acc (%)
CNN	85	0.35	84.51
BCNN-MC Dropout	84	0.26	83.3
BCNN-VI	85	0.40	84.50

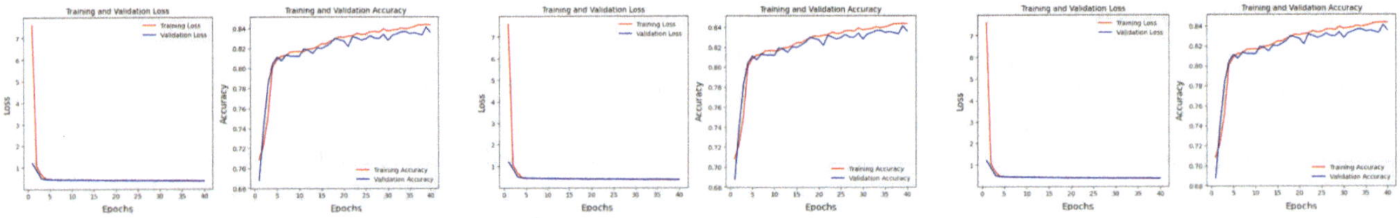

(a) CNN (b) BCNN MC Dropout (c) BCNN VI

Fig. 5 Graphiques d'exactitude et de perte pendant l'entraînement et la validation pour CNN, BCNN MC dropout et BCNN VI

ROC.png droupout.png droupout.png

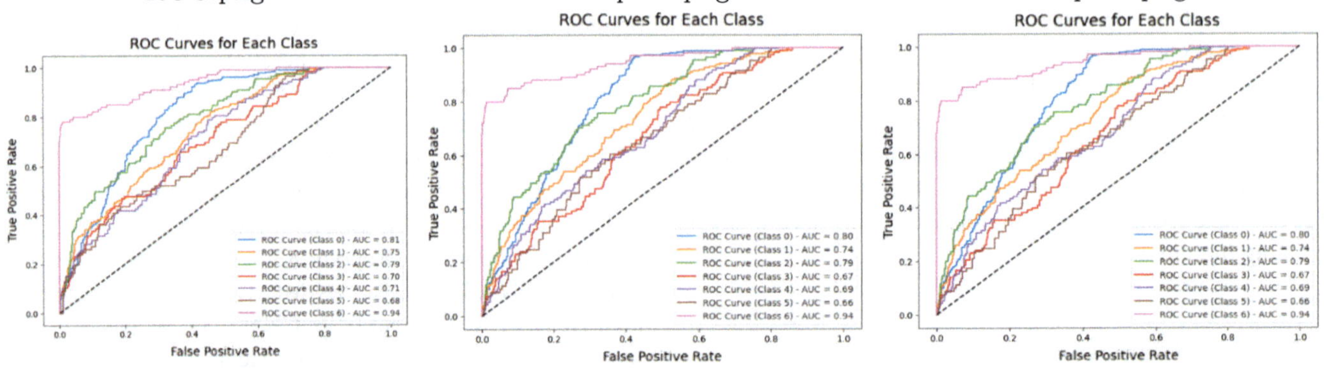

(a) (b) (c)

Fig. 6 Accuracy and loss graphs during validation and training for BCNN MC Dropout (**a**), CNN (**b**), BCNN VI (**c**)

classes show poorer performance. The "diabetic retinopathy," "opacity," and "glaucoma" classes have moderate to good performance. 8 shows class uncertainties, with class 5 (purple) having the least. Our analysis highlights significant

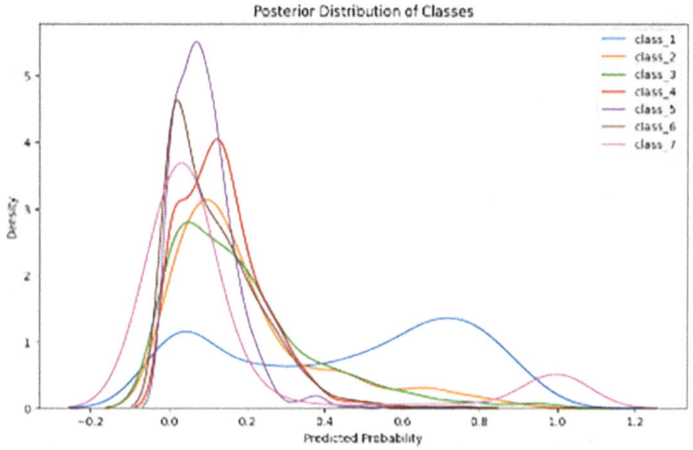

Fig. 7 Predictive posterior distribution for each class during test data

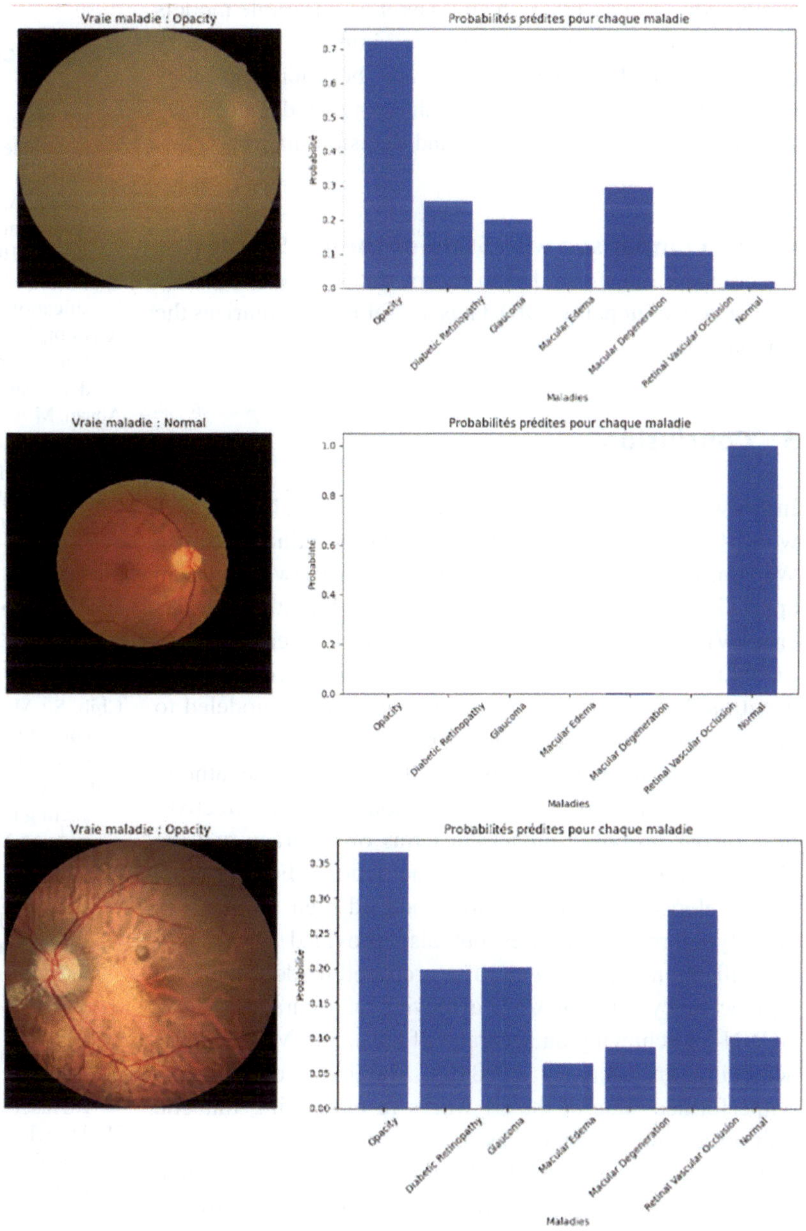

Fig. 8 Probabilistic model predictions with uncertainty for single test samples

Table 4 Comparison of recent state-of-the-art studies on diabetic retinopathy dataset

Reference	Year	Method	Accuracy (%)
Islam et al. [2022]	2022	SCL (supervised contrastive learning)	84
Alyoubi et al. [2021]	2021	CNN512 + Dropout	84
Gangwar and Ravi [2021]	2021	Hybrid inception-ResNet-v2	82.18
Qaddour et al. [2023]	2023	Retina_Net	39.65
Proposed model		CNN	85
		BCNN (VI)	85
		BCNN (MC-Dropout)	84

model uncertainty, which our methods quantify effectively. Reducing this uncertainty is crucial, especially in medical applications. Future research could explore ensemble models or advanced Bayesian inference to improve uncertainty quantification. Bayesian models are beneficial as they identify unreliable predictions, enhancing model accuracy essential for medical diagnosis and decision-making (Figs. 5, 7 and 8).

4.4.2 Comparison with State-of-the-Art Studies

We compare our method with recent state-of-the-art studies on diabetic retinopathy (DR) datasets. Table 4 summarizes the comparison.

5 Conclusion

In this work, we aimed to develop a Convolutional Neural Network (CNN) model for the classification of retinal diseases. We combined the developed CNN model with two Bayesian approximation methods, MC-Dropout and Variational Inference (VI), to create Bayesian CNN models that can quantify model uncertainty by analyzing the posterior predictive distribution. We discussed how uncertainty can be modeled to detect and classify diseases in retinal images.

Our experimental results on the 'Derbi_Hackathon_Retinal_Fundus_Image Dataset' demonstrate the effectiveness of our proposed models in terms of accuracy and the ability to provide uncertainty in predictions. Furthermore, the developed models not only produced high accuracy in classifying retinal diseases but also provided uncertainty quantification in predictions. The proposed models achieved high accuracy, with the CNN achieving an accuracy of 84%, BCNN-VI achieving an accuracy of 85%, and MC-Dropout achieving an accuracy of 84%. Moreover, our models outperformed state-of-the-art works in classifying diabetic retinopathy and retinal diseases.

Our study contributes to the growing body of research on Bayesian models in medical image analysis and provides insights into the use of uncertainty quantification for improved disease diagnosis and treatment.

References

Abdar, M., et al.: A review of uncertainty quantification in deep learning: techniques, applications and challenges. Inform Fusion **76**, 243–297 (2021)

Abdullah, A.A., Hassan, M.M., Mustafa, Y.T.: A review on Bayesian deep learning in healthcare: applications and challenges. IEEE Access **10**, 36538–36562 (2022)

Alahmadi, M.D.: Texture attention network for diabetic retinopathy classification. IEEE Access **10**, 55522–55532 (2022)

Alyoubi, W.L., Abulkhair, M.F., Shalash, W.M.: Diabetic retinopathy fundus image classification and lesions localization system using deep learning. Sensors **21**, 3704 (2021)

Ahsan, M.A., Qayyum, A., Razi, A., Qadir, J.: An active learning method for diabetic retinopathy classification with uncertainty quantification. arXiv preprint arXiv:2012.13325 (2020)

Band, N., Rudner, T.G.J., Feng, Q., Filos, A., Nado, Z., Dusenberry, M.W., Jerfel, G., Tran, D., Gal, Y.: Benchmarking Bayesian deep learning on diabetic retinopathy detection tasks. In: arXiv preprint arXiv:2211.12717 (2022)

Blundell, C., Cornebise, J., Kavukcuoglu, K., Wierstra, D.: Weight uncertainty in neural network. In: Proceedings of the 32nd International Conference on Machine Learning, pp. 1613–1622. PMLR (2015)

Chib, S.: Markov chain Monte Carlo methods: computation and inference. In: Heckman, J.J., Leamer, E. (eds.) Handbook of Econometrics, vol. 5, pp. 3569–3649. Elsevier (2001)

Gal, Y., Ghahramani, Z.: Dropout as a Bayesian approximation: representing model uncertainty in deep learning. In: International Conference on Machine Learning, pp. 1050–1059 (2016)

Gal, Y., Hron, J., Kendall, A.: Concrete dropout. In: Advances in Neural Information Processing Systems, vol. 30 (2017)

Gangwar, A.K., Ravi, V.: Diabetic retinopathy detection using transfer learning and deep learning. In: Bhateja, V., Peng, S.-L., Satapathy, S.C., Zhang, Y.-D. (eds.) Evolution in Computational Intelligence, vol. 1176, pp. 679–689. Springer Singapore, Singapore (2021)

Ghahramani, Z.: Probabilistic machine learning and artificial intelligence. Nature **521**(7553), 452–459 (2015)

Graves, A.: Practical variational inference for neural networks. In: Advances in Neural Information Processing Systems, vol. 24 (2011)

Hüllermeier, E., Waegeman, W.: Aleatoric and epistemic uncertainty in machine learning: an introduction to concepts and methods. Mach. Learn. **110**, 457–506 (2021)

Hüllermeier, E., Waegeman, W.: Aleatoric and epistemic uncertainty in machine learning: an introduction to concepts and methods. Mach. Learn. **110**(3), 457–506 (2021)

Islam, M.R., et al.: Applying supervised contrastive learning for the detection of diabetic retinopathy and its severity levels from fundus images. Comput. Biol. Med. **146**, 105602 (2022)

Ismael, H.R., Abdulazeez, A.M., Hasan, D.A.: Detection of diabetic retinopathy based on convolutional neural networks: a review. Asian J Res Comput Sci **8** (2021), Article no. 3

LeCun, Y., Bengio, Y., Hinton, G.: Deep learning. Nature **521**(7553), 436–444 (2015)

Majumder, S., Kehtarnavaz, N.: Multitasking deep learning model for detection of five stages of diabetic retinopathy. IEEE Access **9**, 123220–123230 (2021)

Myojin, T., Hashimoto, S., Ishihama, N.: Detecting uncertain BNN outputs on FPGA using Monte Carlo dropout sampling. In: Artificial Neural Networks and Machine Learning–ICANN 2020, Cham, pp. 27–38. Springer (2020)

Paisley, J., Blei, D., Jordan, M.: Variational Bayesian inference with stochastic search (2012). arXiv:1206.6430

Papamarkou, T., Hinkle, J., Young, M.T., Womble, D.: Challenges in Markov chain Monte Carlo for Bayesian neural networks. Stat. Sci. **37**, 425–442 (2022)

Qaddour, M., Touimi, Y.B., Minaoui, K.: Classification of retinal fundus images using convolution neural network (CNN). In: 2023 IEEE International Conference on Advances in Data-Driven Analytics And Intelligent Systems (ADACIS), pp. 1–7. IEEE (2023)

Sadr, M., Gante, J., Champagne, B., Falcao, G., Sousa, L.: Uncertainty estimation via Monte Carlo dropout in CNN-based mmwave MIMO localization. IEEE Signal Process. Lett. **29**, 269–273 (2022)

Salimans, T., Kingma, D.P., Welling, M.: Markov chain Monte Carlo and variational inference: bridging the gap (2013). arXiv:1312.6114

Shridhar, K., Laumann, F., Liwicki, M.: A comprehensive guide to Bayesian convolutional neural network with variational inference (2019). arXiv:1901.02731

Srivastava, N., Hinton, G., Krizhevsky, A., Sutskever, I., Salakhutdinov, R.: Dropout: a simple way to prevent neural networks from overfitting. J. Mach. Learn. Res. **15**, 1929–1958 (2014)

Sungheetha, A., Sharma, R.: Design an early detection and classification for diabetic retinopathy by deep feature extraction based convolution neural network. J. Trends Comput. Sci. Smart Technol. TCSST **3**(02), 81–94 (2021)

Towards Efficient Real Time Object Detection Using Edge AI

Mouad Jabrane, Imane Sebari, and Kenza Ait El Kadi

Abstract

In recent years, UAV Earth Observation (EO) generated a vast amount of data, enabling vital problems to be solved. To keep pace with this exponential growth, new data processing solutions are needed. This includes not only artificial intelligence processing, but also the need for real-time processing. Object detection is one of the main purposes of computer vision (CV), with increasing demand for deployment on resource-constrained embedded systems. This paper investigates the possibilities of real-time Edge AI processing of visual aerial data (images and videos) on such platforms using deep learning (DL) techniques. We compared a range of state-of-the-art lightweight discriminative convolutional neural network (CNN) architectures, including MobileNet, EfficientNet and YOLO variants, assessing their performance against key metrics such as accuracy and inference speed. In addition, we explored the feasibility of hardware optimization and acceleration using embedded GPUs and specialized neural network accelerators to further enhance real-time capabilities. This comprehensive benchmark provides valuable information for developers seeking to optimize object recognition models for deployment in resource-constrained embedded systems, thus contributing to the advancement of intelligent vision applications in fields such as robotics and surveillance.

M. Jabrane (✉) · I. Sebari · K. A. El Kadi
Research Unit of Geospatial Technologies for a Smart Decision, IAV Hassan II, Rabat, Morocco
e-mail: jabranemouad@iav.ac.ma

I. Sebari
e-mail: i.sebari@iav.ac.ma

K. A. El Kadi
e-mail: k.aitelkadi@iav.ac.ma

Keywords

Earth observation · UAV · Computer vision · CNN · Edge AI · Embedded systems

1 Introduction

Technological progress continues to develop, with a digital revolution followed by an intelligent one invading various sectors of activity and attracting the attention of the research community. The 4.0 industrial revolution has accelerated the need to exploit robotics, which requires the development of a number of approaches and techniques. The area of artificial intelligence has prompted the main players to draw inspiration from cognitive sciences, making it possible to imitate human behavior by ensuring that machines have at least one of these human senses.

This imitation is mainly reflected in the need to carry out real-time processing of the data collected, which is why the Edge Computing approach coupled with artificial intelligence is gaining ground, given the advantages it offers.

In geomatics, the processing of geospatial information in real time can already be seen, in real-time GNSS surveys, one of the direct survey techniques of spatial geodesy, but can also be considered revolutionary in other disciplines of indirect acquisition of geo data, such as passive aerial remote sensing, known for its work process that follows the following logic: first data collection in the field, then digital post-processing of the images on the desktop, and then analysis and interpretation of the results, something that can have a negative impact on the recovery time of the desired products. For this purpose, on-board systems or nano-computers can be fitted to aerial robotic devices (autonomous or remotely piloted) such as UAVs, to carry out processing during aerial photography missions, exploiting the power of machine vision for a variety of applications. Given the importance of the clean

B. El Bhiri et al. (eds.), *AI and Digital Transformation: Innovations in Supply Chain, Education, and Energy Systems*, Sustainable Artificial Intelligence-Powered Applications, https://doi.org/10.1007/978-3-031-86837-5_11

energy sector in recent years, aerial surveys to inspect solar farms and detect potential faults in the solar panels in real time, particularly hot spots, using thermal imaging, have proved very useful.

The purpose of this document is to present the key components in the development of solutions and systems that guarantee real-time processing of aerial images by Edge AI.

1.1 Earth Observation Applications

With a purpose to solve problems associated with different applications having a geospatial dimension, remote sensing comes to the fore as one of the acquisition techniques that can be exploited in various fields. These applications, which use real-world geospatial data, can be located on different scales, in the atmosphere, on land or even in the oceans. In this context, the application that has attracted our attention is the investigation of solar farms.

1.2 Solar Farms Inspection

The climate change that is sweeping the globe calls for new practices, such as the adoption of green energies to guarantee sustainable development. To this end, the use of solar panels has invaded several sectors of activity. In this sense, these solar panels undergo several failures what requires an obligation of their detection, their follow-up and the prediction of potential problems. To this end, our study focuses mainly on resolving the problem of detecting hot spots on photovoltaic panels in real time.

The nature of the information to be extracted and the geographical extent of these installations mean that specific sensors and platforms need to be used, as we will specify in the next section.

1.3 UAV Image/Video Acquisition

As earth observation specialists, one of the aspects we are interested in is the acquisition of geospatial information, via different sensors and using different platforms. In this context, and since we want to extract spectral information from a vast area with a high spatial resolution, the most appropriate tool is the optical and thermal camera on board a UAV, generating an output in the form of aerial photos or videos.

1.4 Real-Time Edge AI Processing

The processing of the visual data recovered after acquisition is generally carried out in post-processing, on a local machine (on-premise) or in the Cloud, using either conventional or intelligent image processing algorithms.

In recent years, the notion of real time has become increasingly important, given the excessive amount of data produced by users and the need for rapid, instantaneous decision-making. In our case, this gives the UAV autonomy in terms of processing and decision-making.

The key to this is Edge Computing, which consists of bringing computing and storage to the place where the data is produced. This is a distributed information technology architecture that moves certain storage and computing resources away from the central data center and closer to the source of the data they contain. This architecture reduces latency, bandwidth and increases the data sovereignty.

Edge AI moves artificial intelligence processing from the cloud to edge devices, where the data is generated. This is a key element in building distributed and scalable AI systems for resource-intensive applications (Boesch 2024).

- **Edge CV object detection models**

In this section, we will focus on the Tiny ML concept, and we will mention a number of embedded (lightweight or optimised) models of object detection based on different deep learning architectures for computer vision.

These deep learning models represent a spectrum of architectures designed for efficient real-time object detection. **MobileNet**, with its depth-separable convolutions, paved the way for lightweight models. Its variants, combined with frameworks such as **Tiny-SSD**, offer a balance between precision and speed. **Tiny-YOLO**, as its name suggests, favours fast detection, ideal for applications requiring instant responses. R^2**CNN** introduces a new recursive feature pyramid to improve accuracy, while **SqueezeNet** and **ShuffleNet** use architectural tricks to minimise computing costs. **EfficientNet**, thanks to a meticulous approach to scaling, achieves remarkable accuracy for its size. **PeleeNet**, in turn, built with dense connections, aims for performance on various platforms. Finally, **CenterNet** reframes object detection as key point estimation, enabling it to achieve high speed and accuracy. These models, each with its strengths and weaknesses, illustrate the dynamism of object detection research, constantly pushing back the boundaries of efficiency and

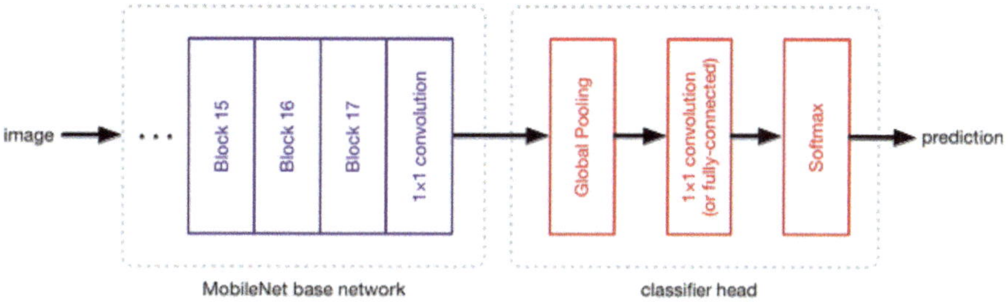

Fig. 1 MobileNetSSDv2 components and architecture

performance. One of the most widely used Edge CNN models for object detection is the **MobileNetSSD** model, due to its light weight, speed and accuracy.

– *MobileNetSSD*

MobileNetSSDv2 (MobileNet Single Shot Detector) is an object detection model with 267 layers and 15 million parameters. It enables real-time inference in the event of computational constraints in devices such as smartphones. Once trained, MobileNetSSDv2 can be stored with 63 MB, making it ideal for small devices.

Architecture De MobileNetSSDv2

The MobileNetSSDv2 model is essentially a two-part model. The first part consists of the base MobileNetV2 network with an SSD layer that classifies the detected image. Essentially, the base MobileNet network acts as a feature extractor for the SSD layer which then classifies the object of interest (Fig. 1).

Comparing Edge CV models

The Microsoft COCO dataset is the most popular benchmark to evaluate models for object detection. Usual metric to assess models performance is **mean average precision**, or mAP. Here, we will make a comparison among top real-time object detection algorithms. The choice of the algorithm would thus be based on the use case and application; different algorithms perform better for different tasks—for instance, Beta R-CNN achieves the best results in pedestrian detection. Ranked by AP on the MS COCO dataset, real-time object detection is led by YOLOv7, followed by vision transformers Swin and DualSwin, PP-YOLOE, YOLOR, then YOLOv4, and EfficientDet (Fig. 2).

An important benchmark measure is **inference time** (ms/frame, the lower the better) or frames per second (FPS, the higher the better). The rapid advances in computer vision technology are clearly visible when inference times are compared. Based on current inference times, YOLOv7 achieves 3.5 ms/frame, compared with 12 ms for YOLOv4 or 29 ms for the popular YOLOv3. Note that the introduction of YOLO has led to significantly faster inference times than previously established methods, such as the two-step Mask R-CNN method (333 ms). From a technical point of view, it is quite complex to compare different architectures and model versions in a meaningful way. And as Edge AI becomes an integral part of scalable AI solutions, the most recent algorithms are accompanied by a lightweight version optimised for the Edge (as is the case with YOLOv7- lite or TensorFlow Lite) (Boesch 2024) (Fig. 3).

- **Edge Frameworks and Accelerators**

All devices share Tensorflow (TF) as a common framework for providing optimised model inference, although they are

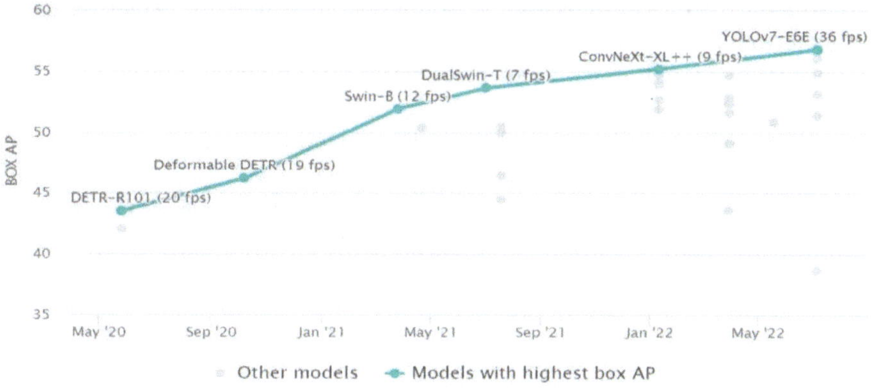

Fig. 2 Real-Time object detection on the COCO Benchmark: Leading algorithms based on average precision

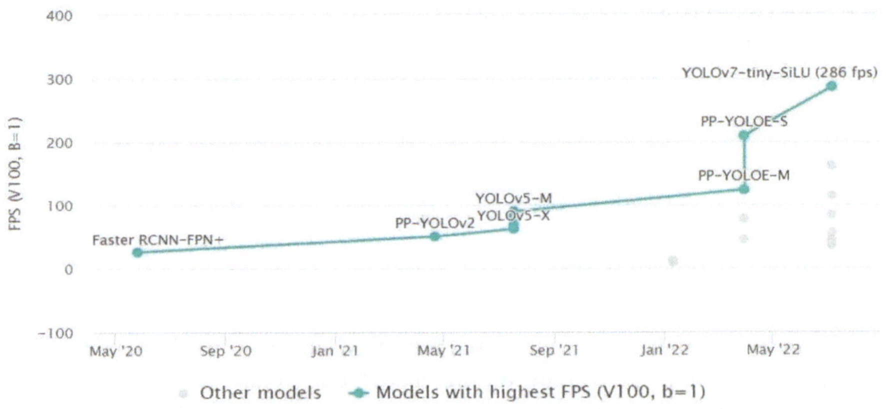

Fig. 3 The state of the art in terms of frames per second (FPS)

only compatible with certain models and variants of Tensorflow. The Google Coral Dev Board's Tensor Processing Unit (TPU) can only be used with optimised Tensorflow Lite models, RKNN-Toolkit for Tinker Edge R only supports conversion of Frozen Graph and Tensorflow Lite models, and TensorRT for the Nvidia Jetson Nano does not support Tensorflow Lite. In addition, there are further restrictions on converting Tensorflow models so that they are optimised for the device in question (Baller et al., 2021).

1. **Tensorflow**: Tensorflow is a free software library for numerical computation using data flow graphs. The nodes of the graph represent the mathematical operations, while the edges of the graph represent the arrays of multidimensional data (tensors) that flow between them. This flexible architecture means that computation can be deployed on one or more CPUs or GPUs without rewriting the code. Tensorflow is the main framework we have used to export pre-trained models for inference on devices. As a framework, it is already optimised for systems equipped with Nvidia graphics cards in combination with their CUDNN neural network library. Nvidia's Jetson Nano is therefore capable of running networks on its embedded GPU, giving it an advantage over other devices that perform inference only on the central processing unit (CPU).
2. **TensorRT**: NVIDIA TensorRT is a C++ library for running inference on NVIDIA GPUs efficiently. It takes a trained network and creates an optimized runtime to run the inference efficiently. TensorRT does this by reordering the graph to combine similar operations together and selecting the fastest implementation of an operation using optimized kernels. In addition, it allows model analysis in ONNX, Caffe, and UFF formats. However, in order to convert models from frameworks like TensorFlow, external tools such as tf2onnx or the uff library are needed (Baller et al., 2021).
3. **Tensorflow Lite**: It is a lightweight and efficient library aimed at running machine-learning models on microcontrollers and other IoT devices with only a few kilobytes of memory. In addition to runtime environments for iOS, Android, and embedded Linux, it also creates them for microcontrollers. The APIs are available in Java, Swift, Objective-C, C++, and Python. In contrast to the standard TensorFlow models that leverage Protocol Buffers for data serialization, TensorFlow Lite makes use of FlatBuffers. Models are at first developed in TensorFlow, and the internal converter is able to convert them into the TFLite format if all operations are supported by the model or if one re-implements them by hand in case it is not.

- **On-board systems**

Real-time processing of geospatial data is only possible if the data processing part is linked to the data acquisition part. For this reason, the component that guarantees real-time processing is the nano-computer or microcontroller, which can be embedded in the various cutting-edge devices. In what follows, we will list the different solutions on the market.

In the context of Edge Computing, it is quite interesting to see how small devices (low power consumption and limited resources) can manage the evaluation of DNNs. In this context, we will list the various recent Edge devices based on accelerators and specialized in DNN applications. The most popular and widely used are **Nvidia Jetson Nano, Google Coral Dev Board, Asus Tinker Edge R, Raspberry Pi 4 and Arduino Nano 33 BLE**.

2 Materials and Methods

2.1 General Architecture

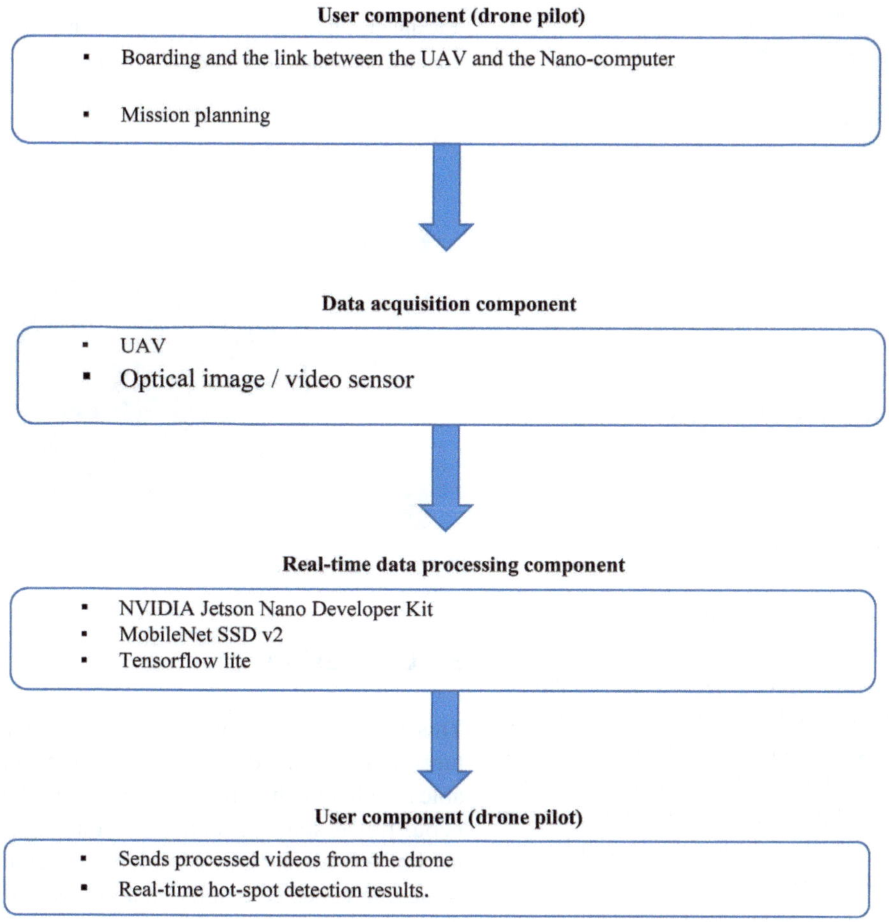

Fig. 4 The M210 V2 UAV

- **UAV**

The UAV used for this project is DJI's M210 V2 (Fig. 4 and Table 1).

- **NVIDIA Jetson Nano Developer Kit**

It is the main computing board, and with sufficient GPU capacity, real-time object detection and higher-end autonomous object detection tasks can be sustained. Networked via an Ethernet switch, it connects to a camera-gimbal system that provides live streams of images and sends commands for adjusting orientations of cameras and zooming. It is also fitted with a 4G module for extended communication with the pilot through which object information can be sent back on detected objects (Fig. 5).

- **Edge CV model**

We present the brain system that leverages MobileNet SSD v2 for object detection. Our approach is a novelli multi-stage deep learning methodology geared toward developing object inference over several consecutive frames and targets an increase in accuracy and real-time processing.

- **Data relay**
  - 4/5G communication: a 4/5G communication module links with the Jetson Nano module for communication back to the pilot. The transmit data includes image frames and metadata like detected objects and their coordinates.
  - Air unit transceiver: This module streams images and UAV telemetry to a ground remote controller via a 2.4 GHz connection.

Table 1 The M210 V2 specifications

Wheelbase	643 mm
Battery	4920 mAh
Maximum takeoff weight	6.14 kg
Maximum flight time	34 min
Limited communication range	8 km

Fig. 5 A Jetson Nano on board the M210 UAV

3 Conclusion

This paper's exploration of various lightweight CNN architectures and associated hardware solutions has revealed a valuable insight into their trade-offs in terms of achieving instantaneous processing. Benchmark results highlight the effectiveness of models such as MobileNet-SSD and Tiny-YOLO in achieving expected results, making them suitable candidates for real-world embedded deployments.

Although promising, this field still presents challenges. Future work will focus on exploiting new advances such as generative AI, exploring more advanced compression techniques (**Pruning**, **Quantization**, **Hyperparameter optimisation**), optimising model architectures for specific hardware platforms and investigating new approaches to further improve accuracy and efficiency for demanding real-time applications. Ultimately, this research paves the way for a new generation of intelligent devices capable of performing complex visual tasks directly at the edge, opening up new possibilities primarily in the field of robotics.

References

Baller, S.P., Jindal, A., Chadha, M., Gerndt, M.: DeepEdgeBench : Benchmarking deep neural networks on edge devices. In: 2021 IEEE International Conference on Cloud Engineering (IC2E), pp. 20–30 (2021). https://doi.org/10.1109/IC2E52221.2021.00016

Boesch, G.: Object detection in 2024 : The definitive guide. Viso.Ai, (2024). https://viso.ai/deep-learning/object-detection/

3D Semantic Segmentation from LiDAR Point Clouds: A Review

Adnan Anouzla, Mohamed Bakali El Mohamadi, Nabila Zrira and Khadija Ouazzani-Touhami

Abstract

In computer vision and computer graphics, segmenting three-dimensional (3D) scenes is an important and difficult subject. The goal of 3D segmentation is to develop computational approaches that anticipate the fine-grained labels of objects in a 3D scene for a variety of applications, such as autonomous driving, industrial control, mobile robots, medical image analysis, and augmented reality. In this study, we explore the most popular approaches and datasets used in 3D semantic segmentation, emphasizing their benefits and limitations.

Keywords

3D LiDAR • Semantic segmentation • Deep learning

1 Introduction

In the larger subject of 3D segmentation, segmentation tasks such as semantic, instance, and part segmentation are crucial. These assignments, which each have different goals and difficulties, entail classifying and defining objects in 3D space. Semantic segmentation is essential for understanding 3D scenes since it assigns class labels to particular points or regions, resulting in high-level comprehension of the scene. Instance segmentation goes a further step, discriminating between several instances of the same class label, allowing for more accurate object delineation in complex scenarios. Part segmentation, on the other hand, provides a finer-grained analysis by dividing objects into their constituent parts, allowing for more comprehension and analysis.

A. Anouzla (✉) · M. Bakali El Mohamadi · N. Zrira ·
K. Ouazzani-Touhami
ADOS Team, LISTD Laboratory, National Superior School of Mines of Rabat - ENSMR, Rabat, Morocco
e-mail: adnan.anouzla@enim.ac.ma

The advancement of deep learning algorithms in recent years has led to a breakthrough in 3D segmentation tasks across multiple modalities and representations. Numerous deep learning-based techniques have been investigated by researchers. These can be broadly categorized as follows: methods based on RGB-D images, methods based on projected images, methods based on voxels, methods based on points, methods based on 3D video, and other representations-based techniques. Because each category addresses particular application scenarios and data properties, it offers a distinct combination of advantages and problems (Zhang et al. 2023).

Among these categories, point-based semantic segmentation algorithms have gained popularity due to their ability to manage the unstructured and irregularly sampled data seen in point clouds. Point-based approaches work directly with point cloud data, describing a scene or object as a collection of 3D points in space. This type of approach is well-suited for tasks like 3D object detection, scene reconstruction, and robotic perception, making it useful in industries such as autonomous driving (Kang et al. 2021), augmented reality (Jin et al. 2020).

We will concentrate on the point-based category of semantic segmentation techniques in this study. Point-based approaches provide distinct advantages in dealing with unstructured point cloud data and have demonstrated promising outcomes in a variety of real-world applications. By focusing on this topic, we try to provide a thorough assessment of cutting-edge methodologies, advantages, and limitations in point-based semantic segmentation. Through this concentrated investigation, we try to contribute to the knowledge and enhancement of point-based semantic segmentation, hence helping progress in computer vision and related domains. The rest of the paper is as follows: The point cloud-based semantic segmentation approaches are introduced in Sect. 2. Section 3 presents the advantages and limitations of each method studied in this paper. Section 4 discusses the majority of publicly available datasets used in 3D semantic segmentation especially S3DIS and ScanNet. The discussion is in Sect. 5. The paper is concluded in Sect. 6.

B. El Bhiri et al. (eds.), *AI and Digital Transformation: Innovations in Supply Chain, Education, and Energy Systems*, Sustainable Artificial Intelligence-Powered Applications, https://doi.org/10.1007/978-3-031-86837-5_12

2 3D Semantic Segmentation

Semantic segmentation in 3D is the process of providing semantic labels to individual points or voxels in a 3D environment, allowing for a more complete comprehension of the scene's geometry and semantics. Unlike typical 2D semantic segmentation, which uses images, 3D semantic segmentation works with volumetric representations like point clouds, voxel grids, or models. This task is critical for a wide range of applications, including robotics, self-driving cars, augmented reality, and urban planning. Point clouds, a sort of 3D object representation, are the most basic output given by 3D sensors. Point clouds, a type of representation of 3D objects, are the most fundamental output that 3D sensors can generate. Point clouds, unlike two-dimensional images, are unstructured and chaotic. Recently, point-based semantic segmentation networks were presented. The methods can be divided into four categories: MLP-based, point convolution-based, graph convolution-based, and transformer-based. An overview of categories based on point clouds is shown in Fig. 1.

To assess the performance of the 3D semantic segmentation model, we look at two essential metrics: mean accuracy (mAcc) and mean intersection over union (mIoU). While other significant metrics include overall accuracy, recall, precision, F1 score, and confusion matrix, we focus on mAcc and mIoU because they provide a thorough and balanced assessment of model performance across all classes.

- Mean Accuracy (mAcc): This measure represents the average accuracy across all classes, which is derived by taking the mean of the accuracy in each class. It provides a balanced measure of model performance, which is particularly relevant in datasets with class imbalance.
- Mean Intersection over Union (mIoU): This metric computes the average overlap between predicted and ground truth segments for all classes. It is calculated by averaging the IoU for each class, resulting in a full assessment of the model's segmentation ability.

2.1 MLP-Based

Shared MLPs are frequently used in MLP-based techniques as the basic building blocks of the network, suggesting that every point in the point cloud has the same properties. Table 1 illustrates the performance evaluation of these approaches.

2.1.1 G+RCU

was presented by Engelmann et al. [2018] as a system for semantic segmentation of unstructured point clouds in three dimensions. In order to compute local or global point features based on the neighborhood's spatial extent, the method employs grouping techniques to establish point neighborhoods in both the initial world space and the learnt feature space. It also uses specific loss functions, such as pairwise distance loss and centroid loss.

2.1.2 PointNet

Qi et al. [2017] operates on 3D points directly. The basic concept is to use a series of MLPs that analyze each point independently in order to extract point features (i.e., point features), and then characterize the points globally using a max-pooling operation (i.e., global features). Before making final label predictions, point and global representations are merged (concatenation + MLP).

2.1.3 ESC

Engelmann et al. [2017] described an adaptation of the PointNet architecture that improves semantic segmentation of 3D point clouds. The authors suggested two techniques for including larger-scale spatial context in the segmentation process: input-level context and output-level context. They demonstrated that their method, which processes multiple scales or nearby regions simultaneously, improves segmentation outcomes for both indoor and outdoor datasets.

2.2 Point Convolution-Based

In 2D image semantic segmentation applications, the essential feature extraction approach is convolution, and in order to extract local contextual information, ordered inputs are needed. Various methods were employed to convert unordered 3D LiDAR data into an ordered feature sequence, which was then fed into convolutional deep networks.

2.2.1 PointwiseCNN

Hua et al. [2018] is a convolutional neural network designed for object detection and semantic segmentation using 3D point clouds. The network is based on a novel convolution operator called pointwise convolution, which can be used to learn pointwise features by applying it to every point in a point cloud. This produces remarkably basic and completely convolutional network designs for scene segmentation and object detection.

2.2.2 RSNet

Huang et al. [2018] proposed by Huang et al. is a 3D segmentation framework for efficiently modeling local features in point clouds. The RSNet's main component is a lightweight local dependence module. Recurrent Neural Network (RNN) layers, a slice unpooling layer, and a new slice pooling layer make up this structure. The slice pooling layer is intended to project features from unordered points onto an ordered set of

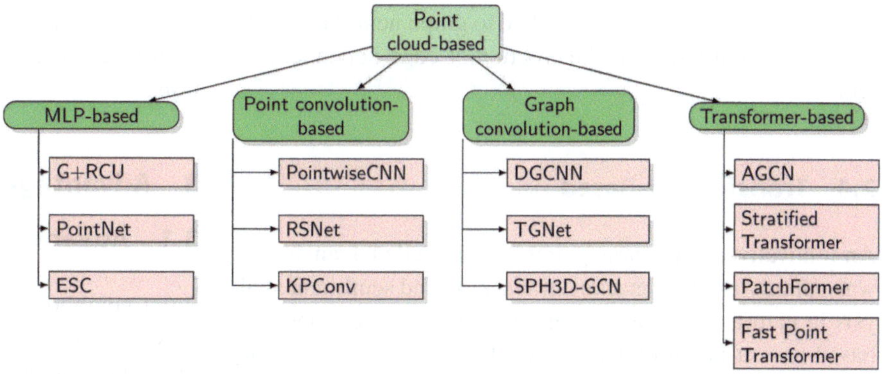

Fig. 1 The taxonomy of point cloud-based approaches of 3D semantic segmentation

Table 1 Performance evaluation of MLP-based semantic segmentation techniques for point clouds on the S3DIS and ScanNet. The symbol "–" indicates unavailable results

Method	S3DIS (%)			ScanNet (%)	
	Area 5		Sixfold	Test set	
	mAcc	mIoU	mIoU	oAcc	mIoU
G+RCU (Engelmann et al. 2018)	54.06	45.14	49.7	75.53	–
PointNet (Qi et al. 2017)	48.98	41.09	47.71	–	14.69
ESC (Engelmann et al. 2017)	59.10	52.17	–	75.53	–

feature vectors so that standard end-to-end learning methods (RNNs) can be used.

2.2.3 KPConv

The Kernel Point Convolution (Thomas et al. 2019) approach transforms point convolution by operating directly on point clouds, eliminating the need for an intermediate representation. By geographically placing convolution weights through kernel points in Euclidean space and applying them to nearby input points, KPConv offers advantages like flexibility in kernel point numbers, adaptability through deformable convolutions, and efficiency in handling variable densities.

2.3 Graph Convolution-Based

In order to broaden the use of graph convolutional networks (GCN) and develop a new paradigm and method for deeper semantic perception, researchers transform the point cloud into a graph structure and develop computational algorithms for nodes and edges. This is accomplished by effectively transmitting learnt information and making full use of the interaction between point pairs.

2.3.1 DGCNN

Wang et al. [2019] is an abbreviation of Dynamic Graph CNN for Learning on Point Clouds. It made use of EdgeConv, a novel fundamental method that preserves permutation invariance while capturing local geometric structure. Instead of providing point features that are obtained straight from their embeddings, EdgeConv offers edge features that describe the connections between a point and its neighbors. EdgeConv is designed to be permutation invariant, meaning that it should be invariant to neighbor ordering. Given that EdgeConv learns edge embeddings and explicitly produces a local graph, it is possible for the model to group points in both semantic and euclidean space.

2.3.2 TGNet

Li et al. [2019] presented a new graph convolution architecture, known as the Taylor Gaussian mixture model (GMM) network for effectively learning compositional and expressive local geometric features from point clouds. The TGNet is composed on fundamental geometric units called TGConv, which are parametrized by a family of filters and perform local convolution on irregular point sets.

2.3.3 SPH3D-GCN

Lei et al. [2020] supports efficient discrete graph convolution by introducing a spherical convolution kernel that divides each neighborhood into bins around a center point. Large point clouds find the suggested kernel computationally appealing because it may be applied to graph neural networks (GNN) without requiring the creation of edge-dependent filters.

2.4 Transformer-Based

An innovative deep learning architecture called Transformer was initially used for machine translation and sentiment analysis in natural language processing (NLP) applications. It is based on self-attention techniques. Recent attempts to apply Transformer to the field of computer vision by academics were motivated by the positive outcomes in NLP, and the results were astounding (Dosovitskiy et al. 2020; Parmar et al. 2018; Strudel et al. 2021).

2.4.1 AGCN

Xie et al. [2020] applied an attention mechanism to study links between the points' local attributes. They also included a global graph structure network to account for the relative information of each individual point in the graph structure network. The suggested network also includes an encoder-decoder structure for segmentation tasks.

2.4.2 Stratified Transformer

Lai et al. [2022] can capture long-range contexts while demonstrating excellent generalization and good performance. They specifically suggested a brand-new key sampling method. In a stratified fashion, they sampled distant points sparsely as keys and adjacent points densely as keys for each query point. This allowed the model to experience long-range contexts at low computational cost and increase its effective receptive field.

2.4.3 PatchFormer

Zhang et al. [2022] is based on Patch ATtention (PAT), which achieves linear complexity in relation to input size by adaptively learning a smaller number of bases for computing attention mappings. Scalability is possible thanks to the PAT mechanism's linear complexity in relation to input size. To further give the model multi-scale capabilities, they introduced a lightweight Multi-Scale Attention (MST) block to establish attention across characteristics of many sizes.

2.4.4 Fast Point Transformer

Park et al. [2022] presented a new local self-attention-based network called Fast Point Transformer, which can process large-scale 3D scenes rapidly. Furthermore, while decreasing space complexity, a lightweight local self-attention module learns continuous positional information from 3D point clouds. The point cloud is processed by the Fast Point Transformer in three phases: (1) Centroid-aware voxelization; (2) lightweight self-attention; and (3) centroid-aware devoxelization.

3 Advantages and Limitations

3.1 Advantages

Point neighborhoods in different feature spaces and certain loss functions make up the majority of the suggested strategy in Engelmann et al. [2018]. These mechanisms allow for the computation of local or global point characteristics based on the neighborhood's spatial extent. The authors demonstrated the efficacy of their technique on indoor and outdoor datasets, delivering cutting-edge results. Overall, their method bridges the gap between 2D and 3D semantic scene comprehension by creating adaptive zones and imposing structure on the learnt feature space.

PointNet (Qi et al. 2017) suggested a revolutionary deep net architecture that does not require rendering or voxelization in order to consume raw point clouds, or groupings of points, avoiding excessive large data and quantization errors. Furthermore, PointNet is robust to small perturbations in input points and can manage missing data or outliers. It is a straightforward, effective, and successful method for a range of 3D recognition applications since it has a unified architecture that learns both local and global point properties.

The authors of Hua et al. [2018] proposed a novel convolution operator that may be used on each point in a point cloud, enabling the learning of pointwise characteristics. Additionally, they reduced implementation while maintaining competitive accuracy for semantic segmentation and object identification tasks.

Features from unstructured input points are projected onto an ordered series of feature vectors by ResNet's slice pooling layer (Huang et al. 2018).

KPConv (Thomas et al. 2019) can be applied to deformable convolutions that adjust kernel points to local geometry. The regular subsampling approach employed by KPConv ensures efficiency and robustness in different densities.

The DGCNN model (Wang et al. 2019) can semantically categorize points by updating a graph of relationships layer by layer. In addition, they exploited local geometric patterns by generating a local neighborhood graph and executing convolution-like operations on the edges linking neighboring pairs of points, similar to GNN.

To increase TGNet's scale invariance (Li et al. 2019), it adopts a multi-scale hierarchical design that runs TGConv on neighborhoods at various scales, allowing it to extract semantic deep features ranging from coarse to fine. Additionally, an extra conditional random field (CRF) layer is applied at the finest scale for feature refining in order to compensate for the loss of feature fidelity caused by the multiple graph pooling and feature interpolation layers.

The proposed kernel in Lei et al. [2020] works for graph neural networks without edge-dependent filter construction, resulting in computational efficiency for huge point clouds. Moreover, traits such as translation invariance and asymmetry allow for consistent weight distribution and the ability to learn accurate geometric details.

In AGCCN (Xie et al. 2020), the authors enhanced the learning process by using a global graph structure to supplement individual node relative location information. They enhanced the point attention layer in order to construct an encoder-decoder attention network for segmentation tasks. Furthermore, by stacking many point attention layers, they can learn from local to global aspects in the network.

The authors of Lai et al. [2022] introduced first-layer point embedding for aggregating local information, which improves performance and convergence. To capture more detailed position information, they also used efficient relative position encoding.

PatchFormer (Zhang et al. 2022) outperforms its baselines while simultaneously having the lowest compute and memory needs. PatchFormer's qualities make it suited for use on an edge device.

The Fast Point Transformer (Park et al. 2022) is optimized for efficiency and speed, with an amazing average inference time of 0.14 s per scene. It is 129 times faster than the Point Transformer (Armeni et al. 2016) for 3D semantic segmentation on the S3DIS dataset. Its architecture is designed for scalability, allowing it to handle large-scale 3D scenes efficiently. Furthermore, it ensures forecast coherency under rigorous transformations, as demonstrated by the introduction of a novel consistency score measure known as CScore.

3.2 Limitations

PointNet (Qi et al. 2017) lacks the ability to understand local structures resulting from the metric space in which points reside, which limits its ability to identify fine-grained patterns and generalize to complicated scenarios.

The order in which the points are entered affects the final global feature vector used for item category prediction. Furthermore, the implementation of pointwise convolution (Hua et al. 2018) is inefficient, taking around two times longer than competing algorithms such as PointNet (Qi et al. 2017) and VoxNet (Maturana and Scherer 2015).

The "vanishing gradient" phenomenon might hinder the performance of RSNets (Huang et al. 2018) by preventing RNN units from accurately modeling dependencies in extended sequences. In addition, RSNets may face a difficult trade-off between slicing resolutions due to larger block sizes, which might result in more slices when a fixed slicing resolution is used or increased slicing resolution.

Deformable kernels can improve descriptive power but increase network complexity, potentially disrupting convergence and leading to overfitting (Thomas et al. 2019).

One limitation remains for TGNet (Li et al. 2019) in multiobject linked area labeling, mainly because of TGConv's limited receptive field.

The Fast Point Transformer (Park et al. 2022) significantly improves processing speed for large-scale 3D scenes, providing a significant advantage over current approaches. However, this speed comes at the cost of accuracy, albeit a good one. Furthermore, the system incorporates novel ways to eliminate quantization artifacts, yet these can still appear, affecting the detection of small geometric patterns.

4 Public Datasets

As artificial intelligence has grown, computer vision tasks now require deep neural networks with higher parameter sizes and more intricate designs in order to extract high-level feature information. Reliable point cloud datasets are a vital assurance for training networks effectively and evaluating the efficacy of proposed segmentation techniques. However, the collection and labeling of vast data necessitates not only a large amount of labor, financial resources, and material but also the direction of domain experts and professional capabilities in relevant industrial software. Some research institutions offer semantically informative and trustworthy public datasets, such as The Stanford 3D Indoor Spaces (S3DIS) (Armeni et al. 2016), ScanNet (Dai et al. 2017), ShapeNet (Chang et al. 2015), Semantic3D (Hackel et al. 2017), and SemanticKitti (Behley et al. 2019), to foster the growth of point cloud semantic segmentation-related research. In addition to guaranteeing the validity and fairness of comparisons with other networks, using these widely available public datasets for network training and validation offers an affordable and workable way to create deep networks with exceptional performance. This section focuses on S3DIS (Armeni et al. 2016) and ScanNet (Dai et al. 2017), two datasets commonly used for point cloud semantic segmentation.

4.1 S3DIS

Armeni et al. [2016] dataset is a massive interior scene dataset created with the use of a Matterport 3D laser scanner. The dataset covers a total area of over 6,000 m^2 and consists of 6 indoor regions with over 215 million points, 70,496 regular RGB images, 1,413 equirectangular RGB images, and 272 indoor scenes with instance-level semantic annotations. Each point in the dataset has surface normals, coordinates, semantic

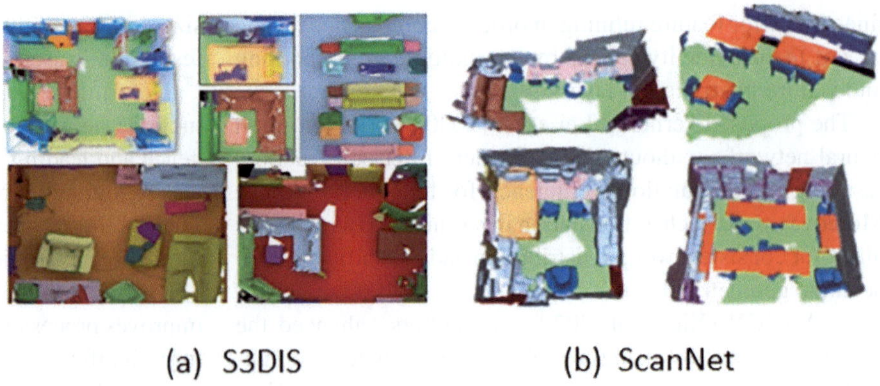

Fig. 2 The annotation samples of **a** S3DIS and **b** ScanNet datasets. *Source* http://www.scan-net.org/

Table 2 Performance evaluation of point convolution-based semantic segmentation techniques for point clouds on the S3DIS and ScanNet

Method	S3DIS (%)			ScanNet (%)	
	Area 5		Sixfold	Test set	
	mAcc	mIoU	mIoU	mAcc	mIoU
PointwiseCNN (Hua et al. 2018)	56.5	–	–	–	–
RSNet (Huang et al. 2018)	59.42	51.93	56.47	–	39.35
KPConv (Thomas et al. 2019)	72.8	67.1	70.6	–	66.6

annotations, and other attributes. This dataset is crucial for 3D vision training interior scene aspects.

4.2 ScanNet

Dai et al. [2017] is an RGB-D video clip dataset of interior situations. The dataset includes 1513 scans of 707 indoor spaces, yielding 2.5 million RGB-D views across 21 categories. The properties include both pre-calibration settings, textures, and locations, as well as instance-level semantic annotations. This dataset makes a significant contribution to the development of 3D scene perception (Fig. 2).

5 Discussion

Different approaches for 3D semantic segmentation have demonstrated variable levels of efficacy on benchmark datasets such as S3DIS and ScanNet, especially when measured by metrics such as mean Accuracy (mAcc) and mean Intersection over Union (mIoU). Tables 1, 2, 3, and 4 depict some results from different point cloud-based approaches.

ESC (Engelmann et al. 2017) outperforms all other MLP-based approaches on the S3DIS and ScanNet datasets. This impressive performance is due to its novel method of using contextual information and efficient feature learning.

KPConv (Thomas et al. 2019) is at the forefront of point-based convolution algorithms. KPConv's ability to dynamically modify convolution kernels to the local geometry of point clouds yields improved segmentation accuracy, raising the bar for mIoU on both datasets.

TGNet (Li et al. 2019) outperforms other graph-based algorithms on the ScanNet dataset, with the greatest mIoU. SPH3D-GCN (Lei et al. 2020) outperforms previous approaches on the S3DIS dataset by using spherical convolutions to better capture geometric structures.

Transformer-based approaches are gaining popularity, with the Stratified Transformer (Lai et al. 2022) emerging as the best performer. It has the highest mIoU on both Area 5 of the S3DIS and the ScanNet datasets, exceeding all other approaches (Lai et al. 2022). Transformers' inherent strengths allow them to mimic long-range relationships and global context through self-attention mechanisms, which MLPs, Point Convolutions, and Graph Convolutional Networks do not do as well. Transformers' ability to capture intricate interactions within point clouds results in more accurate and robust segmentation.

Furthermore, in the S3DIS dataset's sixfold cross-validation, AGCN (Xie et al. 2020) leads. AGCN's architecture, which includes adaptive graph convolutional networks, excels in capturing complicated geometric relationships within point clouds, resulting in higher mIoU scores throughout the dataset's multiple folds.

Table 3 Performance evaluation of graph convolution-based semantic segmentation techniques for point clouds on the S3DIS and ScanNet

Method	S3DIS (%)			ScanNet (%)	
	Area 5		Sixfold	Test set	
	mAcc	mIoU	mIoU	mAcc	mIoU
DGCNN (Wang et al. 2019)	–	51.5	56.1	–	–
TGNet (Li et al. 2019)	–	57.85	–	–	66.2
SPH3D-GCN (Lei et al. 2020)	65.9	59.5	68.9	–	61.0

Table 4 Performance evaluation of transformer-based semantic segmentation techniques for point clouds on the S3DIS and ScanNet

Method	S3DIS (%)			ScanNet (%)	
	Area 5		Sixfold	Test set	
	mAcc	mIoU	mIoU	mAcc	mIoU
AGCN (Xie et al. 2020)	–	–	56.63	–	–
Stratified Transformer (Lai et al. 2022)	78.1	72.0	–	–	73.7
PatchFormer (Zhang et al. 2022)	–	68.1	–	–	–
Fast Point Transformer (Park et al. 2022)	77.3	70.1	–	–	–

These findings show the diversity and specialization of 3D semantic segmentation approaches, as well as the merits of each strategy in dealing with the particular obstacles presented by various datasets. The ongoing evolution of these approaches, particularly the rise of Transformer-based models, promises even greater improvements in the accuracy and reliability of 3D semantic segmentation in the future.

6 Conclusion

3D semantic segmentation is a basic task in the field of 3D segmentation that allows for detailed scene interpretation and item recognition in three dimensions. Despite the difficulties presented by irregular and unstructured data representations, recent advances in deep learning have resulted in significant success in this field. Researchers continue to push the frontiers of 3D semantic segmentation by utilizing approaches such as point-based algorithms, resulting in new potential for applications in computer vision and related domains.

References

Armeni, I., Sener, O., Zamir, A.R., Jiang, H., Brilakis, I., Fischer, M., Savarese, S.: 3d semantic parsing of large-scale indoor spaces. In: Proceedings of the IEEE Conference on Computer Vision and Pattern Recognition, pp. 1534–1543 (2016)

Behley, J., Garbade, M., Milioto, A., Quenzel, J., Behnke, S., Stachniss, C., Gall, J.: Semantickitti: a dataset for semantic scene understanding of lidar sequences. In: Proceedings of the IEEE/CVF International Conference on Computer Vision, pp. 9297–9307 (2019)

Chang, A.X., Funkhouser, T., Guibas, L., Hanrahan, P., Huang, Q., Li, Z., Savarese, S., Savva, M., Song, S., Su, H., et al.: ShapeNet: an information-rich 3d model repository (2015). arXiv:1512.03012

Dai, A., Chang, A.X., Savva, M., Halber, M., Funkhouser, T., Nießner, M.: ScanNet: Richly-annotated 3d reconstructions of indoor scenes. In: Proceedings of the IEEE Conference on Computer Vision and Pattern Recognition, pp. 5828–5839 (2017)

Dosovitskiy, A., Beyer, L., Kolesnikov, A., Weissenborn, D., Zhai, X., Unterthiner, T., Dehghani, M., Minderer, M., Heigold, G., Gelly, S., et al.: An image is worth 16×16 words: transformers for image recognition at scale (2020). arXiv:2010.11929

Engelmann, F., Kontogianni, T., Hermans, A., Leibe, B.: Exploring spatial context for 3d semantic segmentation of point clouds. In: Proceedings of the IEEE International Conference on Computer Vision Workshops, pp. 716–724 (2017)

Engelmann, F., Kontogianni, T., Schult, J., Leibe, B.: Know what your neighbors do: 3d semantic segmentation of point clouds. In: Proceedings of the European Conference on Computer Vision (ECCV) Workshops (2018)

Hackel, T., Savinov, N., Ladicky, L., Wegner, J.D., Schindler, K., Pollefeys, M.: Semantic3d. net: a new large-scale point cloud classification benchmark (2017). arXiv:1704.03847

Hua, B.-S., Tran, M.-K., Yeung, S.-K.: Pointwise convolutional neural networks. In: Proceedings of the IEEE Conference on Computer Vision and Pattern Recognition, pp. 984–993 (2018)

Huang, Q., Wang, W., Neumann, U.: Recurrent slice networks for 3d segmentation of point clouds. In: Proceedings of the IEEE Conference on Computer Vision and Pattern Recognition, pp. 2626–2635 (2018)

Jin, Y.-H., Hwang, I.-T., Lee, W.-H.: A mobile augmented reality system for the real-time visualization of pipes in point cloud data with a depth sensor. Electronics **9**(5), 836 (2020)

Kang, D., Wong, A., Lee, B., Kim, J.: Real-time semantic segmentation of 3d point cloud for autonomous driving. Electronics **10**(16), 1960 (2021)

Lai, X., Liu, J., Jiang, L., Wang, L., Zhao, H., Liu, S., Qi, X., Jia, J.: Stratified transformer for 3d point cloud segmentation. In: Proceedings of the IEEE/CVF Conference on Computer Vision and Pattern Recognition, pp. 8500–8509 (2022)

Lei, H., Akhtar, N., Mian, A.: Spherical kernel for efficient graph convolution on 3d point clouds. IEEE Trans. Pattern Anal. Mach. Intell. **43**(10), 3664–3680 (2020)

Li, Y., Ma, L., Zhong, Z., Cao, D., Li, J.: TGNet: GEOMEtric graph CNN on 3-d point cloud segmentation. IEEE Trans. Geosci. Remote Sens. **58**(5), 3588–3600 (2019)

Maturana, D., Scherer, S.: Voxnet: a 3d convolutional neural network for real-time object recognition. In: 2015 IEEE/RSJ International Conference on Intelligent Robots and systems (IROS), pp. 922–928. IEEE (2015)

Park, C., Jeong, Y., Cho, M., Park, J.: Fast point transformer. In: Proceedings of the IEEE/CVF Conference on Computer Vision and Pattern Recognition, pp. 16949–16958 (2022)

Parmar, N., Vaswani, A., Uszkoreit, J., Kaiser, L., Shazeer, N., Ku, A., Tran, D.: Image transformer. In: International Conference on Machine Learning, pp. 4055–4064. PMLR (2018)

Qi, C.R., Su, H., Mo, K., Guibas, L.J.: Pointnet: deep learning on point sets for 3d classification and segmentation. In: Proceedings of the IEEE Conference on Computer Vision and Pattern Recognition, pp. 652–660 (2017)

Strudel, R., Garcia, R., Laptev, I., Schmid, C.: Segmenter: transformer for semantic segmentation. In: Proceedings of the IEEE/CVF International Conference on Computer Vision, pp. 7262–7272 (2021)

Thomas, H., Qi, C.R., Deschaud, J.-E., Marcotegui, B., Goulette, F., Guibas, L.J.: KPConv: flexible and deformable convolution for point clouds. In: Proceedings of the IEEE/CVF International Conference on Computer Vision, pp. 6411–6420 (2019)

Wang, Y., Sun, Y., Liu, Z., Sarma, S.E., Bronstein, M.M., Solomon, J.M.: Dynamic graph CNN for learning on point clouds. ACM Trans. Graph. (ToG) **38**(5), 1–12 (2019)

Xie, Z., Chen, J., Peng, B.: Point clouds learning with attention-based graph convolution networks. Neurocomputing **402**, 245–255 (2020)

Zhang, C., Wan, H., Shen, X., Wu, Z.: Patchformer: an efficient point transformer with patch attention. In: Proceedings of the IEEE/CVF Conference on Computer Vision and Pattern Recognition, pp. 11799–11808 (2022)

Zhang, R., Yichao, W., Jin, W., Meng, X.: Deep-learning-based point cloud semantic segmentation: a survey. Electronics **12**(17), 3642 (2023)

Active and Reactive Optimal Power Flow Management in a Low-Voltage Microgrid Integrating Renewable Energy Sources and Energy Storage: A Genetic Algorithm-Based Approach for Improved Smart Grid Efficiency

FatimaZahra Zahraoui, Houssam Eddine Chakir, Mehdi Et-taoussi, and Hamid Ouadi

Abstract

This paper presents an optimal power flow management (OPFM) optimization approach for managing active and reactive energy in a low-voltage microgrid (MG) connected to the main grid that incorporates photovoltaic (PV) systems, battery storage (ESS), a gas turbine (GT), and residential loads. The proposed method employs genetic algorithms to solve the optimization problem, addressing active and reactive power scheduling for distributed sources over a 24-h period. Each source contributes to either active or reactive energy mixes; the goal is to respect the operational constraints of each energy source while aiming to (i) minimize greenhouse gas (GHG) emissions and (ii) reduce expenses. The management approach integrates weather and tariff forecasts to improve decision-making. Results demonstrate the effectiveness of the proposed approach in reducing emissions and maintaining grid stability. Simulations are performed in the MATLAB testing environment to assess the accuracy of the suggested optimization methodology.

Keywords

Optimal power flow management · Micro-grid · Operational planning · Photovoltaic system · Hybrid system · Energy storage system · Genetic algorithm · Reactive power

1 Introduction

In order to achieve sustainable energy goals and lower greenhouse gas emissions, power systems must increasingly integrate renewable energy sources. Photovoltaic (PV) systems, which convert solar energy into electrical power, are widely adopted as key renewable energy sources. However, maintaining grid stability and reliability is severely hampered by the intermittent and unpredictable nature of PV generation. Using complex management tactics and combining various distributed energy resources, smart grids (SGs) have emerged as an advanced solution to address these issues. This study focuses on a hybrid distributed power system linked to the main grid, comprising: (i) a PV system as a regenerative source; (ii) lithium-ion batteries as energy storage systems (ESS); and (iii) gas turbines (GT) as auxiliary power generators. The selection of these DS technologies aims to ensure reliable power supply and effective energy management throughout the operating day. In microgrids (MGs), optimal power flow management (OPFM) is essential for balancing the production of both reactive and active power to satisfy power demand while preserving grid stability. Efficient OPFM maximizes the operation of DS energy resources, thereby reducing operating expenses and environmental impacts. Given the complexity of SGs, sophisticated optimization methods are needed to manage the intricate and ever-changing nature of smart grids.

F. Zahraoui (✉) · H. Ouadi
Equipe de Recherche en Electrotechnique, Robotique et Automatique, ENSAM, Mohammed V University in Rabat, Rabat, Morocco
e-mail: fatimazahra.zahraoui@gmail.com

F. Zahraoui
SmartiLAB EMSI-Rabat, Honoris United Universities, Rabat, Morocco

H. E. Chakir
EEIS-Lab, ENSET Mohammedia, Hassan II University, Casablanca, Morocco

M. Et-taoussi
Laboratory of Advanced Research in Industrial and Logistics Engineering (LARILE), Higher National School of Electricity and Mechanics (ENSEM), HassanII University of Casablanca, Casablanca, Morocco

B. El Bhiri et al. (eds.), *AI and Digital Transformation: Innovations in Supply Chain, Education, and Energy Systems*, Sustainable Artificial Intelligence-Powered Applications, https://doi.org/10.1007/978-3-031-86837-5_13

Numerous methods are available in the literature to address multi-source system energy optimization concerns. To reduce costs and emissions, Mohammed et al. (Mohammed et al. 2019) suggested an optimization technique for hybrid PV-wind-tidal turbine-battery systems, applying the Particle Swarm Optimization (PSO) technique. Similarly to this, Chakir et al. (2022) created an optimization model with a cost focus for a grid-connected PV-wind-battery hybrid system and also used PSO. The energy management on the source side is modeled according to linear programming, while the demand side is optimized with mixed-integer GA programming. In a different study, Gheouany et al. (2023) used multi-objective particle swarm optimization (MOPSO) in a grid connected with PV and battery storage to address both management and prediction issues. Moreover, a thorough optimization technique for a wind-PV-MGT-battery hybrid configuration was provided by Zahraoui et al. (2023) to reduce emissions and running costs while taking the main grid fees, weather prediction, and variable load into account. In (Syamala et al. 2024), Syamala et al. looked into how to best utilize Multiple-Agent Modelling (M.A.M.) and blockchain-based technologies to improve the reliability and long-term sustainability of energy-efficient grid energy management. Mixed-Integer Linear Programming (MILP) optimization was studied by Haider et al. (2023) in an effort to lower day-ahead costs and boost the effectiveness of a hybrid PV-wind turbine-ESS system. The Lagrange multiplier method was utilized in Chreim et al. (2023) by Chreim et al.. It has introduced Energy Management in residential Communities with shared storage based on multi-agent systems (EMICO) to determine the best energy trading activities between homes and reduce total energy losses. To minimize the total energy losses, they used the Lagrange multiplier method. To optimize and minimize the operating cost of the microgrid (MG), a PV-battery, Kumar et al. (2024) used intelligent Golden Jackal Optimization (GJO) for distributed-generation energy management. Last but not least, Álvarez-Arroyo et al. (2024) minimized the anticipated microgrid system cost while finding the optimal capacity of the ESS to fulfill the constraint on energy balance for controlling a PV-wind-battery-diesel-turbine hybrid microgrid. The importance of environmental factors in today's energy management strategies is growing. Only two of the examined articles directly address reducing greenhouse gas (GHG) emissions: Mohammed et al. (2019) and Zahraoui et al. (2023). This underscores the need for more comprehensive techniques that incorporate environmental consequences.

The above-mentioned studies don't take into account reactive power, which is crucial for preserving voltage levels and power quality in the grid, even if they all discuss smart grid management and guarantee active power balance. For the power system to be stable and efficient overall, reactive power balance is just as crucial as active power balance.

In contrast to previous works, it will discuss power energy management in a smart grid while taking into account both active and reactive power and the reduction of GHG emissions. In order to stay out of trouble with the energy operator, we will also concentrate on lowering the power factor ($\cos(\varphi)$). The three main objectives of this all-encompassing strategy are to increase power quality, stabilize the system, and support environmental sustainability.

The following are the aims of this work:

- Generates the best power references to verify the power balances of both active and reactive powers.
- Incorporate weather and tariff forecasts into the management plan to improve decision-making.
- Optimize the operating costs by scheduling, in a profitable way, the use of available DS and energy flows.
- Reduce the amount of GHG emissions released by the auxiliary DS.

The remaining document is structured as follows: Section 2 details the system model, the smart grid components, and their constraints. Section 3 presents the mathematical formulation of the optimization problem. Section 4 describes the genetic algorithm approach used to solve the problem. A set of simulation results are discussed in Sect. 5. At the end, the paper is summarized with a conclusion and suggests directions for future research.

2 System Description and Components

Several energy sources are integrated into the smart grid system taken into consideration in this study to guarantee a dependable and effective power supply. Photovoltaic (PV) panels, battery storage systems (ESS), gas turbines (GT), and a grid link are the system's primary parts (Fig. 1). Every one of these elements is essential to the overall plan for energy management.

2.1 Power Balance Constraints

The inherent variability of renewable energy sources, coupled with the uncertainty of predicted power demand, can lead to imbalances in voltage and frequency (Barone et al. 2023; Et-taoussi et al. 2019). Therefore, the main constraints of this paper are maintaining power grid equilibrium.

$$P_t^{pv} + P_t^{PG} + P_t^{MGT} + P_t^{ESS} - P_t^{LD} = 0 \forall t 1..24h \quad (1)$$

$$Q_t^{pv} + Q_t^{PG} + Q_t^{MGT} + Q_t^{ESS} - Q_t^{LD} = 0 \forall t 1..24h \quad (2)$$

Fig. 1 Micro grid architecture

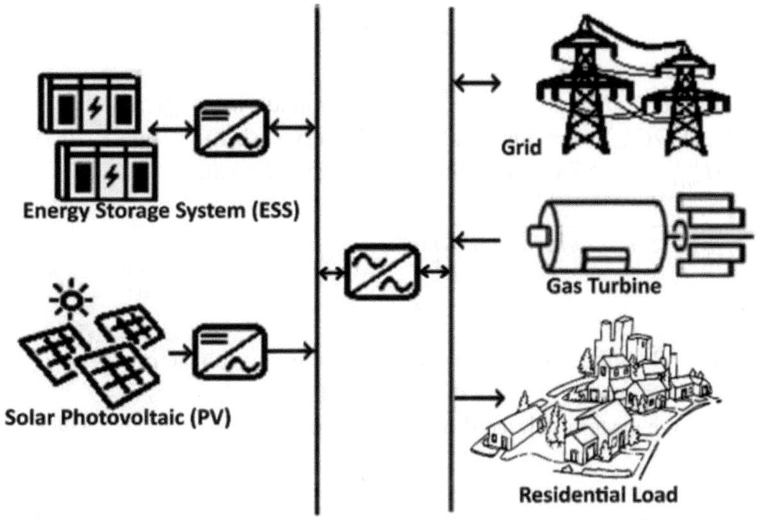

where P_t^i and Q_t^i represent the active and reactive power for the PV cells, grid, ESS, and predicted load, respectively, every hour throughout the day. These constraints ensure that at every moment, the power supplied matches the power demanded by the load. This balance is crucial for maintaining grid stability and preventing power outages or overloading of the system components.

2.2 Modeling and Constraints of the Main Grid

In times of high renewable generation from PV cells, the main grid and the smart grid can exchange electricity. This exchange may involve exporting (selling) excess power or purchasing power when local generation is insufficient.

$$P_t^{PG} \leq 0.9 \times P^{CC} \tag{3}$$

$$\left| Q_t^{PG} \right| \leq Q^{PG\_max} \tag{4}$$

P^{CC} is the contractual power capacity and Q^{PG_max} maximum injected or absorbed reactive power.

2.3 Photovoltaic (PV) Modeling and Constraints

PV systems use semiconductor materials to directly turn sunlight into electricity. The formula in (Gheouany et al. 2023; Bakali et al. 2024) was used to estimate the PV system power generation. It used the PV power to derate the solar irradiance of the PV panel, its degradation factor, and the ambient temperature.

The PV cells must respect these constraints:

$$\sqrt{(P_t^{pv})^2 + (Q_t^{pv})^2} \leq 90\% * P_{MPP_t}^{pv} \forall t 1..24\text{h} \tag{5}$$

Ten percent of each source's nominal (apparent) power is reserved as a buffer in order to meet microgrid integration and commissioning requirements (Eqs. 3, 5 and 11).

2.4 Modeling and Constraints of Energy Storage System (ESS)

When the amount of solar energy produced by PV systems is greater than what is needed, battery storage systems release the excess energy they have stored. By helping to control the supply and demand for electricity, the batteries increase the smart grid's stability and dependability (Hossain et al. 2018). The batteries' efficiency, capacity, and level of charge control how they work.

The available power of the BESS is determined by its state of charge (SOC).

$$SOC_t^{ESS} = \frac{C_t^{ESS}}{C_{ref\ t-1}^{ESS}}; \forall t = 1..24 \tag{6}$$

where $C_t^{ESS} and C_t^{ESS}$ represent, respectively, the battery capacity at time "t" and the ESS reference capacity.

Also (Zahraoui et al. 2023), the ESS may respect those inequalities (Zahraoui et al. 2021) $\forall t 1..24$h:

$$SOC_{min}^{ESS} \leq SOC_t^{ESS} \leq SOC_{max}^{ESS} \tag{7}$$

$$\Delta SOC_t^{ESS} \leq \Delta SOC_{max}^{ESS} \tag{8}$$

$$SOH_{min}^{ESS} \leq SOH_t^{ESS} \tag{9}$$

where SOC_{min}^{ESS} and SOC_{max}^{ESS} are the boundaries of SOC and ΔSOC_{max}^{ESS} the maximal SOC battery's shifting, and SOH_{min}^{ESS} the ESS minimal limit of the state of health.

$$-\sqrt{(S_t^{ESS})^2 - (P_t^{ESS})^2} \leq Q_t^{ESS} \leq \sqrt{(S_t^{ESS})^2 - (P_t^{ESS})^2} \tag{10}$$

where S_t^{ESS} is the apparent power of the storage system at time 't', and P_t^{ESS} *and* Q_t^{ESS} are the active and reactive powers of the storage system delivered or stored by the battery.

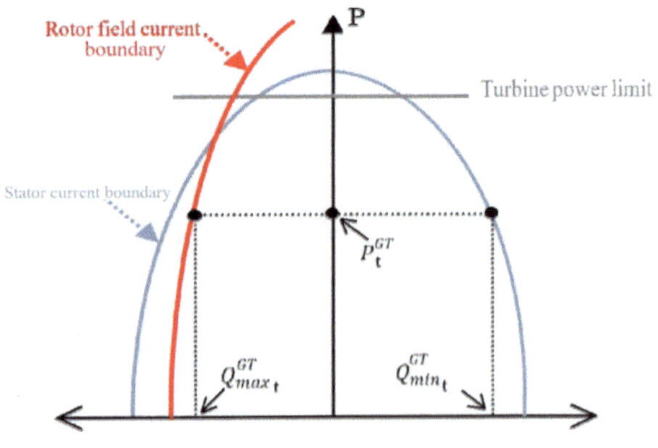

Fig. 3 PQ diagram of PMSG

2.5 Modeling and Operational Limits of the Auxiliary Gas Turbine Source

As a backup power source, the gas turbine can swiftly increase power production to meet peak demand or offset PV generating variability. Compared to renewable energy sources, gas turbines are more dependable and adaptable, but they also emit CO_2. For the MGT to have the highest possible output and the lowest possible CO_2 emissions, it cannot be used to generate power that is less than 50% of its standard power (Boulal et al. 2020). (rf. Fig. 2) The efficiency of the gas turbine has been plotted as a function of the power generated, as shown in Fig. 3.

$$50\% P_{sdt}^{MGT} \leq P_t^{MGT} \leq 90\% P_{sdt}^{MGT} \tag{11}$$

where P_{sdt}^{MGT} is the MGT standard power.

The efficiency of the gas turbine has been plotted as a function of the power generated, as shown in Fig. 2.

The PQ diagram Fig. 3 of the synchronous generator (usually a PMSG) reveals the limitations on the gas turbine's ability to contribute to both active and reactive power production.

$$Q_{min\ t}^{MGT} \leq Q_t^{MGT} \leq Q_{max\ t}^{MGT} \forall t 1..24\text{h} \tag{12}$$

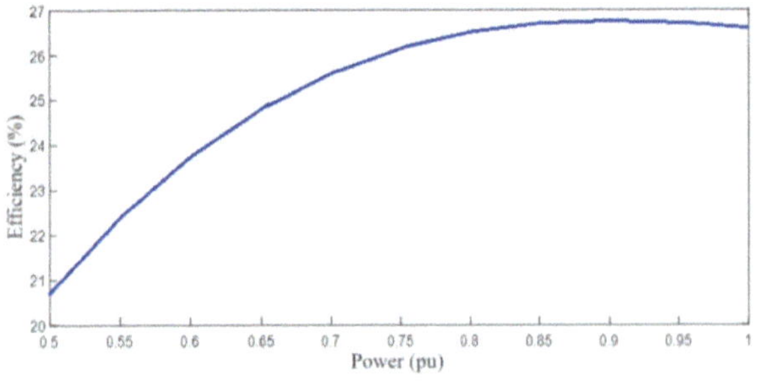

Fig. 2 Gas turbine efficiency

3 Formulation of the Optimization Management Problem

3.1 Objective Function

The primary objective of our optimization management problem is to lower total CO_2 emissions and operating expenses throughout the day. This can be expressed mathematically as follows:

The objective function f_{obj} is designed to minimize the total cost associated with active P_t^i and reactive Q_t^i power, as well as the GHG emissions penalty CO_2^{MGT} from the GT. The function is evaluated over a 24-h period.

- The cost terms ensure that both active and reactive power generation are economically optimized.
- CO_2^{MGT}: represents the cost associated with the GHG emissions from the gas turbine (GT). Incorporating this penalty into the objective function aims to minimize the environmental impact of the power generation process.

$$f_{obj} = \sum_{t=1}^{24} \Big(\underbrace{\sum_i cost\ P_t^i + cost\ Q_t^i}_{\text{active and reactive power cost}} + \underbrace{cost\ CO_2^{MGT}}_{\text{GHG release penality}} \Big) \tag{13}$$

With i = ESS, PG, and MGT.

3.2 Cost of the Main Grid Exchange

The industrial electricity tariff includes both acquisition and sales costs. Acquisition costs are presented in the donut chart Fig. 4. The power sales cost is 0.1176 €/*h*.

The purchase cost corresponds to the price of active energy drawn from the main grid. This price generally varies depending on consumption hours. From 6 AM to 10 PM, the

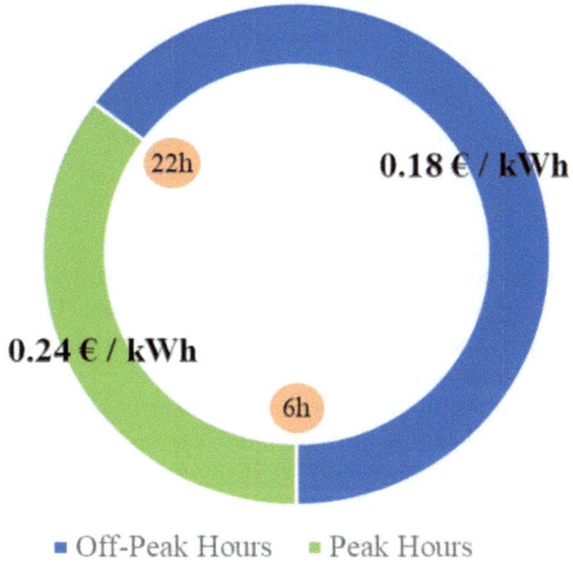

Fig. 4 Purchased grid energy cost

price is 0.24 €/kWh, and at other times, the price is 0.18 €/kWh.

$$C_t^{pen} = 2 \times \left(0{,}8 - \cos \varphi_t^{PG}\right) \times C_t^{PG} \tag{14}$$

where C_t^{PG} is the Acquisition cost.

When the power factor ($\cos \varphi_t^{PG}$) at the main grid connection point is reported as being less than 0.8, this cost is equivalent to the billing for reactive power taken from the main grid. This expense, C_t^{pen}, is determined by "Official website of ONEE". [En ligne] (n.d)

3.3 Cost of PV System

We aim to maximize the use of the generated PV power, and to achieve this, the power generated by the PV cells is considered cost-free.

3.4 Cost of Energy Storage System

Battery cost is associated with the aging degradation charge every hour and is stated as

$$cost_t^{ESS} = \frac{1}{1 - SOH_{min}^{ESS}} \times \left(cost_{ini}^{ESS} \times \Delta SOH_t^{ESS}\right) \tag{15}$$

where $cost_{ini}^{ESS}$ is the first purchase cost of the battery (Zahraoui et al. 2023).

3.5 Cost of the Micro Turbine

The cost of the MGT auxiliary source is made up of:

- Start-up and shutdown expenses for one hour (Zahraoui et al. 2021; Kanchev 2014).

$$\text{cost\_}P_{S/S_t}^{TG} = \begin{cases} 5 \times \text{M}_{\text{gas}} \times C_{ng} \\ 2,5 \times 5 \times \text{M}_{\text{gas}} \times C_{ng} \end{cases} \tag{16}$$

2.5 min for shut down and 5 min for startup mode.

- Active and reactive power expenses.

$$cost\_P_t^{TG} = \text{M}_{\text{gas}} \times C_{ng} \tag{17}$$

where M_{gas} and C_{ng} are, respectively, the natural gas mass and cost.

As seen in Fig. 3, the GT contributes to the generation of active as well as reactive powers, limited to the maximum and minimum (rf. Eq. 12), as mentioned in Boulal et al. (2020). The GT has a maximum volume of active power that it can deliver; if the management system requests reactive power (Q_t^{TG}), the turbine will need to reduce its active power production capacity. This reduction in active production (the opportunity cost) is $\left(P_{nom}^{GT} - P_t^{GT}\left(Q_t^{GT}\right)\right)$. (Boulal et al. 2020; Rider and Paucar 2004)

$$cost\_Q_t^{TG} = \lambda\left[cost\_P_{nom}^{TG} - cost\_P_t^{TG}\left(Q_t^{TG}\right)\right] \tag{18}$$

where λ represents the profit rate from active power production, typically between 5 and 10%.

- CO_2 equivalent emissions penalty cost.

Only the most hazardous and damaging gases are included in the calculation of greenhouse gas emissions, namely NOx, CO, and CO_2 (Kanchev 2014). The penalty cost associated with these emissions is calculated based on the quantity of gas produced.

Figure 5 depicts the variation of CO_2 emission factors with respect to the power provided by the turbine.

$$cost\_CO_{2\ t}^{TG} = \text{CP}_{\text{gas}} \times \text{M}_{\text{CO2}t}^{TG} \tag{19}$$

where CP_{gas} represents the environmental penalty cost, set at thirty euros per ton of CO_2 in this study. And $\text{M}_{\text{CO2}}^{TG}$ is used to describe the carbon dioxide emissions due to the combustion of natural gas for electricity generation.

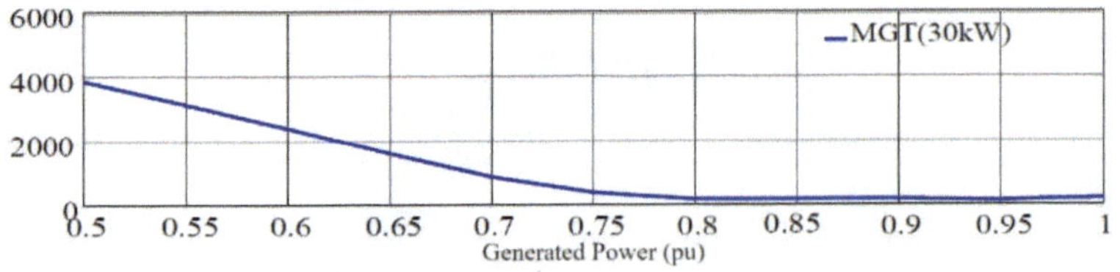

Fig. 5 CO_2 emission factor (mg/kWh)

4 Simulation Technique and Results Simulation Insights and Outcomes

4.1 Genetic Algorithm Approach

Because genetic algorithms are good at handling complicated and non-linear optimization issues, they are used to address the optimization problem (Zahraoui et al. 2023, 2021; Zahraoui et al. xxxx). The following steps are involved in the GA approach (Fig. 6).

- Initialization: Produce a first set of feasible solutions, each of which represents a distinct power generation and consumption schedule for the course of a day.
- Fitness Evaluation: Determine each solution's fitness using the objective function, which takes operational costs and CO_2 emissions into account.
- Selection: The best solutions should be chosen by the parents of the following generation.
- Crossover and Mutation: To create new offspring solutions, apply crossover and mutation operators. This adds diversity and enables the algorithm to explore other regions of the solution space.
- Iteration: Continue doing the assessment, crossover, mutation, and selection processes for a set number of generations, or until convergence is attained.
- Output: The best option from the last generation is chosen as the smart grid's ideal timetable for controlling reactive and active electricity.

4.2 Simulation Conditions

To conduct this simulation, we collected data from the Renewables.ninja website ("Renewables.ninja" 2023) for a moderate PV generation day. Figure 7 presents the maximum extracted power from the PV cells. This power has its maximum at 1 p.m., close to 100 KW.

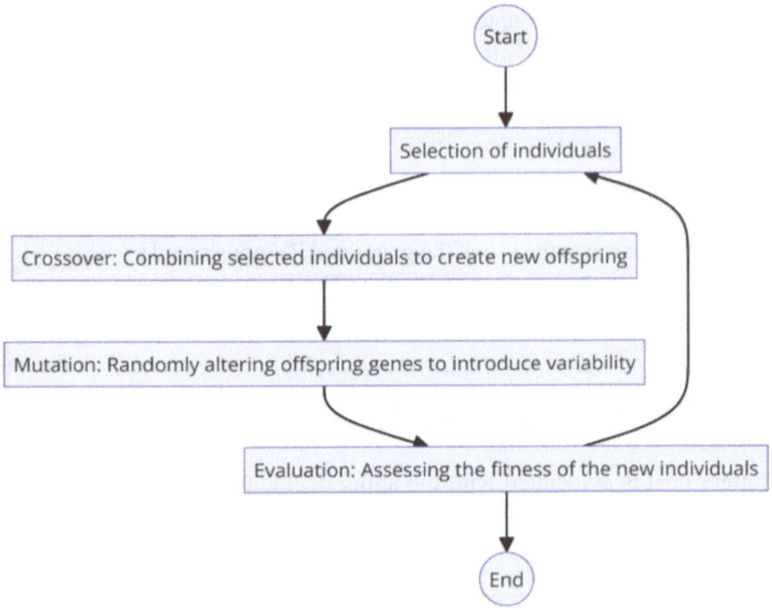

Fig. 6 Genetic algorithm flowchart

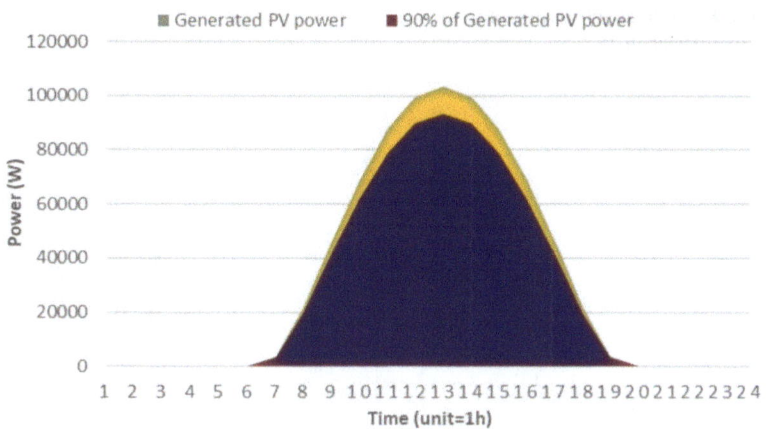

Fig. 7 PV power generated during the studied day

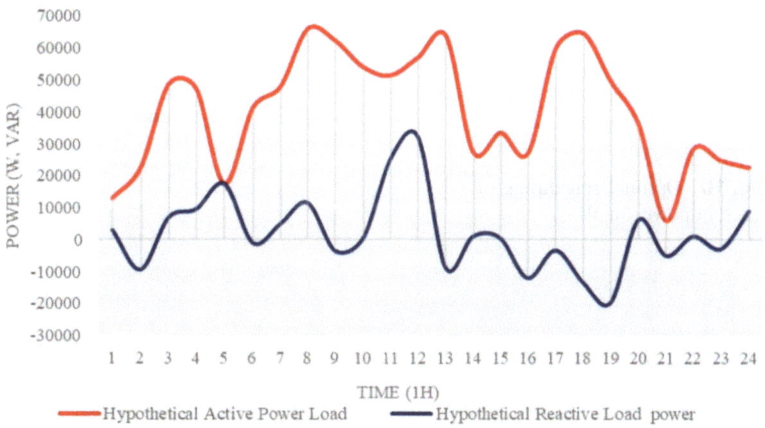

Fig. 8 Forecasted active and reactive consumption profile

Figure 8 depicts a consumption profile characterized by significant fluctuations in both active and reactive power throughout the studied day. The active power ranges from 6 to 66 kW, while the reactive power ranges from -20 kvar to 25 kvar.

In our study, we ran the simulation using a genetic algorithm 400 times with a population size of 10,000. Since genetic algorithms typically approximate rather than provide an exact optimal solution, we averaged the results to obtain a more reliable estimation of the objective function minimization.

4.3 Results and Discussion

In this part, we showcase the outcomes of the simulation of the GA optimization strategy on an active and reactive random load profile. Simulations performed using the MATLAB environment show how effective the suggested genetic algorithm management strategy is. Positive numbers for ESS power show energy being saved for use at a later time when renewable generation is plentiful, whereas negative values show the release of stored energy during times of high demand.

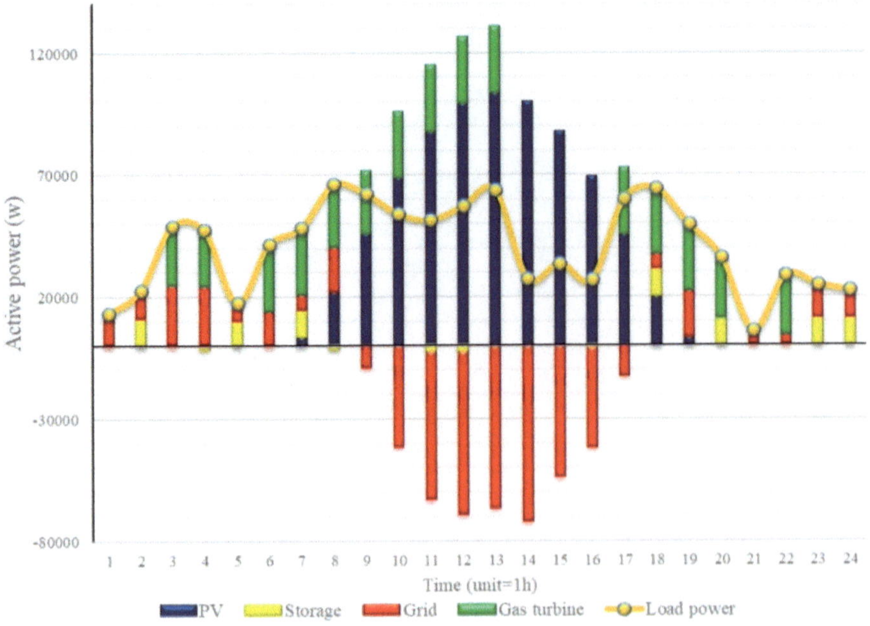

Fig. 9 Optimal scheduling of active power in W

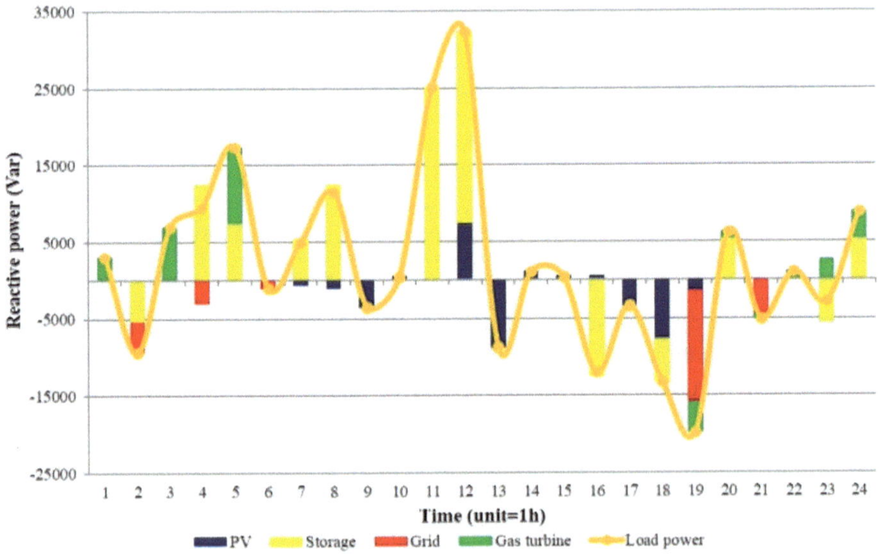

Fig. 10 Optimal scheduling of reactive power in W

Figures 9 and 10, respectively, show the GA distribution of active and reactive power. We can see that it meets the power demand.

In the middle of the day, the genetic algorithm values photovoltaic generation; by authorizing the injection of a large size of active power into the main grid (around 80 kW); while ensuring the full charge of the batteries. This power is mainly produced by the PV generator (around 100 kW).

In this same part of the day, the reactive power demand records its highest value (Figs. 8 and 10). This demand (25 kVAR, 33 kVAR) is largely met by the storage system which acts as a reactive energy compensator (Fig. 10). This latter is charged from the mixed PV/Gas turbine power (see Figs. 9 and 11); where a large part of this power is converted into reactive in the demand side.

The energy storage device is allowed to operate freely during the day, although the SOC is set to begin and end at 50%. Figure 11 demonstrates that this limitation is implemented and allows for efficient battery utilization.

It is observed that at noon, the battery continues its charging cycle, increasing from 40 to 80%. It charges by drawing largely from the combined efforts of PV energy and the gas turbine to meet the reactive power demand of approximately 25 kVAR requested by the load (with the contribution of PV power).

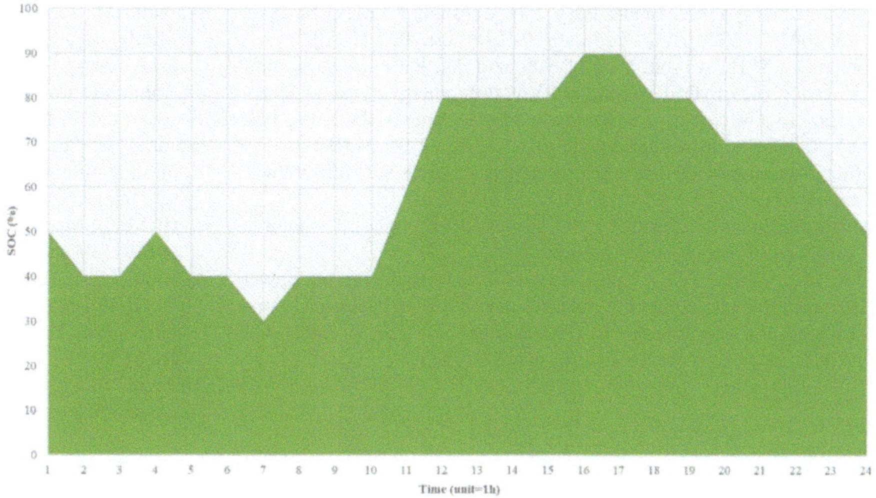

Fig. 11 Profiles of SOC of batteries in studied day conditions

5 Conclusion

This paper offers a successful optimization strategy for managing active and reactive energy in a low-voltage smart grid incorporating PV, battery storage, and a gas turbine. By focusing on minimizing CO_2 emissions and optimizing energy flows, the suggested approach significantly enhances the sustainability and reliability of the power grid. Through intelligent control of power flow within the microgrid, operational cash flow was minimized while maintaining high efficiency in energy conversion from hybrid sources. The genetic algorithm (GA) optimization was tested in various operation modes within the Matlab environment, demonstrating the robustness of the proposed strategy. The primary objectives of this work were met, which include: (i) Generating optimal power references to ensure balance in both active and reactive power. (ii) Incorporating weather and tariff forecasts into the management plan to improve decision-making. (iii) Optimizing operating costs by effectively scheduling the use of available distributed storage (DS) and energy flows. (vi) Reducing greenhouse gas emissions from auxiliary distributed systems.

References

Álvarez-Arroyo, C., Vergine, S., de la Nieta, A. S., Alvarado-Barrios, L., D'Amico, G.: Optimising microgrid energy management: Leveraging flexible storage systems and full integration of renewable energy sources. Renew. Energy **229**, 120701 (2024). https://doi.org/10.1016/j.renene.2024.120701

Barone, G., Buonomano, A., Forzano, C., Palombo, A., Russo, G.: The role of energy communities in electricity grid balancing: A flexible tool for smart grid power distribution optimization. Renew. Sustain. Energy Rev., **187**, 113742 (2023). https://doi.org/10.1016/j.rser.2023.113742

Boulal, A., Chakir, H.E., Drissi, M., Ouadi, H.: Energy bill reduction by optimizing both active and reactive power in an electrical microgrid. IREE **15**(6), Art. n° 6, (2020). https://doi.org/10.15866/iree.v15i6.17909

Chakir, A., Tabaa, M., Moutaouakkil, F., Medromi, H., Alami, K.: Linear programming combined with mixed integer GA for energy management optimization of a PV-wind-battery hybrid system subject to residential demand response. In: présenté à technologies and materials for renewable energy, environment and sustainability: TMREES21Gr, Athens, Greece, p. 020044 (2022). https://doi.org/10.1063/5.0092333

Chreim, B., Esseghir, M., Merghem-Boulahia, L.: Energy management in residential communities with shared storage based on multi-agent systems: Application to smart grids. Eng. Appl. Artif. Intell., **126**, 106886 (2023). https://doi.org/10.1016/j.engappai.2023.106886

El Bakali, S., Ouadi, H., Gheouany, S.: Efficient real-time cost optimization of a two-layer electric water heater system under model uncertainties. Energy Convers. Manag., **304**, 118190 (2024). https://doi.org/10.1016/j.enconman.2024.118190

Et-Taoussi, M., Ouadi, H., Chakir, H.E.: Hybrid optimal management of active and reactive power flow in a smart microgrid with photovoltaic generation. Microsyst. Technol., **25**, (2019). https://doi.org/10.1007/s00542-019-04354-x

Gheouany, S., Ouadi, H., Giri, F., El Bakali, S.: Experimental validation of multi-stage optimal energy management for a smart microgrid system under forecasting uncertainties. Energy Convers. Manag., **291**, 117309 (2023). https://doi.org/10.1016/j.enconman.2023.117309

Haider, H., Jun, Y., Rashed, G. I., Peixiao, F., Kamel, S., Li, Y.: A robust optimization model for microgrid considering hybrid renewable energy sources under uncertainties. Environ. Sci. Pollut. Res., **30**(34), 82470–82484 (2023). https://doi.org/10.1007/s11356-023-27913-2

Hossain, E., Perez, R., Nasiri, A., Bayindir, R.: Stability improvement of microgrids in the presence of constant power loads. Int. J. Electr. Power & Energy Syst., **96**, 442–456 (2018). https://doi.org/10.1016/j.ijepes.2017.10.016

Kanchev, H.: Gestion des flux énergétiques dans un système hybride de sources d'énergie renouvelable : Optimisation de la planification opérationnelle et ajustement d'un micro réseau électrique urbain, Thesis. Technical University of Sofia, Central School of Lille (2014)

Kumar, R.P., Karthikeyan, G.: A multi-objective optimization solution for distributed generation energy management in microgrids with hybrid energy sources and battery storage system. J. Energy Storage, **75**, 109702 (2024). https://doi.org/10.1016/j.est.2023.109702

Mohammed, O.H., Amirat, Y. and Benbouzid, M.: Particle swarm optimization of a hybrid wind/tidal/PV/battery energy system. Application to a remote area in Bretagne, France. Energy Procedia, **162**, 87–96 (2019). https://doi.org/10.1016/j.egypro.2019.04.010

"Official website of ONEE".: [En ligne]. Disponible sur: http://www.one.org.ma/

"Renewables.ninja".: Consulté le: 11 mai 2023. [En ligne]. Disponible sur: https://www.renewables.ninja/

Rider, M.J., Paucar, V.L.: Application of a nonlinear reactive power pricing model for competitive electric markets. In: Generation, transmission and distribution, IEE Proceedings, vol. 151, pp. 407–414 (2004). https://doi.org/10.1049/ip-gtd:20040385

Syamala, M., Gowri, U., Babu, D.V., Nisha, A.S.A., Ahmed, M.A., Muniyandy, E.: Transactive energy management system for smart grids using Multi-Agent Modeling and Blockchain. Sustain. Comput.: Inform. Syst., **43**, 101001 (2024). https://doi.org/10.1016/j.suscom.2024.101001

Zahraoui, F., Chakir, H.E., Et-Taoussi, M., Ouadi, H.: Smart grid cost optimization: Comparing Bellman and genetic algorithms. In: 2021 9th International Renewable and Sustainable Energy Conference (IRSEC), Morocco. IEEE, pp. 1–6 (2021). https://doi.org/10.1109/IRSEC53969.2021.9741131

Zahraoui, F.Z., Et-taoussi, M., Chakir, H.E., Ouadi, H., Elbhiri, B.: Bellman–Genetic hybrid algorithm optimization in rural area microgrids. Energies **16**(19), 6897 (2023). https://doi.org/10.3390/en16196897

Zahraoui, F.Z., Chakir, H.E., Ouadi, H.: Cost reduction in smart grid considering greenhouse gas emissions using genetic algorithm, p 10

A Comprehensive Overview of Green Supply Chain Management Practices in Renewable Energy Systems

Rania Beniaich, Bilal En-Nouaary, and Aziz Hmioui

Abstract

Renewable energy technologies are crucial to address climate change and meet the continuously rising global energy demand. However, their supply chain might pose serious environmental threats for stakeholders to consider, potentially offsetting their key objective of mitigating the environmental issues associated with fossil energy–coal, oil, and natural gas. This paper aims to comprehensively overview green supply chain practices within the literature of renewable energy civil systems. Initially, we introduce the potential of green supply chain management to minimise the ecological impact of renewable energy technologies. Next, we explore green supply chain management practices through applications in the civil engineering field and focus on the renewable energy supply chain, encompassing associated challenges with its three main life stages, namely: manufacturing, operational life, and end-of-life. As such, our contribution provides valuable insights for stakeholders and researchers keen on refining the sustainability of the supply chain for renewable energy civil systems.

Keywords

Sustainability · Green supply chain management · Civil engineering · Renewable energy

R. Beniaich (✉) · B. En-Nouaary · A. Hmioui
Ecole Nationale de Commerce Et de Gestion de Fès, Université Sidi Mohamed Ben Abdellah, Fez, Morocco
e-mail: rania.beniaich@usmba.ac.ma

B. En-Nouaary
e-mail: bilal.ennouaary@usmba.ac.ma

A. Hmioui
e-mail: aziz.hmioui@usmba.ac.ma

1 Introduction

On a global scale, it is undeniable that energy consumption, in whatever form, continues to grow despite international efforts to hinder it (Ritchie et al. 2024). This expansion in energy use leaves behind a heavy carbon footprint, resulting in a series of pressing environmental scars. By far, climate change and water scarcity are agreed upon as the most urgent to handle. Faced with these issues, the world is seeking to reduce its dependence on fossil energy, largely responsible for 90% of greenhouse gas (GHG) emissions in the whole energy sector (Ritchie and Roser 2024), in favour of more sustainable alternatives. That is said, renewable energies (REs) allow to follow increasing world energy needs by simultaneously preserving the environment. Nowadays, various RE sources are harnessed, such as solar, wind, hydro, biomass, geothermal, tidal, and wave power, offering a glimmer of hope in the transition toward a cleaner energy future. However, behind this renewable-dominated future, unveils a complex and challenging supply chain (SC) that could potentially counterbalance their primary reason of existence i.e. environment protection.

Although RE systems boast low to no GHG emissions during their operational life, similar to any industrial process, the manufacturing phase often results in intensive energy consumption leading to high global warming emissions. Unfortunately, the environmental impacts of RE technologies are beyond GHG emissions. Large land requirements, visual and noise pollution, high water consumption, land and water acidification, effects on biodiversity, use of toxic materials, and large volumes of waste after decommissioning represent the main environmental downsides within the SC of RE projects, needed to be handled to ensure their sustainability (Olabi et al. 2023).

B. El Bhiri et al. (eds.), *AI and Digital Transformation: Innovations in Supply Chain, Education, and Energy Systems*, Sustainable Artificial Intelligence-Powered Applications, https://doi.org/10.1007/978-3-031-86837-5_14

Green supply chain management (GSCM) stands out as a solution, aiming to minimise the ecological impact of the entire SC process of RE civil systems, from the manufacturing stage to the end-of-life (EoL) disposal and to ensure their environmentally safe implementation. Throughout the life cycle of RE technologies, GSCM practices focus on optimising resource utilisation, minimising pollution and waste generation, and protecting all the forms of life in our ecosystem, with human beings as a priority. In literature, numerous works reviewed SC sustainability of RE systems laying focus on either GSC practices of a single technology, such as: (Ghosh 2023; Hiloidhari et al. 2023) or on the challenges and performance assessment of RE SC, such as: (Olabi et al. 2023; Wee et al. 2012; Hmouda et al. 2024). Intriguingly, to the extent of our current knowledge, there has not been yet a paper to comprehensively cover the GSC practices of distinct RE sources. As such, our work focused on the most popular sources of RE, namely: solar photovoltaic (PV), wind, and hydropower, since they represent more than 90% of global RE installed capacity (IRENA: Capacity_Statistics_2023 2023), offering a comprehensive overview of GSC practices within the literature of RE civil systems. Green practices were investigated for each of these three technologies, encompassing the three life stages, i.e. manufacturing, operational life, and end-of-life. Our aim is to identify Res' main environmental challenges and describe the appropriate GSC practices to address them.

Subsequent to this introduction, the remainder of this paper is organised as follows: in Sect. 2, we discuss GSCM's role in sustainability and explore their applications in different fields of civil engineering, while in Sect. 3, we shed the light on the green practices within the RE supply chain, outlining the main environmental challenges in the different stages of RE projects lifetime. Finally, in Sect. 4, we summarise the key practices of GSCM within the RE field and draw insightful conclusions.

2 GSCM in Civil Engineering

In civil engineering projects, there is a growing emphasis on implementing sustainable practices to minimise environmental impact and promote long-term resilience. Within this context, GSCM has emerged, referring to the integration of environmental thinking into the entire process of supply chain management, from product design to disposal and reverse logistics, aiming to close the loop and address the environmental impacts within the supply chain (Srivastava 2023; Zhu et al. 2007). Several practices of GSCM were discerned within the field of civil engineering, mainly: internal environmental management (IEM), eco-design (ED), green purchasing (GP), customer cooperation (CC), and investment recovery (IR) (Zhu et al. 2007).

IEM refers to an organisation's creation of its environmental management sustainability measures and goals, the integration of green strategies, the standardisation of processes to ensure environmental protection, and the development of an evaluation framework to track environmental impacts. It can be categorised into three sub-groups; environmental management system (EMS), reduction of resource consumption, and reduction of pollutant emissions (Sarkis 2018). Within civil engineering projects, IEM involves the implementation of robust environmental policies and procedures in organisational structures to reduce risk, minimise environmental impact, and pave the way for a greener and more sustainable future. For instance, building green infrastructures (GIs) such as green parks, greenways, green roofs, rain gardens, permeable pavement, swales, woodlands, and stormwater curb extensions, can sustainably manage stormwater and mitigate urban flood risk. These features work by absorbing, retaining, and filtering rainwater through vegetation and soils. As a result, they contribute to lowering the average urban temperature, reducing air and noise pollution, promoting biodiversity, and enhancing social, physical and mental health (Wang et al. 2024).

ED or design for the environment, and GP are closely related and go often hand in hand. ED is more focused on designing and developing environmentally friendly products from the outset, to minimise environmental impact throughout their lifecycle, it is considered the ultimate pollution prevention tool, since product materials, and up to 80% of environmental impacts are determined during the design phase (Sarkis 2018). On the other hand, GP concentrates on cooperating with manufacturers to develop environmentally friendly goods (Sahoo and Vijayvargy 2023). This involves encouraging manufacturers to adopt green manufacturing practices, such as minimising energy consumption, reducing waste generation, and prioritising recycling throughout the entire lifecycle of the products. In civil engineering practice, using recycled concrete aggregates as a raw material to produce green concrete, as illustrated by Zhang et al. (2023), exemplifies the integration of both practices. On one hand, ED is satisfied by adopting recycled concrete aggregates from demolished structures, effectively reducing natural resource consumption, minimising waste generation, and creating a circular economy approach. On the other side, GP is fulfilled via the use of recycled aggregates, which diminishes the need for new raw materials, thereby lowering the environmental footprint associated with concrete manufacturing.

In civil engineering, CC plays a pivotal role in driving sustainability initiatives within the construction industry. It refers to the cooperation endeavors between the constructing firms and their downstream clients in, for instance, creating an environmentally friendly design, greener production, and energy-efficient transportation to minimise the adverse ecological effect of its supply chain operations (Guo et al. 2021). Existing literature often underscores CC as the most significant component of GSCM since it represents a whole system, illustrating a company's capacity to collaborate with customers in jointly planning for GSCM activities. Government entities, often the largest clients of the construction sector, wield significant influence in advancing sustainability targets (Lenderink et al. 2022). As clients and regulators, governments have the power to shape purchasing initiatives by reducing obstacles to innovation and encouraging supportive frameworks towards environmentally responsible construction practices.

In the seeking for a green SC, IR refers to the financial gain achieved through the management of surplus ressources, scraps, used and recycled components, as well as additional capital assets arising from equipment replacement (Zhu et al. 2007). Take for instance as previously mentioned, the use of recycled concrete aggregates (RCA) to produce green concrete. This practice not only offers environmental benefits but also significant cost savings, as studies have shown that the long-term cost of producing one tonne of coarse RCA was around 40% lower than the cost for coarse natural aggregate (Ohemeng and Ekolu 2020).

Across industries and civil systems, GSCM has revolutionised the traditional supply chain by promoting sustainability while mitigating environmental impact. With its roots in civil engineering projects, GSCM principles extend into the renewable energy sector, offering new avenues for application and innovation.

3 Environmental Challenges and Green Practices in the Renewable Energy Supply Chains

In the RE sector, GSCM plays an essential role in driving sustainable practices throughout the entire supply chain. In this section, we shed light on the green practices spanning the lifecycle of three prominent RE technologies: photovoltaic (PV) systems, wind turbines, and hydropower plants. These practices are examined across three main life stages; manufacturing, operational life, and end-of-life, outlining the environmental challenges encountered in each stage. Figure 1 summarises these challenges and green SC practices for each life cycle stage and RE technology.

3.1 Solar Photovoltaic Projects

Solar Photovoltaic converts the sun's light into electricity via solar cells. It stands out as the most popular and widely adopted source of RE (IRENA: Capacity_Statistics_2023 2023). The manufacturing stage of PV though, poses significant environmental concerns, particularly in the energy-intensive and carbon-emitting processes involved in manufacturing PV panels. (Akulenko et al. 2023) emphasise the necessity of integrating eco-design principles and considering recycling during the research and design phase to enable efficient separation and recovery of valuable components. Silicon, a key component in PV panel production,

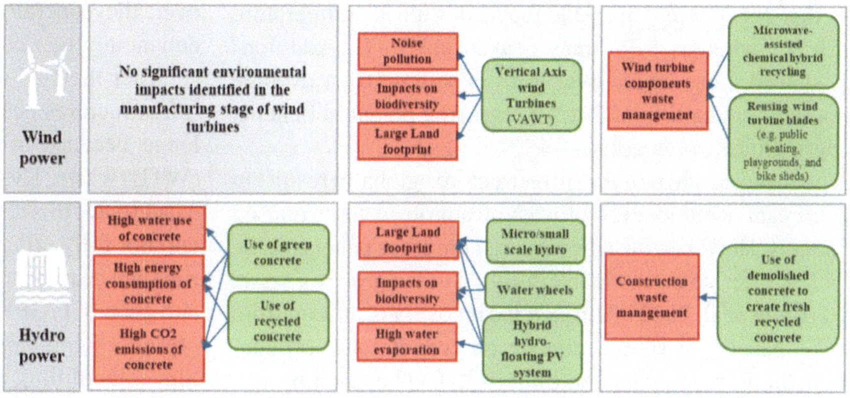

Fig. 1 Main environmental challenges and associated green practices per life cycle stage for each RE system supply chain

is obtained through methods that contribute heavily to CO_2 emissions. Recognising this challenge, authors in Riahi et al. (2023) proposed using recycled silicon as a raw material to produce high-quality silicon, which could significantly reduce energy consumption and cut down CO2 emissions by up to 75% compared to conventional methods of silicon production. On the flip side, authors in Sheikholeslami and Khalili (2024) address the challenge of efficiently cooling PV panels, as heat reduces the panel's efficiency. Despite various cooling systems being explored, such as thermoelectric coolers, evaporative cooling, air cooling, and the use of nanofluids, selecting an optimal cooling method that balances energy output with environmental considerations remains another challenge. Authors in Sheikholeslami and Khalili (2024) suggest a novel design for the heat transfer tube's cross-section, incorporating non-toxic and eco-friendly additives.

Regarding the operational life of PV systems, land availability poses a challenge. Given their considerable land footprint, PV installations can lead to conflicts with other land use types, particularly agriculture. Authors in Tölgyesi et al. (2023) examine the dual land use of Agrivoltaics PV systems and agriculture, known as Agrophotovoltaic (APV) i.e., continuing agricultural production under and between solar panels. Moreover, (Ghosh 2023) delves into water-based PV (WPV) systems as a means of leveraging untapped surfaces for green energy generation. This includes floating PV (FPV) or Floatovoltaic plants, underwater or submerged PV, offshore PV, and canal top PV. Beyond avoiding conflicts with agricultural land, these innovative approaches contribute to the cooling of the PV panels as well as, minimise water evaporation rates to ensure water can still be utilised for farming purposes. In addressing the same problem statement, new models of Solar Photovoltaic (SPV) trees have emerged as potential solutions. These last, have been compared with a conventional flat PV by Almadhhachi et al. (2024), whose empirical findings revealed that SPV trees afford 85% land savings while generating 16–23% more energy compared to the flat PV. Furthermore, temperature assessments indicate that the flat PV module reached a peak temperature of 51 °C, whereas SPV trees peaked at 41 °C. Additionally, shaded areas under solar trees experienced an average temperature reduction of 3 °C, showcasing potential benefits for agricultural applications.

The disposal stage of PV infrastructures globally results in a significant accumulation of waste from obsolete or unused PV panels. However, the silicon present in these panels can be recycled, offering a sustainable solution to the growing waste issue (Riahi et al. 2023). Through the recovery of silicon from old PV panels for reuse, reliance on conventional production methods, significant contributors to CO2 emissions, can be eliminated. Compared to PV, wind turbines are a much cleaner technology (Asdrubali et al. 2015). Still, a number of environmental impacts lay on their way toward absolute sustainability.

3.2 Wind Energy Projects

Wind energy use the kinetic power of wind to generate electricity. It stands out as the renewable technology with the lowest overall environmental impact, characterised by the lowest CO_2 emissions and embodied energy values, i.e. the amount of energy to produce a material, during the manufacturing stage (Asdrubali et al. 2015). However, the operational life of wind parks causes significant effects on neighbouring ecosystems. These include the necessity for ample spacing between turbines in wind farms, where each turbine requires a distance of 7 times the rotor diameter from its adjacent counterpart, hence, requiring larger land areas for installation (Gemayel et al. 2023). Nevertheless, concerns over noise pollution and its negative effects on both humans and wildlife arise. The constant noise emitted by these systems can lead to sleep disturbances, stress, and annoyance in humans. While in wildlife, it can alter the ability of animals to communicate and locate the prey or predators. Also, issues related to biodiversity conservation, particularly for birds and bats, have been underscored, as they face the risk of colliding with the high-speed rotating blades of these giants. To address these challenges, authors in Gemayel et al. (2023) investigate a new design of wind energy system through Vertical Axis Wind Turbines (VAWTs). In contrast to conventional Horizontal Axis Wind Turbines (HAWTs), VAWTs have a unique design that allows for closer packing in wind farms, optimising land use, and operating effectively closer to the ground, where wind speeds tend to be lower, facilitating easier maintenance and installation on tall structures like chimneys. Their omnidirectional design enables efficient capture of wind energy from any direction, making them suitable for areas with variable wind patterns. As VAWTs lack large spinning blades, they mitigate collision threats to bird and bat populations. Moreover, they operate more quietly than conventional HAWTs, enhancing their compatibility with residential areas.

After their 20 to 25-year lifespan, decommissioned wind infrastructures become a heavy environmental obstacle challenge because of the accumulation of wind turbine blade (WTB) waste. The intricate nature of wind power components exacerbates this challenge, as the recycling process is complex and not fully efficient. While 80–85% of wind turbine materials can be recycled, the remaining 15%, consisting of carbon and glass (CF/GF) fiber-reinforced polymer (FRP) composite, presents significant recycling difficulties (Khalid et al. 2023). Since, landfilling and incinerating WTBs might have serious health and environmental consequences, many end-of-life

(EOL) approaches have been proposed for composite wastes to recover CF/GF from FRPs such as mechanical, chemical, thermal, and hybrid recycling techniques (Wang et al. 2023). Recently, researchers have shifted their focus to hybrid recycling technologies, and precisely microwave-assisted chemical hybrid recycling (Wang et al. 2023), a highly efficient technique to recover high-quality fibers. Another viable solution for managing WTB waste involves reusing them as public seating, playgrounds, or bike sheds. These green practices not only reduce waste but also contribute to the circular economy by extending the lifespan and utility of wind turbine components. This concept of circularity is also leveraged in hydropower systems, to green their SCs.

3.3 Hydropower Projects

Hydropower, as the oldest technology for harnessing renewable energy, converts the kinetic energy stored in moving waters into electricity. Nevertheless, the environmental impacts associated with the supply chain of hydropower pose serious challenges, primarily stemming from the extensive use of concrete in dam construction. Concrete is known for its high energy consumption, leading to substantial GHG emissions. However, innovative green practices offer solutions to address the environmental effects of construction. On one hand, authors in Fakhri and Dawood (2023) have introduced a novel concrete design, termed green concrete, which exhibits commendable performance while maintaining environmental and sustainable attributes within the construction industry. This is achieved by incorporating sustainable materials like superplasticizers to reduce water requirements and mitigate CO_2 emissions by up to 37% (Fakhri and Dawood 2023). On the other hand, recycled concrete represents an effective solution to hinder the carbon footprint of hydropower plants. Authors in Zhang et al. (2023) conducted a life cycle assessment to evaluate the engineering material footprint (EMF) and the fossil fuel material footprint (FMF) in closed-loop recycling of concrete. The findings underscore that the concrete recycling system may decrease 99% of the EMF and 66–93% of the FMF when compared to the baseline system, in which concrete waste is landfilled.

Upon the commissioning of a hydropower dam, challenges related mainly to significant land occupation, water evaporation, and biodiversity loss become prominent. Various GSC practices were introduced by literature to address such issues. The synergy of hydropower and PV emerges as a sustainable solution to mitigate the negative impacts of implementation on the ecosystem. Floating photovoltaic efficiently combines both technologies, offering a smart configuration to conserve water through mitigation of evaporation losses and improve the aquatic environment and water quality, as PV panels provide shading and limit the evaporation from wind (Ghosh 2023; Kakoulaki et al. 2023). Remarkably, covering 10% of the hydropower basin surfaces with FPV systems has been shown to increase hydropower plant energy production by 65% (Cazzaniga et al. 2019). Additionally, combining FPV and hydropower plants can share existing hydropower equipment, therefore achieving investment recovery (Kakoulaki et al. 2023). Furthermore, water wheels and micro-hydropower stations provide a way to produce renewable electricity with reduced surface occupation, thereby addressing the large-scale land requirements of traditional hydro dams. Water wheels boast their fish-friendly characteristics operating with free surface flow, low rotational speeds, and large buckets, exerting minimal impacts on fish populations. They are particularly suitable for remote localities and rural areas, due to their low costs, low environmental impacts, and good performance (Quaranta and Wolter 2021). Micro-hydropower plants, known also as small-scale hydropower with a production capacity of no more than 10 MW, exhibit land occupation per MWh that is two times smaller than that of large-scale hydropower projects (Bakken et al. 2014).

After a lifetime often exceeding a century, decommissioned hydropower plants leave behind large infrastructures with enormous volumes of concrete waste. As discussed in the previous section, concrete waste can be recycled to create a fresh form of concrete. The recycling process entails the crushing of demolished concrete to obtain recycled concrete aggregates, which serve as an essential raw material in green concrete production (Zhang et al. 2023). The overarching goal of this approach is to form a circular material flow and to preserve the limited sources of natural aggregates.

4 Conclusion

The transition towards RE technologies is essential for mitigating climate change and reducing our reliance on fossil fuels. However, their supply chain presents significant environmental challenges across its lifecycle stages. This paper contributes to the growing body of knowledge on GSCM's role in addressing these challenges and enhancing the sustainability of RE systems, through eco-friendly practices, such as eco-design, green purchasing, green manufacturing, investment recovery, reuse, and recycling. In this work, we have focused on the most popular RE sources, namely solar PV, wind, and hydropower, encompassing the environmental concerns associated with each technology, particularly CO_2 emissions, land footprint, resource optimisation, waste management, and biodiversity conservation. We shed light on innovative green solutions across three lifecycle stages: manufacturing, operational life, and end-of-life.

Regarding solar PV systems, our key findings emphasised the importance of using recycled silicon to reduce

CO2 emissions, adopting eco-friendly cooling additives to address cooling system inefficiencies, using alternative PV types such as water-based PV, agri-PV, and solar PV trees to mitigate large land footprint, also, implementing recycling practices for panel components, particularly silicon, to repurpose them as raw materials and minimise waste. For wind turbines, we outlined the absence of any significant environmental impacts in the manufacturing stage. However, to mitigate concerns arising during operational life, such as noise pollution, impacts on biodiversity, and large land footprint, we depicted the adoption of a vertical axis turbine design. Also, at the end-of-life stage of wind turbines, we underlined the adoption of microwave-assisted chemical hybrid recycling techniques or repurposing wind turbine blades for alternative uses such as public seating, playgrounds, bike sheds, and more. As for hydropower, our research highlighted the necessity of using green or recycled concrete to reduce CO_2 emissions, water, and energy consumption at the manufacturing stage. While, the adoption of micro and small-scale hydro, water wheels, and hybrid hydro floating PV systems helps to address operational-related issues. Furthermore, the repurposing of demolished concrete to create fresh recycled concrete was underlined as a sustainable practice for waste management at the end-of-life of hydro infrastructures.

We believe that our overview provided valuable insights for stakeholders and researchers looking for solutions to refine the sustainability of the specific SC of RE civil systems. Still, our analysis did not cover other emerging RE sources such as solar thermal, biomass, and geothermal technologies. Having said that, future research will focus on extending the scope of this study toward a more comprehensive review of GSCM practices for RE systems.

Acknowledgements The two first authors would like to acknowledge the financial support provided by the CNRST under the "PhD-Associate Scholarship—PASS" program, which has enabled us to pursue our research.

References

Akulenko, E.S., Hadadian, M., Santasalo-Aarnio, A., Miettunen, K.: Eco-design for perovskite solar cells to address future waste challenges and recover valuable materials. Heliyon. **9**, e13584 (2023). https://doi.org/10.1016/j.heliyon.2023.e13584

Almadhhachi, M., Seres, I., Farkas, I.: Sunflower solar tree vs. flat PV module: A comprehensive analysis of performance, efficiency, and land savings in urban solar integration. Results Eng. **21**, 101742 (2024). https://doi.org/10.1016/j.rineng.2023.101742

Asdrubali, F., Baldinelli, G., D'Alessandro, F., Scrucca, F.: Life cycle assessment of electricity production from renewable energies: Review and results harmonization. Renew. Sustain. Energy Rev. **42**, 1113–1122 (2015). https://doi.org/10.1016/j.rser.2014.10.082

Bakken, T.H., Aase, A.G., Hagen, D., Sundt, H., Barton, D.N., Lujala, P.: Demonstrating a new framework for the comparison of environmental impacts from small- and large-scale hydropower and wind power projects. J. Environ. Manage. **140**, 93–101 (2014). https://doi.org/10.1016/j.jenvman.2014.01.050

Cazzaniga, R., Rosa-Clot, M., Rosa-Clot, P., Tina, G.M.: Integration of PV floating with hydroelectric power plants. Heliyon. **5**, e01918 (2019). https://doi.org/10.1016/j.heliyon.2019.e01918

Gemayel, D., Abdelwahab, M., Ghazal, T., Aboshosha, H.: Modelling of vertical axis wind turbine using large eddy simulations. Results Eng. **18**, 101226 (2023). https://doi.org/10.1016/j.rineng.2023.101226

Ghosh, A.: A comprehensive review of water-based PV: Flotavoltaics, underwater, offshore & canal top. Ocean Eng. **281**, 115044 (2023). https://doi.org/10.1016/j.oceaneng.2023.115044

Guo, Y., Yen, D.A., Geng, R., Azar, G.: Drivers of green cooperation between Chinese manufacturers and their customers: An empirical analysis. Ind. Mark. Manage. **93**, 137–146 (2021). https://doi.org/10.1016/j.indmarman.2021.01.004

Hiloidhari, M., Sharno, M.A., Baruah, D.C., Bezbaruah, A.N.: Green and sustainable biomass supply chain for environmental, social and economic benefits. Biomass Bioenerg. **175**, 106893 (2023). https://doi.org/10.1016/j.biombioe.2023.106893

Hmouda, A.M.O., Orzes, G., Sauer, P.C.: Sustainable supply chain management in energy production: A literature review. Renew. Sustain. Energy Rev. **191**, 114085 (2024). https://doi.org/10.1016/j.rser.2023.114085

IRENA.: Capacity_Statistics_2023. (2023)

Kakoulaki, G., Gonzalez Sanchez, R., Gracia Amillo, A., Szabo, S., De Felice, M., Farinosi, F., De Felice, L., Bisselink, B., Seliger, R., Kougias, I., Jaeger-Waldau, A.: Benefits of pairing floating solar photovoltaics with hydropower reservoirs in Europe. Renew. Sustain. Energy Rev. **171**, 112989 (2023). https://doi.org/10.1016/j.rser.2022.112989

Khalid, M.Y., Arif, Z.U., Hossain, M., Umer, R.: Recycling of wind turbine blades through modern recycling technologies: A road to zero waste. Renew. Energy Focus. **44**, 373–389 (2023). https://doi.org/10.1016/j.ref.2023.02.001

Lenderink, B., Halman, J.I.M., Boes, J., Voordijk, H., Dorée, A.G.: Procurement and innovation risk management: How a public client managed to realize a radical green innovation in a civil engineering project. J. Purch. Supply Manag. **28**, 100747 (2022). https://doi.org/10.1016/j.pursup.2022.100747

Ohemeng, E.A., Ekolu, S.O.: Comparative analysis on costs and benefits of producing natural and recycled concrete aggregates: A South African case study. Case Stud. Constr. Materials. **13**, e00450 (2020). https://doi.org/10.1016/j.cscm.2020.e00450

Olabi, A.G., Elsaid, K., Obaideen, K., Abdelkareem, M.A., Rezk, H., Wilberforce, T., Maghrabie, H.M., Sayed, E.T.: Renewable energy systems: Comparisons, challenges and barriers, sustainability indicators, and the contribution to UN sustainable development goals. Int. J. Thermofluids. **20**, 100498 (2023). https://doi.org/10.1016/j.ijft.2023.100498

Quaranta, E., Wolter, C.: Sustainability assessment of hydropower water wheels with downstream migrating fish and blade strike modelling. Sustainable Energy Technol. Assess. **43**, 100943 (2021). https://doi.org/10.1016/j.seta.2020.100943

Riahi, S., Mckenzie, J.A., Sandhu, S., Majewski, P.: Towards net zero emissions, recovered silicon from recycling PV waste panels for silicon carbide crystal production. Sustain. Mater. Technol. **36**, e00646 (2023). https://doi.org/10.1016/j.susmat.2023.e00646

Ritchie, H., Roser, M.: CO_2 emissions. Our World in Data. (2024)

Ritchie, H., Rosado, P., Roser, M.: Energy production and consumption. Our World in Data. (2024)

Sahoo, S., Vijayvargy, L.: Green supply chain management practices and its impact on organisational performance: evidence from Indian manufacturers. https://www.scopus.com/record/display.uri?eid=2-s2.0-85095416900&origin=inward. Last accessed 18 Nov 2023

Sarkis, J., Dou, Y.: Green supply chain management : A concise introduction. (2018)

Shamseldeen Fakhri, R., Thanon Dawood, E.: Limestone powder, calcined clay and slag as quaternary blended cement used for green concrete production. J. Build. Engineering. **79**, 107644 (2023). https://doi.org/10.1016/j.jobe.2023.107644

Sheikholeslami, M., Khalili, Z.: Solar photovoltaic-thermal system with novel design of tube containing eco-friendly nanofluid. Renewable Energy **222**, 119862 (2024). https://doi.org/10.1016/j.renene.2023.119862

Srivastava.: Green supply-chain management: A state-of-the-art literature review. Int. J. Manag. Rev. https://onlinelibrary.wiley.com/doi/abs/10.1111/j.1468-2370.2007.00202.x. Last accessed 17 Nov 2023

Tölgyesi, C., Bátori, Z., Pascarella, J., Erdős, L., Török, P., Batáry, P., Birkhofer, K., Scherer, L., Michalko, R., Košulič, O., Zaller, J.G., Gallé, R.: Ecovoltaics: Framework and future research directions to reconcile land-based solar power development with ecosystem conservation. Biol. Cons. **285**, 110242 (2023). https://doi.org/10.1016/j.biocon.2023.110242

Wang, Y., Li, A., Zhang, S., Guo, B., Niu, D.: A review on new methods of recycling waste carbon fiber and its application in construction and industry. Constr. Build. Mater. **367**, 130301 (2023). https://doi.org/10.1016/j.conbuildmat.2023.130301

Wang, Z., Li, Z., Wang, Y., Zheng, X., Deng, X.: Building green infrastructure for mitigating urban flood risk in Beijing, China. Urban For. & Urban Greening. **93**, 128218 (2024). https://doi.org/10.1016/j.ufug.2024.128218

Wee, H.-M., Yang, W.-H., Chou, C.-W., Padilan, M.V.: Renewable energy supply chains, performance, application barriers, and strategies for further development. Renew. Sustain. Energy Rev. **16**, 5451–5465 (2012). https://doi.org/10.1016/j.rser.2012.06.006

Zhang, C., Hu, M., van der Meide, M., Di Maio, F., Yang, X., Gao, X., Li, K., Zhao, H., Li, C.: Life cycle assessment of material footprint in recycling: A case of concrete recycling. Waste Manage. **155**, 311–319 (2023). https://doi.org/10.1016/j.wasman.2022.10.035

Zhu, Q., Sarkis, J., Lai, K.: Green supply chain management: pressures, practices and performance within the Chinese automobile industry. J. Clean. Prod. **15**, 1041–1052 (2007). https://doi.org/10.1016/j.jclepro.2006.05.021

Incentives Allocation for Renewable Energy Projects Based on Machine Learning

Bilal En-Nouaary and Yassine Rami

Abstract

Widely recognized is the crucial role of renewable energy (RE) in promoting sustainability, addressing climate change, resource conservation, energy security, and green economic development. Despite these imperatives, financial challenges persist, notably in regions with fossil fuel subsidies. Incentives are essential for driving investments towards a low-carbon future. However, policymakers often encounter difficulties in selecting the optimal projects to incentivise. This difficulty arises partially from the sheer volume of projects and the extensive data associated with them, making the decision-making process complex. To cope with these hurdles, artificial intelligence, particularly machine learning, offers the right tools indispensable when working on large datasets. In this paper, we overview the challenges and main categories of incentives, we present recent applications of artificial intelligence in the RE field, and we introduce an innovative framework inspired by a machine learning segmentation model adopted in the marketing domain. By leveraging state of art machine learning algorithms and data processing techniques, we propose a new model for incentives allocation capable of identifying various clusters of projects, based on their production and financial performance parameters. Aiming at enhancing the incentives allocation process, our developed model ensures that targeted segments of RE projects receive appropriate and effective incentives. Our contribution will serve as a blueprint for policymakers seeking to optimise their incentives allocation strategies and to efficiently stimulate production in the renewable energy market.

Keywords

Policymakers strategies · Incentives allocation · Machine learning · Data processing · Renewable energy

1 Introduction

The energy sector is the source of around three-quarters of greenhouse gas emissions today, with fossil fuels accounting for 90% of this share. Renewable energy (RE) sources, such as: solar, wind, hydro, biomass and geothermal, hold the key to mitigating climate change effects and environmental pollution caused by conventional energy sources (oil, coal and natural gas) (Our World in Data–CO_2 emissions 2024). However, many barriers hinder the development of sustainable energy projects in favour of conventional sources. These include insufficient financing, inadequate grid connectivity, high output intermittency, and low-capacity factor, resulting in fragile cost competitiveness compared to fossil fuel alternatives (Qadir et al. 2021).

Targeted incentives have thus far proven to be effective in expediting RE projects deployment. In a recent literature review by Qadir et al. (2021), four categories of incentives were discerned, namely: research and development (R&D), fiscal and tax, market development, and tariff incentives. R&D incentives are provided to improve RE technologies via R&D facilities. Hence production costs are decreased by enabling local production of RE technologies (Qadir et al. 2021). Fiscal and tax incentives work towards improving the cost competitiveness of RE technologies, by waiving imports taxes on RE equipment, taxing fossil fuel generation, and providing various investment incentives, such as: loans at preferential rates, grants, and production tax credit (Qadir et al. 2021;

B. En-Nouaary (✉)
Ecole Nationale de Commerce et de Gestion de Fès, Université Sidi Mohamed Ben Abdellah, Fez, Morocco
e-mail: bilal.ennouaary@usmba.ac.ma

Y. Rami
Ecole Supérieure de Technologie de Fès, Université Sidi Mohamed Ben Abdellah, Fez, Morocco

B. El Bhiri et al. (eds.), *AI and Digital Transformation: Innovations in Supply Chain, Education, and Energy Systems*, Sustainable Artificial Intelligence-Powered Applications, https://doi.org/10.1007/978-3-031-86837-5_15

Simsek and Simsek 2013; Zhao et al. 2016). Market development is facilitated through establishing standard tariffs for RE investors, implementing standardized testing and certification for small power producers, and setting obligatory RE quotas (Qadir et al. 2021; Simsek and Simsek 2013; Zhao et al. 2016). Feed in tariffs (FiT) represent the most common form of grid connection incentive, where the policymakers set a fixed guaranteed price for selling green electricity into the power network. Different FiTs mechanisms are used: fixed priced FiT, premium FiT and fixed premiums (Qadir et al. 2021; Zhao et al. 2016).

However, there are many challenges to consider prior to incentives allocation. Policymakers, faced with a significant number of heterogenous projects, translating to a huge volume of data, often grapple with the challenge of identifying the most promising technologies for incentives. That once initiated, they become entrenched, making their withdrawal a difficult process. Poorly allocated incentives may inadvertently favour projects that would have succeeded independently, or even worse drive-up technology costs as demand rises, exacerbating challenges rather than mitigating them (Duffy et al. 2015). Recognizing these challenges, our study introduced a novel approach, leveraging state of art machine learning (ML) algorithms and advanced data processing. Drawing inspiration from an innovative model for predicting customer segment changes to optimize advertisement campaigns (Abidar et al. 2023). Through our innovative framework, we aimed to address the complex problem of incentives allocation in the RE sector, ensuring appropriate segmentation of these projects, so that targeted categories receive effective incentives. This paper serves as an insightful road map to enhance the effectiveness of incentives programs, maximize the impact on the RE market, and contribute to the broader goals of sustainability.

Subsequent to this introduction, the remainder of this paper is organized as follows: in Sect. 2, we explore artificial intelligence (AI) applications in the RE field, while in Sect. 3, we describe our adapted framework outlining the step-by-step process of RE projects segmentation changes. Finally, in Sect. 4, we summarize the key aspects of our adapted model and draw insightful conclusions.

2 Literature Review

In the transition to renewable and low carbon energy systems using ML and data analytics can enhance the performance of energy systems on multiple levels mainly predictive maintenance, energy storage, grid management, and hybrid RE systems optimisation leading to improved energy efficiency, cost-effectiveness, and environmental sustainability.

AI, particularly ML, proved to be useful in predictive maintenance in RE infrastructure. (Afridi et al. 2021) offered an overview of predictive/prognostic maintenance frameworks found in the literature. In fact, (Paolanti et al. 2018) presented a ML architecture for predictive maintenance utilizing a random forest approach, demonstrating high accuracy in predicting various machine states. Expanding on this application, predictive maintenance was extended by (Bouabdallaoui et al. 2021) to the building sector, providing guidelines for implementation, especially in heating, ventilation, and air conditioning installations.

Also, in the energy storage and grid management fields, AI plays a crucial role. Authors such as (Alimi et al. 2020) focused on developing ML models for power system security and stability, emphasizing aspects such as cyberattack detection, power quality disturbance studies, and dynamic security assessment. While (Omitaomu and Niu 2021) conducted a structured review of prevalent AI techniques applied in the smart grid domain. These techniques encompassed load forecasting, power grid stability assessment, fault detection, and security concerns within smart grid and power systems. Moreover, other papers addressed the intermittency challenge in RE by predicting these systems for enhanced stability; predictive models provided an effective alternative to traditional electricity storage and standby capacity (Sharifzadeh et al. 2019).

Moreover, to optimize hybrid RE systems, a range of studies have explored the potential of using Authors in Cai et al. (2022) provided a comprehensive review of these systems, discussing their configurations, energy storage options, and the use of AI for simulation and optimization. While (Mercado et al. 2016) emphasized the benefits of AI techniques in optimizing hybrid energy sources, and demonstrated that the use of Artificial Neural Networks yielded superior cost efficiency results, underscoring AI versatility in enhancing the performance and cost-effectiveness of hybrid RE systems (Dellosa and Palconit 2021).

Intriguingly, to the extent of our current knowledge, there has not been yet a development of AI machine learning models specifically tailored for the allocation of incentives to RE projects. The global perspective encompassed by AI machine learning models presents a significant opportunity for their application in the strategic allocation of incentives, potentially optimizing their impact in the RE sector.

3 Workflow Model

In our adapted framework, a variety of cutting-edge ML algorithms and data analysis tools were employed to reach our set objective: data preprocessing, including cleaning, reduction, scaling, transformation, and partitioning, to ensure dataset quality and interpretability; computing PCE and UNMI parameters addresses projects' productivity, performance, and financial viability; clustering which employs the

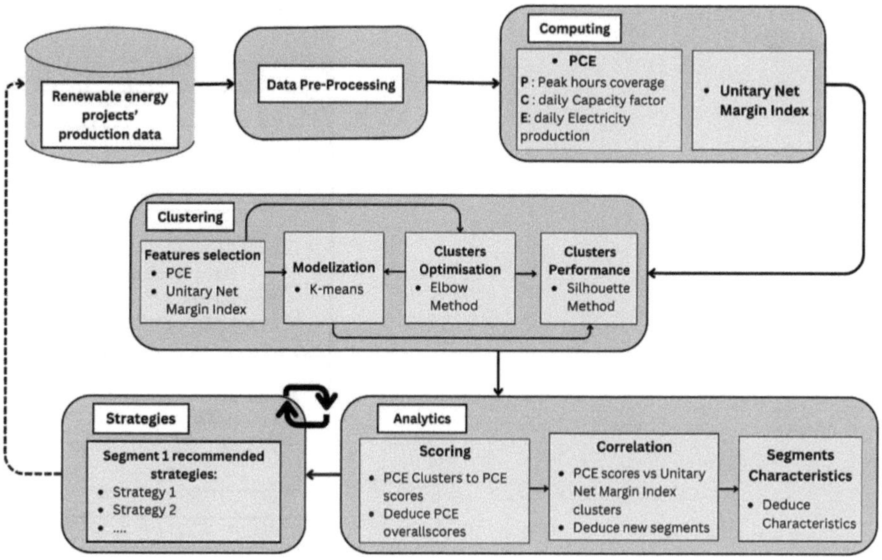

Fig. 1 Main framework workflow

K-means algorithm, using PCE and UNMI as the selected features, the elbow method for cluster optimisation, and the silhouette method for performance evaluation; analytics and segmentation to categorize the projects based on clustering results, guiding the formulation of tailored incentive strategies. The resultant framework's workflow is depicted in Fig. 1.

The initial step of the framework involves gathering the necessary projects daily production data, to be injected into the clustering stage after being pre-processed.

3.1 Data Preprocessing

The following are the crucial steps to be undertaken before initiating the analysis with the actual data for model training. Figure 2 displays the data preprocessing main stages.

Initiating with data cleaning, a crucial step in ensuring dataset quality and reliability, involves identifying and correcting incorrect or noisy data, as well as handling missing values. It also encompasses outlier detection and treatment, given their potential impact on model performance (Maharana et al. 2022; Abidar et al. 2020).

Subsequently, data reduction addresses challenges associated with high-dimensional datasets. Beyond reducing computational costs, it enhances model interpretability and mitigates the risk of overfitting (Fan et al. 2021). Data, organized in a two-dimensional table with rows representing data samples and columns representing variables, undergoes both row and column reduction in this process.

Following data reduction is data scaling, also known as data integration, which involves combining multiple heterogeneous data sources into a cohesive data store and unified data view. The primary objective is to merge diverse datasets into a standardized format for seamless analysis, facilitating effective predictive modelling (Fan et al. 2021; Abidar et al. 2022).

The next step is data transformation, encompassing various techniques to prepare the data for effective modelling. The goal is to ensure that the data aligns with the assumptions and requirements of the chosen ML algorithms (Zhang and Chen 2012).

The final step is data partitioning, involving the division of the entire data table into several subsets, each containing a specific number of rows (Fan and Xiao 2018). Typically, the dataset is split into training, validation, and test sets. The training set is utilized to train the model, the validation set aids in tuning hyperparameters, and the test set evaluates the model's performance on entirely new data.

Data preprocessing holds utmost significance as it greatly influences the performance of ML Algorithms (Maharana et al. 2022; Abidar et al. 2020).

3.2 PCE and UNMI Computing

Following the designated workflow, PCE (Peak hours coverage, daily Capacity factor, and daily Electricity production) index and the UNMI (Unitary Net Margin Index) need to be computed before going into clustering.

PCE index. Our PCE index employs three chosen parameters to evaluate a project's productivity and performance: Peak hours coverage (P), daily Capacity factor (C), and daily Electricity production (E).

Peak hours coverage addresses a significant challenge faced by RE technologies—intermittency. By nature RE sources might, in some cases, be unable to produce electricity during periods of high demand. For example, solar power

Fig. 2 Data preprocessing steps (Fan et al. 2021)

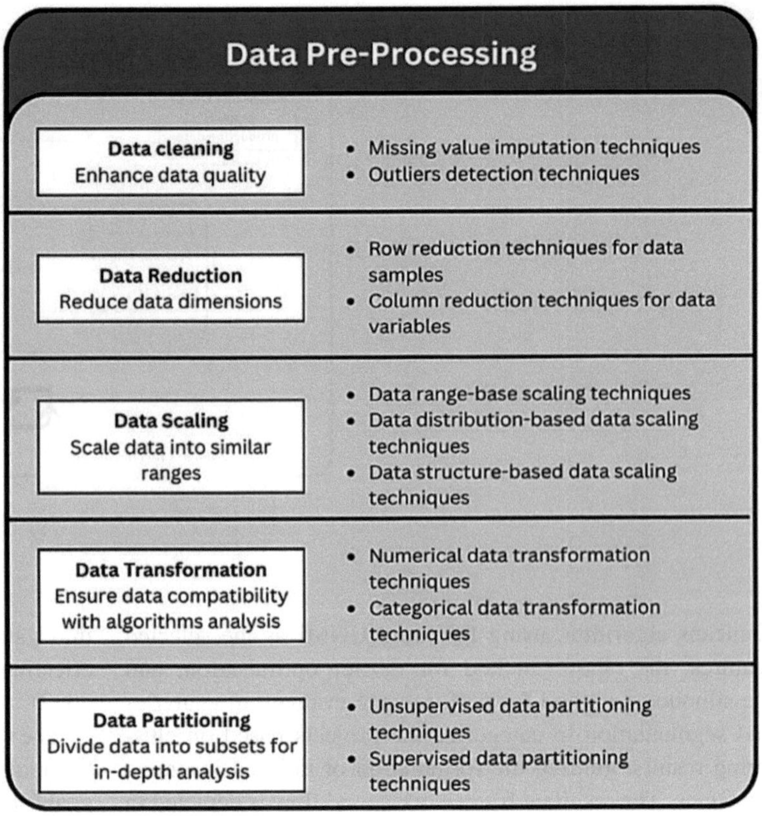

is produced exclusively under daylight, whereas electricity demand spikes around 8 pm as people usually come back home from work (Ferrando et al. 2023). The daily capacity factor measures the daily fraction of time, expressed as a percentage, during which an RE system operates at its rated power. Thus, our model assigned more importance to projects that operate most of the day with minimal maintenance and repair time. Daily electricity production serves as a fundamental indicator, representing the amount of electricity generated by a project. In essence, projects with high electricity output, extended operating hours, and substantial coverage of peak hours are deemed performant, receiving high scores in our model.

UNMI (Unitary Net Margin Index). The UNMI is a synthetic indicator of the financial viability of a project based on the total net cash flows from a project over its rated lifetime, while accounting for the time value of money. The index was computed as the difference between the discounted selling prices of one unit of electricity and the levelized cost of energy (LCOE). The UNMI is calculated using the following formulas:

$$UNMI = \sum_{i=1}^{n}\left(\frac{SP_i - C_i}{(1+d)^i}\right) \quad or \quad UNMI = \sum_{i=1}^{n}\left(\frac{SP_i}{(1+d)^i}\right) - LCOE \tag{1}$$

where:

- SPi represents the average selling price of one unit of generated electricity in period i;
- Ci the total costs of generating electricity per unit in period i;
- d the discount rate;
- n the estimated lifetime of the project;
- $LCOE$ the average cost of one unit of electricity output discounted by the d rate, over the lifetime period n of the project.

The resulting calculations of the PCE and UNMI indices represents the selected features to inject in the clustering stage.

3.3 Clustering

In the field of ML, a well-orchestrated clustering is essential for extracting meaningful insights. The starting point in this process involves feature selection, where the optimal set of features was meticulously chosen to construct models that effectively capture the intricacies of the analysed phenomena (Abidar et al. 2023). In our model, we set the CPE index and the UNMI as pivotal features for input in the subsequent clustering process.

The subsequent phase in our workflow encompasses modelling, a process characterized by the selection of an appropriate algorithm and the training of a model on a dataset for predictive or decision-making purposes. Within our framework, the K-means algorithm takes centre stage, chosen for its widespread popularity, simplicity and low computational complexity. This algorithm is adept at clustering data points into distinct groups, aiming to group similar points while maintaining separation between clusters by minimizing the sum of squared error for each of the k clusters (Abidar et al. 2023; Ikotun et al. 2023). Furthermore, the cluster optimisation phase emerges as paramount to the performance of the clustering; the choice of an appropriate value for k (number of clusters) represents a crucial aspect of the K-means algorithm (Ikotun et al. 2023). This decision is often intricate and was addressed, in our model, through the elbow method, a widely recognized technique for determining the optimal k number of clusters within a given dataset (Abidar et al. 2023). Following the cluster optimisation, the evaluation of clustering performance emerges as a critical consideration. A typical clustering system must address the fundamental question of accuracy and reliability. Here, the Silhouette method comes into action, offering a quantitative metric to assess the distance between each point in a cluster and points in other clusters. This metric provides a visual examination of similarities and differences between the clusters (Abidar et al. 2023).

In conclusion, this carefully crafted clustering process, encompassing feature selection, modelling, cluster optimisation, and clustering performance evaluation, ensured a systematic and robust approach to culminate meaningful data insights. The analytics phase focused on shedding light on the clustering outcomes as detailed in the next sub-section.

3.4 Analytics and Segmentation

The PCE parameters represent the angular stone of the analytics stage. Based on peak hours coverage, daily capacity factor and daily electricity production trilogy, an overall score is calculated for each project in the dataset. The organisation of clusters followed a hierarchical order, with '1' signifying the least favourable cluster and'k' representing the most favourable one. The assigned cluster numbers, derived from the clustering arrangement, acted as scores for each corresponding PCE parameter. After that, three levels of PCE performance were formed, visually depicted by the vertical blue outlined rectangles in Fig. 3. This classification stratified the projects into three discernible levels: the leftmost rectangle representing the low-performance tier, followed by the middle rectangle denoting average performance, and the rightmost rectangle indicating the high-performance level.

Subsequently, the analysis proceeded to an integrative stage, with a comprehensive visualisation of all components. Feature sets are combined, as depicted in Fig. 3, illustrating the juxtaposition of the Unitary Net Margin Index (UNMI) against PCE overall scores. UNMI clusters are represented by red, yellow, and green horizontal floors. Consequently, each project is categorized into one of the eight segments proposed in the figure. As depicted, the projects are segmented based on UNMI clustering colours, serving to rank them according to their financial viability. Projects within the green floor have positive UNMI, indicating financial viability. Projects below this threshold generate a negative UNMI and are further divided into two categories: those above the threshold (T) falling in the yellow band and those below the threshold within the red zone. The ranking is based on the idea that, even though a project may not achieve financial viability, if it attends a certain threshold set by the policymakers, incentives

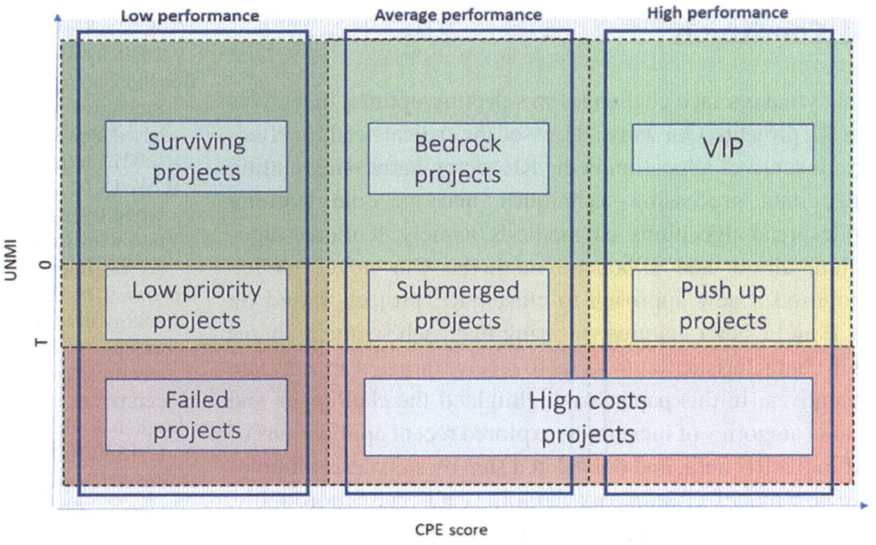

Fig. 3 UNMI versus CPE scores

could potentially propel it to success. As a result, 8 segments of projects were outlined.

The red floor includes failed and high-cost projects, where the former have low production metrics and the latter boast relatively high PCE scores, both generating a significant margin deficit. The yellow band comprises projects without a positive UNMI, but surpassing the threshold value (T). This level is crucial, as incentives for projects in this category could more easily lead to financial viability. It further categorizes projects based on PCE values into low priority, submerged, and push-up projects: the low priority projects are those with low production performance, presenting small attraction for incentives; Submerged projects lack sufficient revenue for achieving viability, despite having average PCE metrics. Push-up projects, more performant than the previous segment, could be easily pushed up towards the green floor. In this upper level, three segments of viable projects are identified, namely: surviving, bed-rock, and Very Important Projects or VIP, translating respectively to low contributors to the network, pyramid base projects needed to insure ongoing injection of electricity, and highly productive and performant projects.

3.5 Targeted Strategies

Within the analytics and segmentation stage, numerous clusters of RE projects exhibiting comparable characteristics are identified. While several incentives strategies and programs exist, it's crucial to acknowledge that not every group reacts uniformly to these measures. Consequently, distinct tailored strategies need to be devised for each segment, predicated on a thorough analysis of their characteristics, to ensure a maximum impact on the market and encourage more investments.

4 Conclusion

Policymakers face challenges in selecting optimal incentives for RE projects. Our work addressed the critical need for effective incentives allocation in the RE sector, harnessing cutting edge data processing techniques, and machine learning widespread algorithms and methods, namely: K-means algorithm, elbow and silhouette methods. Our novel model, proposed a new approach to cluster RE projects based on PCE and UNMI features, ensuring that each segment shares akin characteristics and was set to receive targeted appropriate incentives. In this paper, we highlighted the challenges and main categories of incentives, explored recent applications of AI in the RE field, and delved in a step by step description of our new model's framework, detailing each phase thoroughly, from data preprocessing to targeted strategies encompassing features computing, clustering, and analytics and segmentation stages. We believe that we provided a comprehensive guide for RE policymakers to overcome incentives allocation challenges, through a well-designed projects segmentation approach based on ML.

We are currently working on applying our innovative model to a real-world case dataset in the RE sector. The focus is on using real case data to validate the model's performance, refine its parameters, and further develop incentives allocation strategies.

Acknowledgements The authors would like to express their gratitude to Mr Abdeslam En-Nouaary for his invaluable guidance and advices during the writing of this paper. The authors would also like to acknowledge the financial support provided by the CNRST under the "PhD-Associate Scholarship – PASS" program, which has enabled us to pursue our research.

References

Abidar, L., Zaidouni, D., Ennouaary, A.: Customer segmentation with machine learning: New strategy for targeted actions. In: 13th International conference on intelligent systems: theories and applications (SITA'20), pp. 1–6. ACM, New York (2020). https://doi.org/10.1145/3419604.3419794

Abidar, L., El Asri, I., Zaidouni, D., Ennouaary, A.: A data mining system for enhancing profit growth based on RFM and CLV. In: International conference on future internet of things and cloud–FiCloud 2022. IEEE, pp. 247–253, Rome (2022). https://doi.org/10.1109/FiCloud57274.2022.00041

Abidar, L., Zaidouni, D., EL, I., Ennouaary, A.: Predicting customer segment changes to enhance customer retention: A case study for online retail using machine learning. Int. J. Adv. Comput. Sci. Appl. **14**(7) (2023). https://doi.org/10.14569/IJACSA.2023.0140799

Afridi, Y.S., Ahmad, K., Hassan, L.: Artificial intelligence based prognostic maintenance of renewable energy systems: A review of techniques, challenges, and future research directions. Int. J. Energy Res. **46**(15), 21619–21642 (2021). https://doi.org/10.1002/ER.7100

Alimi, O.A., Ouahada, K., Abu-Mahfouz, A.M.: A review of machine learning approaches to power system security and stability. IEEE Access **8**, 113512–113531 (2020). https://doi.org/10.1109/ACCESS.2020.3003568

Bouabdallaoui, Y., Lafhaj, Z., Yim, P., Ducoulombier, L., Bennadji, B.: Predictive Maintenance in building facilities: A machine learning-based approach. Italian National Conference on Sensors **21**(4), 1–15 (2021). https://doi.org/10.3390/S21041044

Cai, W., Li, C., Agbossou, K., Bénard, P., Xiao, J.: A review of hydrogen-based hybrid renewable energy systems: Simulation and optimization with artificial intelligence. J. Phys. Conf. Ser. **2208**(1), 012012 (2022). https://doi.org/10.1088/1742-6596/2208/1/012012

Dellosa, J.T., Palconit, E.C.: Artificial intelligence (AI) in renewable energy systems: A condensed review of its applications and techniques. In: 21st IEEE International conference on environment and electrical engineering and 2021 5th IEEE Industrial and commercial power system Europe, EEEIC / I and CPS Europe 2021. IEEE, pp. 1–6, Bari (2021). https://doi.org/10.1109/EEEIC/ICPSEUROPE51590.2021.9584587

Duffy, A., Rogers, M., Ayompe, L.: Renewable energy and energy efficiency: assessment of projects and policies. John Wiley & Sons, United Kingdom (2015)

Fan, C., Chen, M., Wang, X., Wang, J., Huang, B.: A review on data preprocessing techniques toward efficient and reliable knowledge discovery from building operational data. Front. Energy Res **9** (2021). https://doi.org/10.3389/fenrg.2021.652801

Fan, C., Xiao, F.: Mining big building operational data for improving building energy efficiency: A case study. Build. Serv. Eng. Res. Technol. **39**(1), 117–128 (2018). https://doi.org/10.1177/014362441 7704977

Ferrando, M., Banfi, A., Causone, F.: Changes in energy use profiles derived from electricity smart meter readings of residential buildings in Milan before, during and after the COVID-19 main lockdown. Sustain. Cities Soc. **99**, 104876 (2023). https://doi.org/10.1016/j.scs. 2023.104876

Ikotun, A.M., Ezugwu, A.E., Abualigah, L., Abuhaija, B., Heming, J.: K-means clustering algorithms: A comprehensive review, variants analysis, and advances in the era of big data. Inf Sci (n y) **622**, 178–210 (2023). https://doi.org/10.1016/j.ins.2022.11.139

Maharana, K., Mondal, S., Nemade, B.: A review: Data pre-processing and data augmentation techniques. Global Transitions Proceedings **3**(1), 91–99 (2022). https://doi.org/10.1016/j.gltp.2022.04.020

Mercado, K.D., Jiménez, J., Quintero, C.G.M.: Hybrid renewable energy system based on intelligent optimization techniques. In: IEEE International conference on renewable energy research and applications–ICRERA 2016. IEEE, pp. 661–666, Birmingham (2016). https://doi.org/10.1109/ICRERA.2016.7884417

Omitaomu, O.A., Niu, H.: Artificial intelligence techniques in smart grid: A survey. Smart Cities **4**(2), 548–568 (2021). https://doi.org/10.3390/SMARTCITIES4020029

Our World in Data–CO_2 emissions. Accessed on Feb 01 2024. https://ourworldindata.org/co2-emissions

Paolanti, M., Romeo, L., Felicetti, A., Mancini, A., Frontoni, E., Loncarski, J.: Machine learning approach for predictive maintenance in industry 4.0'. In: IEEE/ASME International conference on mechatronic and embedded systems and applications. IEEE, pp. 1–6, Oulu (2018). https://doi.org/10.1109/MESA.2018.8449150

Qadir, S.A., Al-Motairi, H., Tahir, F., Al-Fagih, L.: Incentives and strategies for financing the renewable energy transition: a review. Energy Rep. **7**, 3590–3606 (2021). https://doi.org/10.1016/j.egyr. 2021.06.041

Sharifzadeh, M., Sikinioti-Lock, A., Shah, N.: Machine-learning methods for integrated renewable power generation: A comparative study of artificial neural networks, support vector regression, and gaussian process regression. Renew. Sustain. Energy Rev. **108**, 513–538 (2019). https://doi.org/10.1016/J.RSER.2019.03.040

Simsek, H.A., Simsek, N.: Recent incentives for renewable energy in Turkey. Energy Policy **63**, 521–530 (2013). https://doi.org/10.1016/j.enpol.2013.08.036

Zhang, W., Chen, T.: Data preprocessing for web data mining, pp 303–307 (2012). https://doi.org/10.1007/978-3-642-28658-2_47

Zhao, Z.Y., Chen, Y.L., Chang, R.D.: How to stimulate renewable energy power generation effectively?—China's incentive approaches and lessons. Renew. Energy **92**, 147–156 (2016). https://doi.org/10.1016/j.renene.2016.02.001

Life Cycle Energy, Emission, and Costing Footprints for Post-disaster Schools: Case from Damascus Post-conflict

Lina A. Khaddour, Ismail Elhassnaoui, and Rania Obead

Abstract

The past few decades have witnessed a notable escalation in the frequency and severity of natural disasters, significantly impacting the infrastructure, particularly in the realm of building construction. Among the various structures susceptible to such events, educational institutions, notably schools, emerge as particularly vulnerable. The repercussions are profound, with an estimated 175 million students globally finding themselves excluded from educational opportunities annually due to the aftermath of natural and human-induced calamities. Given the vulnerability of children in disaster scenarios, the uninterrupted provision of educational services assumes paramount importance. In Syria, following 11 years of conflict, exacerbated by the COVID-19 pandemic and economic turmoil, the educational landscape for Syrian children and youth has suffered substantial disruption, resulting in over 2.4 million individuals being deprived of schooling. The disruption in educational programming carries immediate and long-term adverse effects for both Syrian children and the nation's future trajectory. To address these challenges, it is imperative for governments and donors to initiate robust measures aimed at fostering a sustainable and high-quality educational response, thereby contributing to Syria's stabilization and recovery endeavors. This research endeavors to scrutinize the impact of post-disaster school initiatives, employing a comprehensive assessment framework encompassing life cycle energy and carbon emissions analyses, alongside a life cycle cost evaluation to provide decision-makers with cost-effective solutions for post-disaster schools design and construction. Utilizing two software platforms, Edupack and DesignBuilder, the study conducts an in-depth analysis of a representative school case study situated in Damascus post-conflict. The assessment encompasses the totality of energy and emissions across the school's life cycle, including embodied and operational aspects. Findings reveal that operational energy overwhelmingly dominates the total energy consumption, accounting for approximately 95%, whereas embodied energy constitutes a mere 5%. Similarly, operational emissions constitute the bulk of total emissions, comprising 96%, while embodied emissions represent a marginal 4%. Integration of renewable energy sources (Solar Photovoltaics) notably diminishes operational energy consumption and carbon emissions, leading to a reduction of 38% and 53%, respectively. Furthermore, the study elucidates the economic dimensions, indicating that maintenance costs exert the most substantial influence on total life costs, constituting 77%, followed by material costs at 21%. Integrating circularity principles for school's building material selection is proposed to directly impact life cycle energy, emissions, and costing footprints. This approach serves as a flexible and effective decision-support tool for post-disaster recovery, with applicability across diverse disaster types and relevance to the reconstruction challenges faced in Syria and comparable regions.

Keywords

Post-disaster · Reconstruction · Netzero · LCA · Decision-making

L. A. Khaddour (✉) · R. Obead
Edinburgh Napier University, Edinburgh, Scotland, UK
e-mail: L.Khaddour@napier.ac.uk

I. Elhassnaoui
Water Witness International, Edinburgh, Scotland, UK

B. El Bhiri et al. (eds.), *AI and Digital Transformation: Innovations in Supply Chain, Education, and Energy Systems*, Sustainable Artificial Intelligence-Powered Applications, https://doi.org/10.1007/978-3-031-86837-5_16

1 Introduction

Globally, more than 80 million people are impacted by a combination of natural and man-made disasters (Eltaweel et al. 2023). The frequency and severity of such events have seen a rapid rise, as evidenced by the World Disasters Report 2016, which reported over 770 thousand fatalities within a decade spanning from 2006 (Singh 2020). This escalating trend places significant pressure on all stakeholders, particularly decision-makers, to formulate and assess effective disaster mitigation strategies (NUISDR 2017).

Following a disaster, the implementation of recovery and reconstruction plans not only serves to rebuild physical structures but also fosters community resilience, as advocated by the United Nations Office for Disaster Risk Reduction (UNDRR) (Toyoda et al. 2021). Embracing the "build-back better" principle, these endeavors aim not only to reconstruct infrastructure but also to fortify societal resilience by considering social, economic, and environmental factors (Toyoda et al. 2021).

The impact of natural disasters, especially on the built environment, has surged in recent decades (Félix et al. 2015). Among the affected structures, schools stand out due to their pivotal role in fostering community cohesion, stability, and the well-being of children (Alisjahbana et al. 2022). Post-disaster, children emerge as particularly vulnerable, facing significant physical and psychological challenges (Li et al. 2016). Disruption to educational services due to damage or destruction of school facilities exacerbates the impact on children and their families, underscoring the urgency of post-disaster school reconstruction efforts (Li et al. 2016). Beyond their educational function, schools serve as vital hubs for immediate shelter and the provision of essential services post-disaster, thus contributing significantly to community recovery and well-being (Parrott et al. 2023). Recognizing the crucial role of post-disaster school reconstruction, various studies have explored different facets of this endeavor. For instance, research by Parrott et al. delves into the extended role of schools in post-disaster contexts, highlighting their positive impact on students' recovery despite some concerns about the school environment (Parrott et al. 2023). Similarly, comparative studies in New Zealand and Japan underscore the pivotal role of post-disaster schools in fostering community resilience and providing a supportive environment for healing and social cohesion (O'Connor and Takahashi 2014).

In Syria post-conflict 2011–2020, the education system in Syria is characterized by being overstretched, underfunded, fragmented, and incapable of delivering safe, equitable, and sustained services to a vast number of children ('After almost ten years of war in Syria, more than half of children continue to be deprived of education' 2024). Within Syria, the number of out-of-school children exceeds 2.4 million, with nearly 40% being girls ('After almost ten years of war in Syria, more than half of children continue to be deprived of education' 2024). This figure is anticipated to have risen in 2020 due to the exacerbating effects of the COVID-19 pandemic on educational disruptions in Syria. A significant proportion of schools, approximately one in three, within Syria have become unusable due to destruction, damage, or repurposing for military activities ('After almost ten years of war in Syria, more than half of children continue to be deprived of education' 2024). For those children who can attend school, learning environments are often marked by overcrowded classrooms and inadequate facilities, including insufficient access to water and sanitation, electricity, heating, and ventilation ('After almost ten years of war in Syria, more than half of children continue to be deprived of education' 2024). Additionally, the Syrian energy sector is not able to meet the built environment demand for electricity, even for a few hours per day (Khaddour 2023). Therefore, the environmental footprint of post-disaster schools warrants attention, with studies advocating for sustainability assessments and the adoption of eco-friendly practices ('After almost ten years of war in Syria, more than half of children continue to be deprived of education' 2024).

Addressing the challenges of post-disaster reconstruction, particularly resource constraints and time pressures, is imperative to avoid unsustainable practices (Yatmo et al. 2021). Managing post-disaster school reconstruction poses distinct challenges, as each disaster event is characterized by its unique set of circumstances and obstacles (Barrett et al. 2019). Therefore, the reconstruction solutions should be tailored to the specific challenges of each disaster and the impacted region. Post-disaster schools design and construction have often relied on the implementation of minimum standards (Ismail et al. 2017; Hosseini et al. 2021). It is therefore incumbent upon states to implement appropriate measures, e.g., comprehensive Life Cycle Assessment (LCA) to meet these standards and to incorporate best practices in addressing the aftermath of each disaster (Khaddour 2022).

In response to these diverse challenges, this research aims to evaluate the life cycle impact of post-disaster schools in Damascus, Syria, focusing on energy, emissions and cost. By leveraging LCA approach, the study seeks to identify opportunities to reduce environmental impacts and enhance sustainability. Additionally, Life Cycle Cost (LCC) analysis will provide insights into the financial implications of energy efficiency measures, aiding decision-making in post-disaster reconstruction efforts (Osunsanmi et al. 2024). Through these efforts, the research endeavors to contribute to informed decision-making and sustainable practices in post-disaster school reconstruction. In this paper, we approached the post-disaster school recovery, through decision-making policies and strategies for reconstruction.

2 Aim and Objectives

This research aims to assess the sustainability and economic viability of post-disaster school reconstruction in Damascus, Syria by integrating Life Cycle Assessment (LCA) of energy consumption and carbon emissions with Life Cycle Costing (LCC), in order to inform sustainable and cost-effective 'build back better' strategies.

- Develop a life cycle energy, emission and cost assessment for post-disaster school design and construction measures.
- Evaluate the total life cycle impact including the operation energy/emissions and embodied energy/emissions for a suggested case study.
- Examine the total life cycle cost for the case study including the costs associated with construction, services, maintenance, and operation costs.

3 Methodology

This research aims to evaluate post-disaster schools in Damascus, Syria based on combining LCA energy/emissions and life cycle cost. As can be seen in Fig. 1, the research methodology can be exemplified in three main stages: stage one: parametric analysis for standardized post-disaster school used in Damascus post-conflict (building details and climatical zone details), stage two: methodological details: planning simplified criteria of LCA (energy and carbon) and LCC, and stage three: the simulation work encludes dupack (embodied energy and emission) and DesignBuilder (operation energy and emission) as explained in the following sections. Furthermore, this method does not depend on a particular disaster and can be applicable to most disasters including hurricanes, floods, and snowstorms, making it a powerful tool for post-disaster decision-making.

3.1 Simplified Life Cycle Energy, Emission, and Costing Assessment

To ensure effective disaster recovery strategies, the utilization of Life Cycle Analysis (LCA) can serve as a valuable tool for assessing environmental impacts and providing decision-making frameworks for comparing alternative emergency buildings (McAvoy et al. 2021). Employing LCA within the built environment offers a comprehensive perspective on the environmental ramifications of the construction process, thereby augmenting sustainability-focused decision-making. LCA facilitates the evaluation of environmental impacts across various stages of construction, encompassing material extraction, transportation, construction, operation, and end-of-life considerations (Barbhuiya and Das 2023). Typically, LCA calculations are conducted throughout the life cycle of a building, spanning from material extraction and manufacturing (A1–A3), construction stages (A4–A5), operational phases (B1–B7), to end-of-life considerations (C1–C4), and post-operational stages (D) (McAvoy et al. 2021).

In light of the substantial influx of migrants following disasters, the demand for reconstruction and utilization of emergency shelters has witnessed a rapid escalation, with projections indicating a further surge in the future. Consequently, enhancing the energy efficiency and reducing carbon emissions of post-disaster buildings, while concurrently minimizing overall costs, emerges as a critical imperative closely aligned with broader global initiatives within the construction sector aimed at curbing energy consumption, carbon emissions, and costs.

As can be seen in Fig. 1, this research constitutes a case study examining the standardized post-disaster school scenario in Syria. The study is structured into three distinct phases. Initially, the investigation involves to calculate the total life energy and emission two different software were used utilizing input data gathered from Damascus during the timeframe spanning June 2023 to September 2023. The software Edupack was used to calculate the case study embodied energy/emissions including the raw material (A1), manufacturing (A3), transport (A4), construction (A5), and end-of-life (C1–C4). Then the DesignBuilder software was used for the simulation to calculate the operation life energy and emission impact. To calculate the LCC for the case study, the total cost for the construction, services, maintenance, and operation were calculated separately and accumulated to find the total LCC. The primary output will address the school energy consumption, carbon emission, and life cycle costing. The final output will elaborate on this simplified approach for optimizing the environmental impact and the cost implications of post-disaster schools.

To calculate the LCC for the suggested school case study, the total cost for the construction, services, maintenance, and operation were calculated separately and accumulated to find the total LCC. The construction cost was calculated by reflecting the construction elements and quantities shown in Table 1. Then the cost of materials and construction were compiled using the Syrian General Company of Housing pricing data. For the services cost, the Central Bureau of Statistics data reveal that for Syria the services

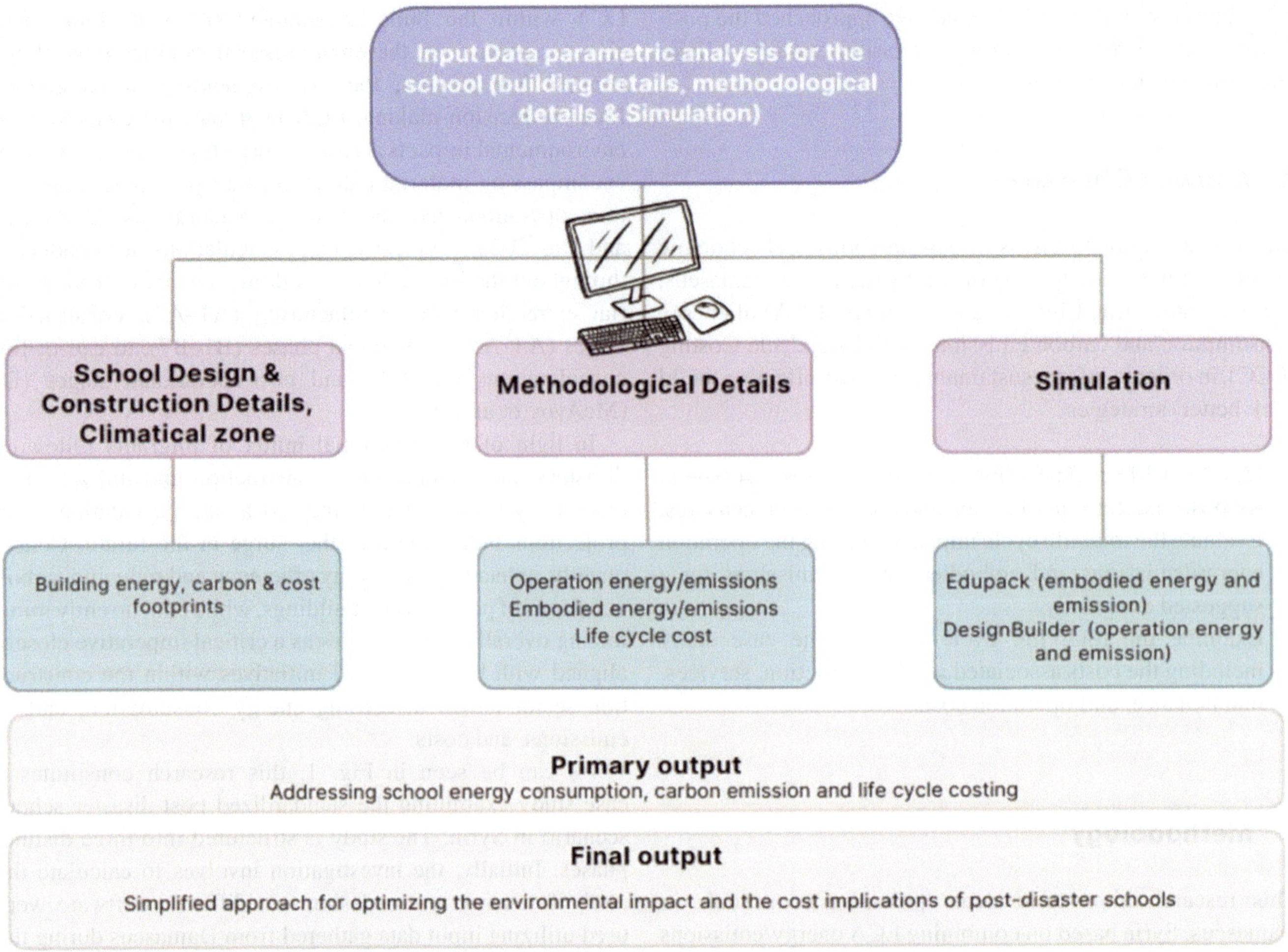

Fig. 1 Framework for evaluating life cycle energy, emission and cost footprints for post-disaster school

cost is estimated as 0.054$ per square meter ('Cost per floor area square meter by stages of construction in governorates centers for privatepublic sectors 2011). For the maintenance cost, according to Kwon et al., this cost is estimated to be 6% of the total construction cost (Kwon et al. 2020). While for the operation cost, the cost of the total operation energy is multiplied by the Syrian energy tariff which equivalent to $0.0000318 per kWh according to Khaddour (Khaddour 2023).

3.2 The Case Study Design and Construction Inputs

The suggested case study is the standard post-disaster high school constructed in Damascus, Syria as shown in Fig. 2. A standardized post-disaster school model, representative of those used in Damascus in the post-conflict period, was developed. This stage involved the definition of architectural, construction, and material parameters based on local building practices, as well as the classification of the building within the specific climatic zone of the region. The school is two stories with a total build area of 1260 m^2. The floor dimensions are 41.15 m × 16.6 m × 6.5 m, and the ground floor has 4 classrooms, 3 offices, a store, and bathrooms. The first floor has 2 classrooms, a library, 2 offices, a computer lab, and 2 science labs as shown in Fig. 3.

The construction of the school is a reinforced concrete frame, walls are concrete blocks, wood doors, and windows are wooden frames with double glazes. The construction materials and quantities are shown in Table 1.

3.3 Simulation Inputs and Assumptions

The software Edupack was used to calculate the case study embodied energy/emissions including the raw material, manufacturing, transport, construction, and end of life. Edupack is a software that uses comprehensive data and tools to enhance sustainable design by providing a life cycle

Table 1 The school construction materials and quantities

Construction element	Material	Illustration	Dimensions Length × width × high (m)	Quantities (m^3)	Density (Kg/ m^3)
The frame	Columns RC		0.5 × 0.25 × 8	48	2300
	Beams RC		0.4 × 0.3 × 0.3	12.5	2300
Foundation	RC		0.7 × 0.7 × 0.5	11.8	2300
Floors	RC slab		41.15 × 16.6 × 0.15	205	2300
Roof	RC flat slab		41.15 × 16.6 × 0.15	102.5	2300
Stairs	RC		2 × 4 × 0.15	2.4	2300
External walls	Concrete blocks (Autoclaved aerated concrete)	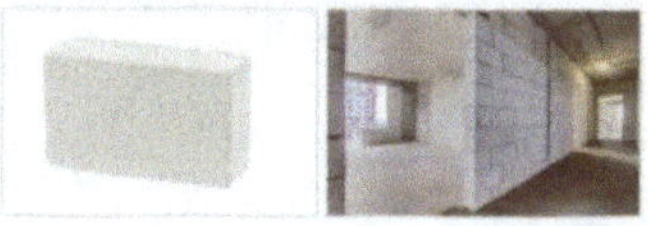	115.5 × 0.3 × 6	208	900
Internal walls	Concrete blocks (Autoclaved aerated concrete)	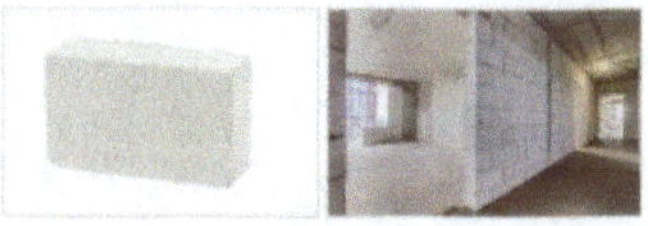	306.6 × 0.25 × 6	460	900
Doors	Wood (Hardwood Oak parallel to the grain)		1.2 × 2.1	50 number	1030
Windows	Wood frame-double gazed		1.2 × 2	30 number	2490

impact for more than 4000 different materials used in the construction industry ('Ansys Granta EduPack|Software for Materials Education' 2024). The software enhances eco-design by allowing the designer to select between different materials and evaluate their environmental impact through their total life cycle ('Ansys Granta EduPack|Software for Materials Education' 2024).

To run the Edupack simulation, the main required to calculate the raw material and the construction impact is the construction materials and qualities data Table 1. To calculate

Fig. 2 School location, situation plan

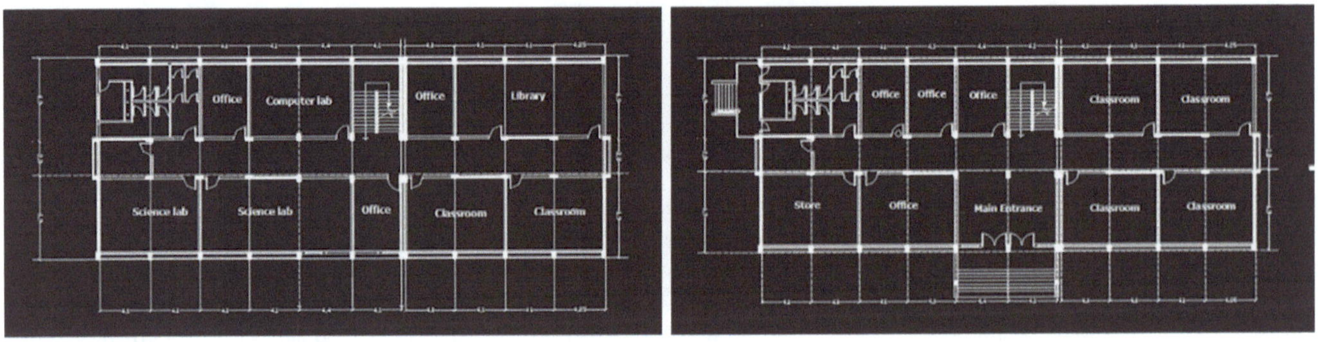

Fig. 3 The school design

the transportation emission, the suggested distance is 40 km to the construction site with the transport mean is 26 ton (3 axle) trucks.

The software DesignBuilder was used for the simulation to calculate the operation life energy and emission impact. Designbuilder is an advanced software that uses building data and geometry to calculate the total building performance as well as calculate the daylight impact, thermal comfort, and natural ventilation ('DesignBuilder Software Ltd—Home' 2024). This software is widely used by architects, sustainability consultants, and researchers to evaluate the buildings' performance and suggest sustainable alternatives to improve the buildings' environmental profile ('DesignBuilder Software Ltd—Home' 2024).

The capital city of Syria, Damascus (36° 13′ N, 33° 29′ E), is selected as the study area. The city is located in the southern region of Syria, roughly 80 km from the Mediterranean Sea's western shore (Khaddour and Yeboah 2022). The summer is from June 1st to September 31st (a total of 120 days in a year) (Khaddour and Yeboah 2022). From November 15 to April 15 (a total of 150 days) is the winter season. Syria's climate zones map indicates that the country's temperature is between

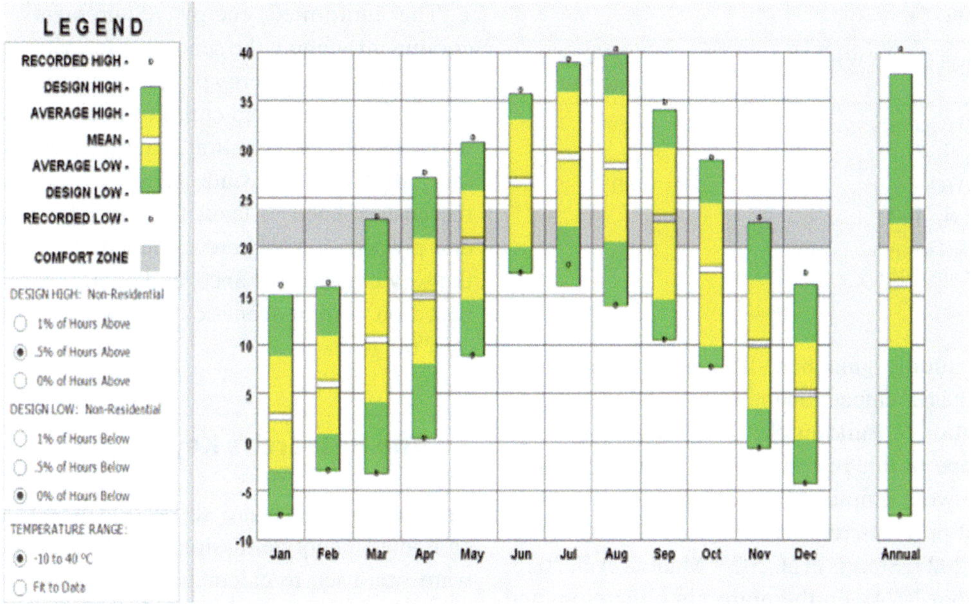

Fig. 4 Temperature range at the school location

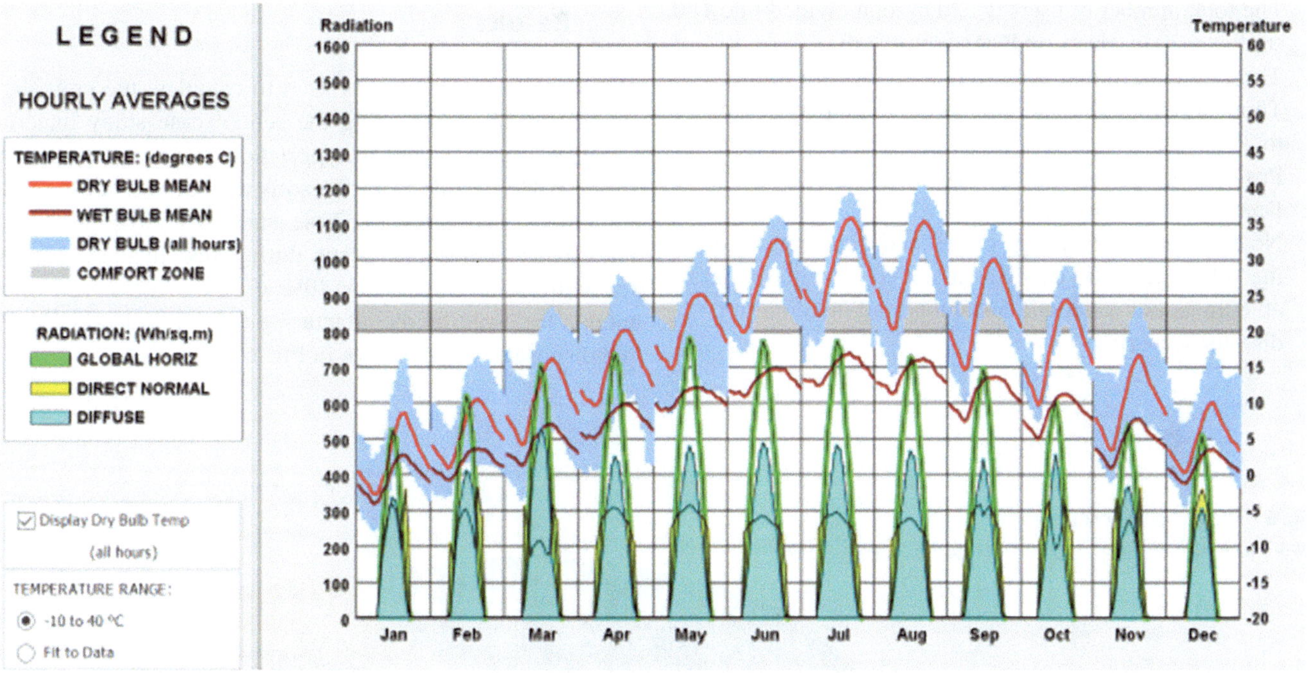

Fig. 5 Range of humidity at the school location

40° the highest to −2 °C the lowest (Khaddour and Yeboah 2022). The city climate profile was used for the climate input for the software simulation.

The majority of the hours in Damascus are out of the comfort zone, with many hours below or above the comfort zone (Fig. 4), therefore, it means the necessity for using an HVAC system to improve thermal comfort. The most important thing about the chart is the thermal lag effect provided between the peak direct normal radiation and the hottest time of the day. Additionally, Fig. 5 indicates that the relative humidity is low for most of the year, as evidenced by the relatively high difference between the dry bulb mean curve and the wet bulb mean curve. Internal heat gain is an important part of the simulation input, for the case study the internal gain is shown in Table 2, for this project the internal heat gain sources are the occupants, computers and fluorescent lamps.

Table 2 Internal gain

Source	Heat Gain (kW)	Number	Total Heat Gains (kW)
Occupants	90 (Georges et al. 2015)	220	19,800
Computers	80 (Georges et al. 2015)	20	1600
Fluorescent Lamps	15 (Georges et al. 2015)	46	690

Buildings will naturally gain heat from their fittings and occupants which has an impact on the simulation calculation (Chartered Institution of Building Services Engineers 1999).

Within the scope of this research, several simplifications and assumptions were employed, drawing from construction practices and previous relevant literature (Mohammadpourkarbasi et al. 2023; Sturgis et al. 2023; Maduta et al. 2023; carbonfootprint.com 2024). Furthermore, simplifications and assumptions were employed encompass:

- The school building life span is 60 years.
- The total number of users is 220 in total divided into 180 students, 20 teachers, and 20 support staff.
- The two educational trimesters are (8–3 pm, Sunday to Thursday) from 15 September to 15 December, 15 January to 30 June.
- Project design specifications were requested from Damascus Governorate.
- No fluctuation in energy prices and consumption over the subsequent 60 years was assumed, despite the probable impact of economic instability and inflation in post-disaster scenarios.

The uniformed energy requirements for heating and cooling throughout the school's life span was assumed, disregarding potential energy poverty the country is facing. Recognizing that energy and emissions calculations are contingent upon design, construction materials, climate, and users are integrated into the study inputs specific to Syria. Hence, the costs related to land, furniture, landscaping, insurance, and property taxes were excluded from this study. Overall, the services, maintenance, and operation costs are calculated based on current scenarios and assumed to remain constant over 60 years.

4 The Research's Key Outcomes

The Edupack software simulation was run to calculate the embodied energy/emissions while the DesignBuilder software was used to calculate the operation energy/emissions. The following are the results from the two soft wares.

4.1 The Embodied Energy/ Emissions Results

The software Edupack was used to calculate the embodied energy/emissions reflecting the school case study material quantities Table 1, and transportation means and distances Table 2. The results from the software come with, the total embodied energy for the case study is 3801338,24 MJ. The highest emissions occur during raw material supply, accounting for more than 40% of the overall embodied energy, followed by manufacturing (40%), transport (11%), and disposal (7%) as shown in Fig. 6.

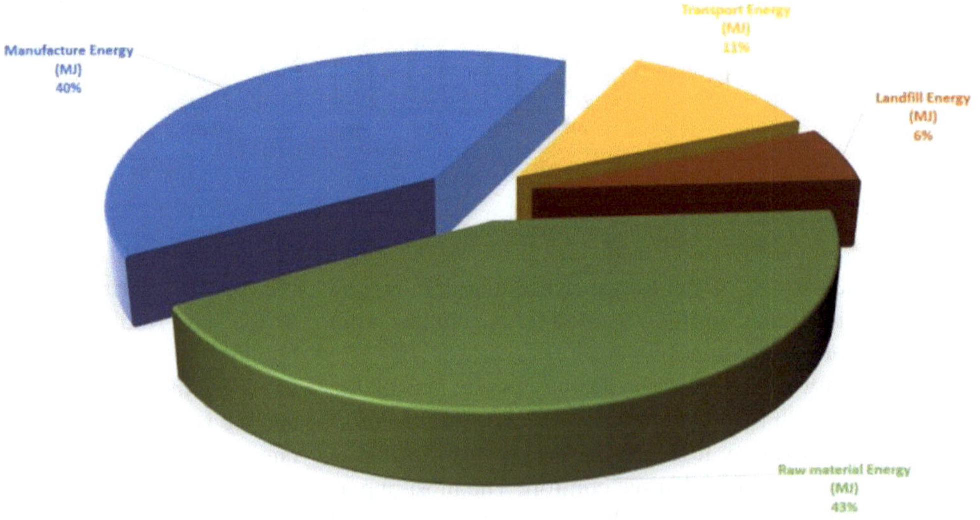

Fig. 6 The embodied energy for the case study

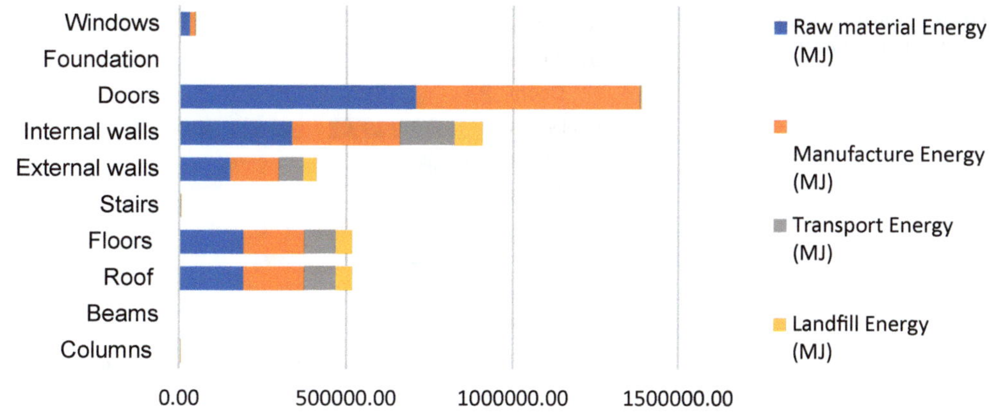

Fig. 7 The embodied energy per construction elements

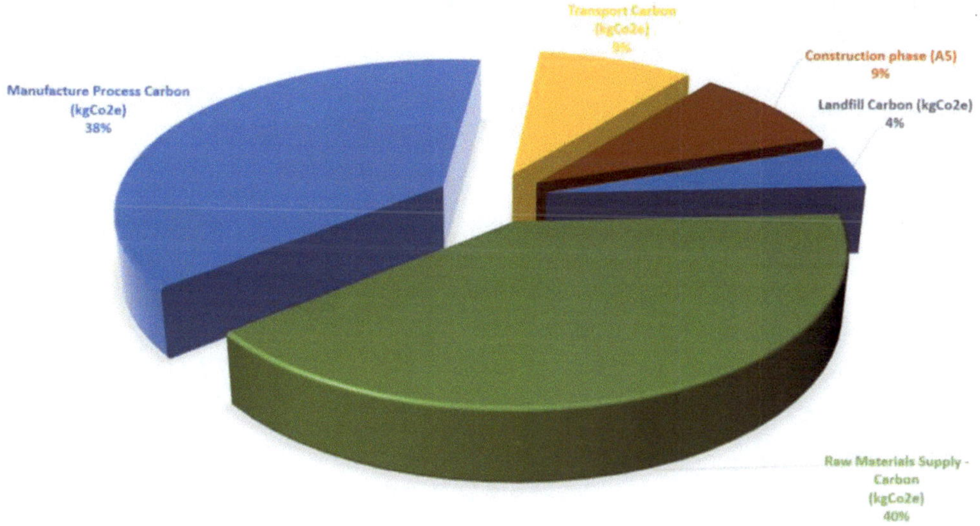

Fig. 8 The embodied carbon for the case study

In the elements sectors, the doors dominate the picture regarding the energy footprint representing 36% followed by external walls representing 24%, the floor and roof representing 14%, and external walls representing 11% as shown in Fig. 7.

For the embodied emissions, the results show that the embodied carbon represents 312868,3 KgCO2e, and the highest emissions occurring during raw material supply, accounting for 40% of the overall embodied carbon, followed by manufacturing (38%), transport (9%), Construction (9%) and disposal (4%) as shown in Fig. 8.

As shown in Fig. 9, for the embodied emissions per element, the internal walls dominate the picture regarding the carbon footprint representing 37% followed by the floors and roof representing 21%, and external walls representing 17%. Doors and windows have less than 5% impact. The high carbon footprint for the structure and the walls is due to the concrete materials which have high carbon intensity.

4.2 The Operation Energy/Emissions

The case study operation energy and emissions sources are assumed to be HVAC system, lighting, computers, and DHW. The software DesignBuilder is used to calculate the total operation energy/emission. The Case Study Geometry Fig. 3 was generated reflecting the construction materials shown in Table 1 as well as the country climate profile including the temperature average and humidity range shown in Figs. 4 and 5.

The simulation results are the total energy consumption for the case study is 377.89 MWh per year and 299.9 KWh/m^2 as shown in Fig. 10. The total operation carbon calculated reflecting the Syrian electricity grid carbon factors is 0.5314 kgCO_2e/kWh, and for the natural gas, the grid intensity is 0.189 kgCO^2e/kWh (Durganjali et al. 2020). The total operational carbon for the case study is 145810 kgCO2e per year.

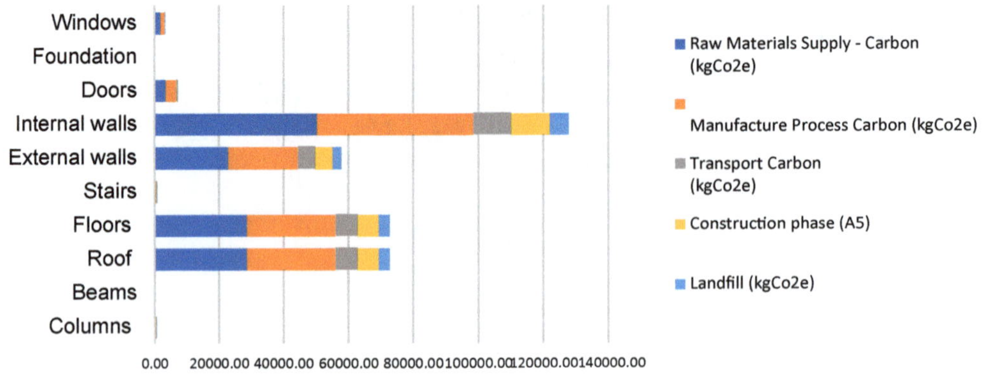

Fig. 9 Embodied emissions per construction elements

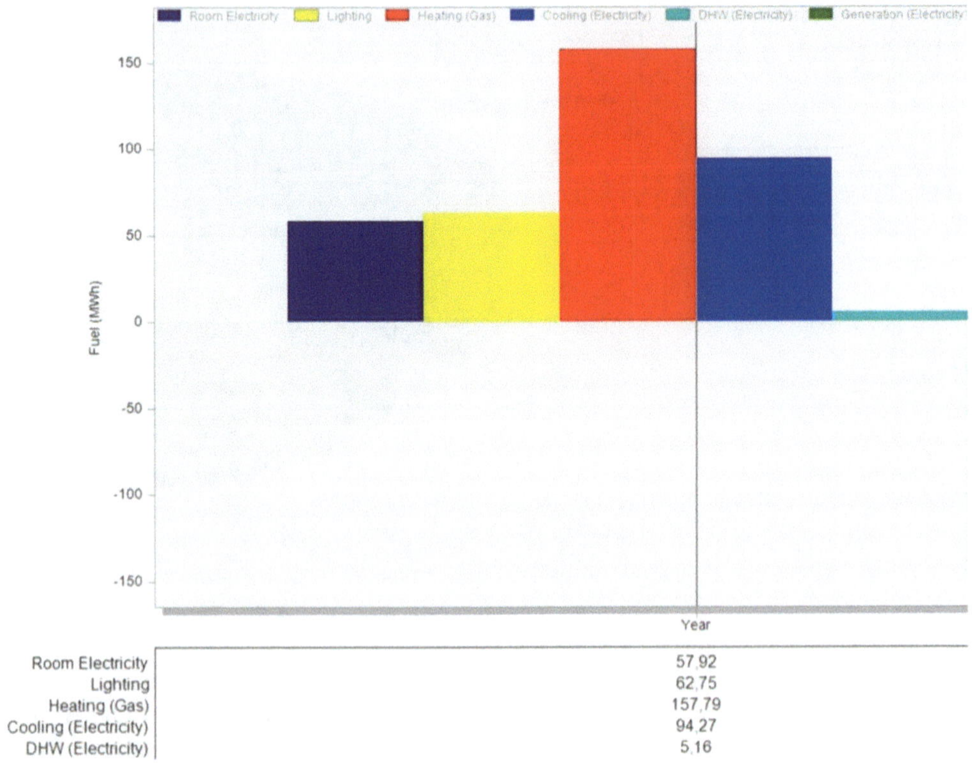

Fig. 10 The case study operation energy

The total operation energy for the school is 377.89 MWh/year. Integrating renewables can reduce the total energy/emissions for the building. Solar panels especially Photovoltaics (SPV) panels are considered one of the most sustainable and environmentally friendly energy resources (Ji et al. 2016). Also, SPV is a suitable investment as they depend on natural cost-free energy sources (the sun) and can provide dependable energy sources (Ji et al. 2016). Therefore, this research suggests adding SPV panels to reduce the total building energy and emissions. Adding solar panels to the case study could generate 149.72 MWh/year the results are shown in Fig. 11.

4.3 The Total Life Energy/Emissions

The total embodied energy for the case study is 3801338,24 MJ equivalent to 1,055 MWh. While the total energy consumption for the case study is 377.89 MWh per year and in 60 years the life span of the building's operation energy is 22,673.4 MWh. The embodied carbon represents 312,868,3 KgCO2e equivalent to 312.9 Tons CO2e. The operation's carbon emissions are 8748.6 Ton CO2e. Figure 12 represents the embodied, operation and the total life energy and emissions. As can be seen, the embodied emissions represent 3.45% of the Total Life emissions.

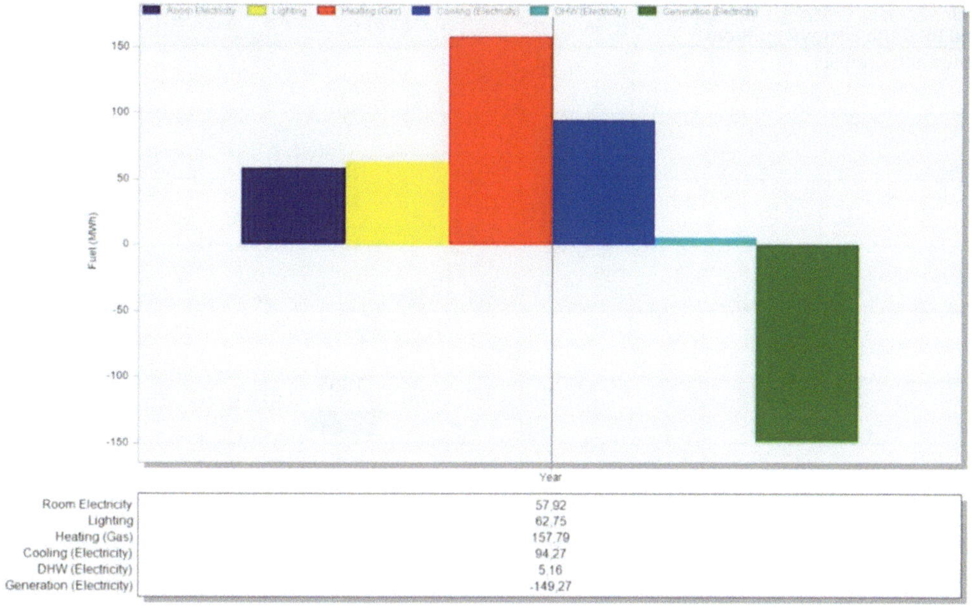

Fig. 11 The renewables impact

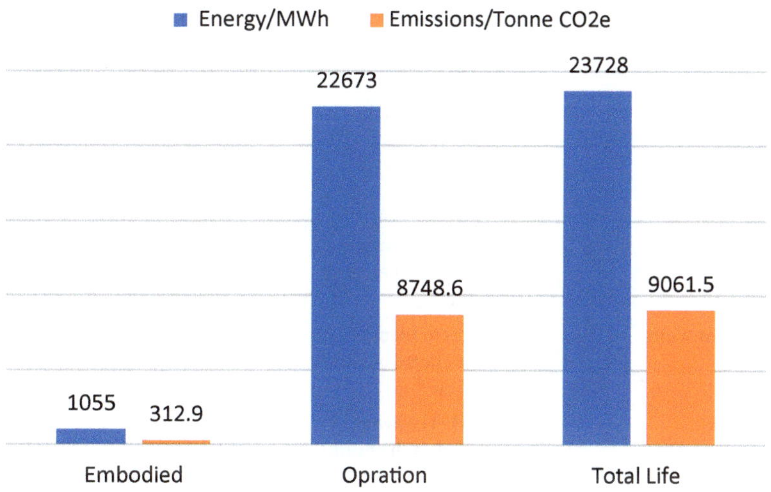

Fig. 12 The embodied operation and total life energy/emissions

Whereas the operational emissions make up the remaining 96.55%. These results emphasize that operational energy and emissions overwhelmingly dominate the building's lifecycle impact, suggesting that efficiency measures targeting the operational phase could yield the most significant improvements in overall sustainability.

Adding renewables reduces the total operation energy to 228.62MWh/year; in 60 years, the total operation energy is 13,717 MWh. After adding the renewable the total operational carbon will drop to 3,996 Ton CO_2e (Fig. 13). Incorporating renewable energy sources significantly mitigates the environmental impact of the operational phase (Khaddour and Deng 2023; Khaddour et al. 2023). Following the integration of renewables, the building's annual operational energy demand is reduced to 228.62 MWh/year, resulting in a 60-year cumulative operational energy consumption of 13,717 MWh. This marks a 39.5% reduction compared to the original operational energy total of 22,673 MWh. In terms of emissions, renewable energy adoption leads to a substantial decrease in operational carbon emissions, lowering the total to 3,996 tonnes CO_2e over 60 years. This represents a 54.3% reduction from the original 8,748.6 tonnes CO_2e (see Fig. 13). These results clearly demonstrate the critical role of renewable energy in enhancing building sustainability, particularly by addressing the dominant operational phase. The integration of clean energy not only decreases dependency on conventional power sources but also significantly reduces the building's long-term carbon footprint.

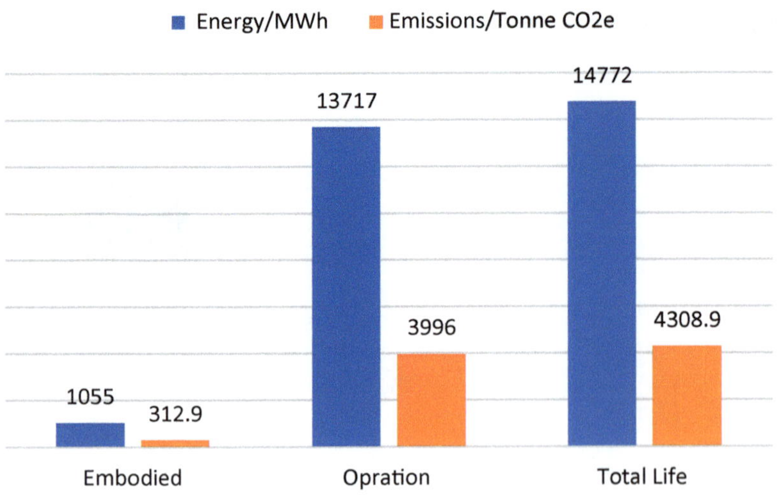

Fig. 13 The embodied operation and total life energy/emissions renewables impact

Table 3 The construction cost for the case study

Construction element	Material	Quantities (m^3)	Construction cost ($) per m^3	Total ($)
The frame	Columns R. concrete	48	238.46	11,446
	Beams R. concrete	12.5	238.46	2,980
Foundation	R. concrete	11.8	238.46	2,813
Floors	R. concrete slab	205	238.46	48,884
Roof	R. concrete slab	102.5	238.46	24,442
Stairs	R. concrete	2.4	238.46	572
External walls	Concrete blocks (Autoclaved aerated concrete)	208	50.77	10,560
Internal walls	Concrete blocks (Autoclaved aerated concrete)	460	50.77	23,354
Doors	Wood (Hardwood Oak parallel to the grain)	50 number	97.69	4,884
Windows	Wood frame-double gazed	30 number	78.46	2,353
Total				**132,288**

4.4 Life Cycle Costing

To calculate the LCC for the suggested school case study, the total cost for the construction, services, maintenance, and operation were calculated separately and accumulated to find the total LCC. Table 3 shows the construction cost calculation. To evaluate the Life Cycle Cost (LCC) of the proposed school case study, costs were categorized and calculated separately across four main components: construction, services, maintenance, and operation. These individual costs were then aggregated to determine the total LCC. The construction cost was estimated based on the materials and quantities outlined in Table 1. Unit prices for materials and construction were obtained from the Syrian General Company of Housing. Table 3 presents the breakdown of construction costs for each building element, according to 2023 exchange rates.

For the services cost the Central Bureau of Statistics data reveal that for Syria the services cost is estimated as 0.054$ per square meter ('Cost per floor area square meter by stages of construction in governorates centers for privatepublic sectors 2011). For the maintenance cost, according to Kwon et al., this cost is estimated to be 6% of the total construction cost (Kwon et al. 2020). While for the operation cost, the cost of the total operation energy is multiplied by the Syrian energy tariff which is equivalent to $0.0000318 per kWh according to 2023 exchange rate (Khaddour 2023) (Table 4). This LCC analysis highlights that while construction constitutes the initial major investment, maintenance

Table 4 The construction, services, maintenance, and operation for the case study

The cost	Cost Per year ($)	Cost for 60 Years ($)
Construction cost	132,288	132,288
Services cost	0.054 × 1260 = 68	4,080
Maintenance cost	6% of 132,288 = 7,937	476,236
Operation cost	377,890 × 0.0000318 = 12	720
Total		613,324

costs dominate the long-term expenditures, emphasizing the importance of durable design and materials to minimize future financial and environmental impacts.

5 Results Analysis and Discussion

Schools play a crucial role in providing children with a thermally comfortable educational environment. This not only enhances students' comfort and achievements but also significantly impacts the energy efficiency and overall sustainability of buildings. Syria's energy supply has traditionally relied on national oil and natural gas resources. However, since the onset of the ongoing conflict, the country's energy sector has been in disarray. Oil and natural gas production have plummeted significantly, primarily due to sanctions imposed by the USA, war-related damage to energy infrastructure, and the government's loss of control over numerous oil fields (Khaddour 2023). In addition to the destruction of major power generation plants and transmission infrastructure during the conflict, the existing power plants are operating beyond their designed capacity and lifespan, incurring exorbitant maintenance and repair costs (Khaddour 2023). Notably, aging power plants such as Banians and Tishreen require more than $15 billion each for maintenance alone, excluding operational expenses (Khaddour and Yeboah 2022). In April 2021, Ghassan al-Zamil, the Minister of Electricity, announced the implementation of an electricity- saving program, rationing power delivery to one hour for every minimum of five hours of power cuts. However, electricity availability remains highly inconsistent across cities and between urban and rural areas, with schedules subject to frequent and unpredictable changes depending on fuel availability.

This research reveals that the total embodied energy for the case study amounts to 3,801,338.24 MJ, with the embodied carbon representing 312,868.3 $KgCO_2e$. Notably, raw materials exert a substantial influence on both embodied energy and carbon, comprising 43% and 40% of the respective totals. Manufacturing processes also contribute significantly, accounting for 40% of total energy and 38% of total carbon emissions. Conversely, transportation, construction, and end-of-life phases have a comparatively minor impact on overall energy and emissions, ranging from 11 to 4% of the total life impact. These findings underscore the significant influence of the production stage, constituting 83% of total embodied energy and 78% of embodied carbon. This aligns with previous studies that have highlighted the predominant role of the production phase in determining the overall environmental impact of buildings (Ji et al. 2016; Alshamrani et al. 2014; Pachta and Giourou 2022).

Regarding operational energy, the case study records an annual consumption of 22,673 MWh, with the heating system being the primary contributor at 157.79 MWh, followed by the cooling system at 94.27 MWh. Lighting, other electricity use, and Domestic Hot Water (DHW) contribute 62.57 MWh, 57.92 MWh, and 5.16 MWh, respectively. Operation emissions total 8,748.6 tons CO_2e, with approximately half attributed to heating, totalling 3,995 tons CO_2e.

In terms of total life energy, the case study registers 23,728 MWh, with operational energy accounting for approximately 95%, while embodied energy comprises approximately 5%. Similarly, total emissions amount to 9,061.5 tons CO_2e, with operational emissions constituting 96% of the total life impact and embodied emissions making up the remaining 4%. These results underscore the significant influence of the operational phase on the overall environmental profile, consistent with previous research indicating the dominant role of operational energy and emissions in determining total life impact (Alshamrani et al. 2014; Zabalza et al. 2013).

To enhance the environmental performance of the building, the integration of Solar Photovoltaic (SPV) systems is recommended to reduce operational energy and emissions. Integration of SPV resulted in a reduction of operational energy to 13,717 MWh and operational carbon emissions to 3,996 tons CO_2e. This translates to a total life impact of 14,772 MWh for energy and 4,308.9 tons for emissions, representing reductions of 38% and 53%, respectively. The incorporation of renewable energy sources in post-disaster construction demonstrates significant potential for mitigating environmental impact.

In the light of lake of availability of SPV technology due to the sanction and economy downturn, further research is recommended on integrating circularity principles in the selection of locally sourced building material with low embodied carbon and high thermal performance to reduce both the overall energy/carbon and costing. Significant impact of material production underscores the importance of utilizing sustainable, natural-based, and recyclable materials to reduce environmental impact in post-disaster construction scenarios (Khaddour 2021).

Additionally, the total life cost for the case study is calculated at $613,324, with maintenance costs comprising the highest proportion at 77%, followed by construction costs at 21%. Conversely, services and operation costs represent minimal percentages of the total life cost due to low energy tariffs and service availability in Syria. To reduce construction costs in post-disaster scenarios, the utilization of local materials and employing durable construction methods are recommended to minimize maintenance expenses.

6 Conclusion

This study emphasizes the necessity of adopting a comprehensive approach to post-disaster school reconstruction, encompassing immediate educational requirements alongside considerations of long-term sustainability, environmental impact, and economic feasibility. Beyond their educational role, schools play a pivotal role in fostering environmental and socio-economic benefits, including optimizing investment efficacy, reinstating societal norms, and enhancing overall well-being. In the aftermath of disasters, schools serve not only as educational centers but also as crucial sources of psychological and emotional support, underscoring the urgency of rapid post-disaster reconstruction endeavors. However, the expeditious resumption of educational services faces multifaceted challenges, including the persistent threat of subsequent disasters, population displacement, logistical constraints in reaching inaccessible areas, and limitations in resources, funding, and time. Despite the acknowledged significance of post-disaster school reconstruction, notable shortcomings endure, such as excessive costs and disregard for cultural and environmental considerations in affected regions.

This research implements a comprehensive life cycle impact assessment for a post-disaster school constructed in Damascus, Syria. The study evaluates the total life energy and emissions, encompassing both embodied and operational energy and emissions. Edupack software is utilized for calculating embodied energy/emissions, while DesignBuilder simulation software is employed for assessing operational energy/emissions, alongside the total life cost, including construction, services, maintenance, and operation.

The findings reveal that the total life energy for the case study amounts to 23,728 MWh, with operational energy constituting approximately 95% and embodied energy approximately 5%. Similarly, total emissions amount to 9,061.5 tons CO2e, with operational emissions representing 96% of the total life impact and embodied emissions constituting 4%. The operational phase significantly influences the overall environmental profile throughout the building's life cycle. Integration of renewable energy sources results in a notable reduction in operational energy and emissions, thereby highlighting the importance of renewable energy adoption in mitigating overall environmental impact.

Regarding cost, the total life cost for the case study is estimated at $613,324, with maintenance costs comprising the highest proportion at 77%, followed by construction costs at 21%. To address the environmental impact of post-disaster reconstruction, prioritizing solutions that mitigate the environmental footprint of materials production is essential. Furthermore, incorporating renewable energy sources is crucial for reducing operational energy/emissions and enhancing overall environmental performance. To mitigate overall costs, the use of durable and long-lasting materials is recommended, given that maintenance costs represent the primary contributor to total life costs.

This study focuses on a standardized school design in the Damascus area, utilizing conventional building materials. Further research is warranted to explore alternative school designs and building materials, providing a broader perspective on post-disaster school construction. Integrating circularity principles for building material selection is proposed to directly impact life cycle energy, emissions, and costing footprints. This approach serves as a flexible and effective decision-support tool for post-disaster recovery, with applicability across diverse disaster types and relevance to the reconstruction challenges faced in Syria and comparable regions.

References

'After almost ten years of war in Syria, more than half of children continue to be deprived of education'. Accessed: May 13, 2024. https://www.unicef.org/press-releases/after-almost-ten-years-war-syria-more-half-children-continue-be-deprived-education

'Ansys Granta EduPack|Software for Materials Education'. Accessed: May 10, 2024. https://www.ansys.com/en-gb/products/materials/granta-edupack#tab1-1

'carbonfootprint.com—Carbon Footprint Calculator'. Accessed: May 11, 2024. https://www.carbonfootprint.com/calculator.aspx

Central Bureau of Statistics: Cost per floor area square meter by stages of construction in governorates centers for private & public sectors dwellings (2011)

'DesignBuilder Software Ltd—Home'. Accessed: May 07, 2024. https://designbuilder.co.uk/

Alisjahbana, I., Graur, A., Lo, I., Kiremidjian, A.: Optimizing strategies for post-disaster reconstruction of school systems. Reliab. Eng. Syst. Saf. **219** (2022). https://doi.org/10.1016/j.ress.2021.108253

Alshamrani, O.S., Galal, K., Alkass, S.: Integrated LCA-LEED sustainability assessment model for structure and envelope systems of school buildings. Build. Environ. **80**, 61–70 (2014). https://doi.org/10.1016/j.buildenv.2014.05.021

Barbhuiya, S., Das, B.B.: Life cycle assessment of construction materials: methodologies, applications and future directions for sustainable decision-making. Case Stud. Constr. Mater. **19** (2023). https://doi.org/10.1016/j.cscm.2023.e02326

Barrett, P., Treves, A., Shmis, T., Ambasz, D., Ustinova, M.: The Impact of School Infrastructure on Learning a Synthesis of the Evidence (2019)

Chartered Institution of Building Services Engineers, Environmental design. CIBSE (1999)

Durganjali, C.S., Bethanabhotla, S., Kasina, S., Radhika, D.S.: Recent developments and future advancements in solar panels technology. J. Phys. Conf. Ser. Inst. Phys. Publ. (2020). https://doi.org/10.1088/1742-6596/1495/1/012018

Eltaweel, A., Saint, R., D'Amico, B., Pomponi, F.: A parametric thermal analysis of refugees' shelters using incremental design and affordable construction material. Energy Build **290** (2023). https://doi.org/10.1016/j.enbuild.2023.113110.

Félix, D., Monteiro, D., Branco, J.M., Bologna, R., Feio, A.: The role of temporary accommodation buildings for post-disaster housing reconstruction. J. Housing Built Environ. **30**(4), 683–699 (2015). https://doi.org/10.1007/s10901-014-9431-4

Georges, L., Haase, M., Houlihan Wiberg, A., Kristjansdottir, T., Risholt, B.: Life cycle emissions analysis of two nZEB concepts. Build. Res. Inf. **43**(1), 82–93 (2015). https://doi.org/10.1080/09613218.2015.955755

Hosseini, S.M.A., Farahzadi, L., Pons, O.: Assessing the sustainability index of different post-disaster temporary housing unit configuration types. J. Build. Eng. **42** (2021). https://doi.org/10.1016/j.jobe.2021.102806

Ismail, F.Z., Halog, A., Smith, C.: How sustainable is disaster resilience?: An overview of sustainable construction approach in post-disaster housing reconstruction. Int. J. Disaster Resil. Built Environ. **8**(5), 555–572 (2017). https://doi.org/10.1108/IJDRBE-07-2016-0028

Ji, C., Hong, T., Jeong, J., Kim, J., Lee, M., Jeong, K.: Establishing environmental benchmarks to determine the environmental performance of elementary school buildings using LCA. Energy Build. **127**, 818–829 (2016). https://doi.org/10.1016/j.enbuild.2016.06.042

Khaddour L.A.: Strategic framework of operational energy performance improvement potential for Damascus post-war social housing. Intell. Build. Int. 1–15 (2021). https://doi.org/10.1080/17508975.2021.1874859

Khaddour, L.A.: Life-cycle sustainability risk management a multi-stakeholder approach: the case of Damascus post-war residential projects. Environ. Dev. Sustain. **24**(11), 12756–12786 (2022). https://doi.org/10.1007/s10668-021-01963-3

Khaddour, L.A.: Comparative analysis of residential building envelopes newly implementing the building insulation code in Damascus. Int. J. Environ. Sci. Technol. (2023). https://doi.org/10.1007/s13762-023-05053-x

Khaddour, L.A., Yeboah, S.K.: Multiple-criteria optimization of residential buildings envelope toward nZEBs: simplified approach for Damascus post-war. In: Energy and Sustainable Futures, Proceedings of the 3rd ICESF, 7–8 September 2022. Springer (2022). https://lnkd.in/eSeKVf4P

Khaddour, L.A., Deng, W.: Multi-criteria sustainability risk management for post-war residential re-construction projects: the case of Damascus. J. Housing Built Environ. **38**(3), 1939–1983 (2023). https://doi.org/10.1007/s10901-023-10024-2

Khaddour, L.A., Yeboah, S.K., Dodoo, J.K.: Ecological and carbon footprints of cities. In: Encyclopedia of Sustainable Technologies, 2nd edn. Elsevier (2023). https://doi.org/10.1016/B978-0-323-90386-8.00044-9

Kwon, N., Song, K., Ahn, Y., Park, M., Jang, Y.: Maintenance cost prediction for aging residential buildings based on case-based reasoning and genetic algorithm. J. Build. Eng. **28** (2020). https://doi.org/10.1016/j.jobe.2019.101006

Li, S.-J., Wu, C.S.T., Wong, H.T.: School safety and children health in a post-disaster community: implications to collaborative care and service learning in school health. J. Acute Dis. **5**(1), 46–50 (2016). https://doi.org/10.1016/j.joad.2015.08.005

Maduta, C., Melica, G., D'agostino, D.: Defining zero-emission buildings support for the revision of the energy performance of buildings directive (2023). https://doi.org/10.2760/107493

McAvoy, S., Grant, T., Smith, C., Bontinck, P.: Combining life cycle assessment and system dynamics to improve impact assessment: a systematic review. J. Clean. Prod. **315**, 128060 (2021). https://doi.org/10.1016/j.jclepro.2021.128060

Mohammadpourkarbasi, H., Riddle, B., Liu, C., Sharples, S.: Life cycle carbon assessment of decarbonising UK's hard-to-treat homes: a comparative study of conventional retrofit vs EnerPHit, heat pump first vs fabric first and ecological vs petrochemical retrofit approaches. Energy Build. **296** (2023). https://doi.org/10.1016/j.enbuild.2023.113353

NUISDR: Build back better in recovery, rehabilitation and reconstruction (2017)

O'Connor, P., Takahashi, N.: From caring about to caring for: case studies of New Zealand and Japanese schools post disaster. Pastor. Care Educ. **32**(1), 42–53 (2014). https://doi.org/10.1080/02643944.2013.875584

Osunsanmi, T.O., Gillies-Denning, E., Khaddour, L.A., Olawumi, T.O., Moobela, C.: Framework for the UK construction industry transition to net zero design: a Delphi techniques and structural equation model approach. Int. J. Constr. Manag. (2024). https://doi.org/10.1080/15623599.2024.2337143

Pachta, V., Giourou, V.: Comparative life cycle assessment of a historic and a modern school building, located in the city of Naoussa, Greece. Sustainability (Switzerland) **14**(7) (2022). https://doi.org/10.3390/su14074216

Parrott, E., Lomeli-Rodriguez, M., Pacheco, E.M., Rahman, A., Direzkia, Y., Joffe, H.: Adolescent girls' representations of the role of schools and teachers post-disaster: "second parents, second homes". Int. J. Adolesc. Youth **28**(1) (2023). https://doi.org/10.1080/02673843.2023.2242450

Singh, Z.: Disasters: implications, mitigation, and preparedness. Indian J. Publ. Health NLM (Medline) **64**(1), 1–3 (2020). https://doi.org/10.4103/ijph.IJPH_40_20.

Sturgis, S., et al.: Whole life carbon assessment for the built environment RICS Professional Standard (2023). www.rics.org

Toyoda, T., Wang, J., Kaneko, Y.: Build Back Better Challenges of Asian Disaster Recovery (2021). https://link.springer.com/bookseries/16115

Yatmo, Y.A., Atmodiwirjo, P., Saginatari, D.P., Harahap, M.M.Y.: Development of modular school design as a permanent solution for post-disaster reconstruction in Indonesia. Int. J. Disaster Resil. Built Environ. **12**(1), 101–113 (2021). https://doi.org/10.1108/IJDRBE-10-2019-0070

Zabalza, I., Scarpellini, S., Aranda, A., Llera, E., Jáñez, A.: Use of LCA as a tool for building ecodesign. A case study of a low energy building in Spain. Energies (Basel) **6**(8), 3901–3921 (2013). https://doi.org/10.3390/en6083901

Preventive and Predictive Maintenance Approach with the Use of Digital Twin Concept

Ihsane Abouzid and Rajaa Saidi

Abstract

The adoption of Digital Twins in automated maintenance represents a pivotal shift in industrial operations, involving the creation of virtual replicas of physical assets or processes to monitor, analyze, and optimize performance. In the context of Smart Factories and Industry 4.0, Digital Twins play a crucial role as models representing the factory, facilitating data analysis and strategic planning to enhance production efficiency. While predictive maintenance predominates in maintenance management studies, preventive maintenance also holds significance, contributing to overall maintenance strategies based on factors like equipment complexity and criticality. Digital Twins further enable remote monitoring of production processes, offering operational flexibility to maintenance teams. The paper is structured to delve into the significance of Digital Twins for both predictive and preventive maintenance in subsequent sections, presenting a proposed approach and an illustrative example of a motor digital twin on the AWS platform, concluding with perspectives on the topic.

Keywords

Preventive maintenance · Predictive maintenance · Digital twins · Smart factories

I. Abouzid (✉) · R. Saidi
SI2M Laboratory, National Institute of Statistics and Applied Economics Institute Rabat, Rabat, Morocco
e-mail: iabouzid@insea.ac.ma

R. Saidi
e-mail: rsaidi@insea.ac.ma

1 Introduction

Automated maintenance with the use of Digital Twins involves creating virtual replicas of physical assets, systems, or processes to monitor, analyze, and optimize their performance. Digital Twins are dynamic models that simulate real-world objects or processes, providing a comprehensive view of their behavior.

Digital Twin collaborates effectively with other associated paradigms, such as IoT systems and Industry 4.0. Analysts have already anticipated that by 2021, approximately half of the major industrial companies will integrate Digital Twin technology, leading to an expected 10% improvement in effectiveness for these organization (D'Amico et al. 2022). As of now, 29% of global manufacturing companies have either fully or partially implemented their Digital Twin strategies.

In the present era, terms such as Smart Factory, Digital Twin, and Industry 4.0 (Errandonea et al. 2020) herald the advent of a new era in intelligent automation. The vision involves computers effectively monitoring an actual plant and autonomously making decisions to optimize production and reduce costs (Rath et al. 2023). This necessitates the analysis of data derived from the physical plant and their integration into a model representing the factory, commonly known as the Digital Twin. While the concept of a Digital Twin is not new, the technology has become closely associated with manufacturing processes in recent years. (Shao and Helu 2020). The Digital Twin provides a comprehensive view of the entire production line, enabling data analysis and strategic planning aimed at maximizing production while considering real constraints from the physical factory.

Numerous studies in maintenance management have concentrated on predictive more than preventive maintenance within production processes. Predictive maintenance aims to predict and address issues just in time, based on real-time data and analysis, while preventive maintenance is

B. El Bhiri et al. (eds.), *AI and Digital Transformation: Innovations in Supply Chain, Education, and Energy Systems*, Sustainable Artificial Intelligence-Powered Applications, https://doi.org/10.1007/978-3-031-86837-5_17

a systematic approach of scheduled tasks to keep equipment in good working condition and prevent potential failures. Both approaches contribute to overall maintenance strategies, and the choice between them depends on factors like equipment complexity, cost considerations, and the criticality of the system being maintained. Moreover, Digital twins enable remote monitoring of production processes, allowing maintenance teams to evaluate equipment status, identify problems, and implement actions from a distance, thereby improving operational flexibility (Tao et al. 2019; Kritzinger et al. 2018).

Considering the significance of the Digital Twin concept, this article provides a review of the utilization of Digital Twins, focusing specifically on their application in maintenance across different industry sectors. To achieve this, a targeted search encompassing both "Digital Twin" and "maintenance" concepts was conducted.

The rest of this paper is structured as follows: In Sect. 2, we provide a summary of key related work regarding the significance of digital twins for predictive maintenance, expanding the discussion to include preventive maintenance. Section 3 delineates predictive and preventive maintenance, highlighting both their similarities and differences. In Sect. 4, we present a proposed approach utilizing digital twin technology. In Sect. 5, we offer an example of a motor digital twin on the AWS platform. Finally, we conclude the paper in Sect.6 by expressing our perspectives.

2 Related Work

The authors have previously published an earlier version of the conceptual model for Digital Twins aimed at monitoring degradation throughout the lifespan, providing theoretical support for Digital Twin's application in maintenance (D'Amico et al. 2019). However, recent studies underscore the enhanced benefits of employing Digital Twin for maintenance purposes. For instance, (Bevilacqua et al. 2020) emphasized how predictive maintenance Digital Twin can prevent high-risk events for operators. Another study by Aivaliotis et al. (2019) implemented models in the Digital Twin for calculating the Remaining Useful Life (RUL) of components, thereby improving decision-making strategies and production planning (Aivaliotis et al. 2019). Generally, the application of Digital Twin in maintenance has been shown to reduce costs (Khajavi et al. 2019), and studies such as (Coraddu et al. 2019) have demonstrated improved prediction accuracy, contributing to a more sustainable, reliable, and efficient manufacturing system (Table 1).

Conversely, dynamic scheduling techniques play a pivotal role in manufacturing systems due to the unpredictability inherent in production processes, including factors like new job additions, order cancellations, employee absenteeism, and equipment failures. The advent of digital twin technology offers a solution by continuously comparing the physical environment with its virtual counterpart, swiftly identifying disruptions, and initiating rescheduling protocols

Table 1 Predictive maintenance and digital twin

Papers	Year	Purpose
Cognitive digital twin: An approach to improve the maintenance management (D'Amico et al. 2022)	2022	A Digital Twin retains information about its physical counterpart throughout its life cycle and utilizes this data for predictive analytics, thereby minimizing unscheduled downtimes. Additionally, it enhances decision-making by providing accurate insights based on reliable information
The design of a digital-twin for predictive maintenance (Centomo et al. 2020)	2020	The entire framework is designed to identify the present health status of equipment. This crucial information is utilized to mirror the detected health status in the model, enabling the creation of the Digital Twin for the equipment
The use of digital twin for predictive maintenance in manufacturing (Aivaliotis et al. 2019)	2019	The approach presented, utilizing Digital Twin for estimating Remaining Useful Life (RUL), yields precise results suitable for predictive maintenance considerations and associated decision-making, including production planning. Consequently, this approach is applicable and predominantly employed for manufacturing resources that do not typically exhibit extended Remaining Useful Life (RUL)
The role of data fusion in predictive maintenance using digital twin (Liu et al. 2018)	2018	The utilization of digital twin technology in aircraft maintenance, repair, and overhaul involves an ecosystem that integrates modeling and simulation. This system capitalizes on the capabilities of the industrial Internet of Things for data acquisition and information processing, seamlessly incorporating cloud computing

Table 2 Preventive maintenance and digital twin

Papers	Year	Purpose
Digital twin-enabled dynamic scheduling with preventive maintenance using a double-layer Q-learning algorithm (Yan and Wu xxxx)	2022	In this research, a hybrid approach combining reinforcement learning and digital twin technology was employed within the scheduling domain to address the complex optimization challenge involving dynamic Flexible Job Shop Scheduling (FJSS) and adaptable maintenance strategies
Real-Time Prediction of Remaining Useful Life and Preventive Maintenance Strategy Based on Digital Twin (Guo et al. xxxx)	2021	Predicting failures in functional components is instrumental in proactively planning maintenance, ensuring the continued stability of manufacturing processes and meeting stringent production quality standards. With the advent of digital twin technology, renowned for its immersive virtual reality features and dynamic real-time mapping capabilities, this study introduces a novel DT-driven approach for real-time prediction of Remaining Useful Life (RUL) and the formulation of effective preventive maintenance strategies

in response. This capability facilitates dynamic scheduling, significantly minimizing discrepancies between planned and actual schedules. Furthermore, the preventive maintenance addresses the intertwined optimization challenge of flexible job shop scheduling and flexible preventive maintenance, incorporating considerations for both machine and personnel resources (Werner et al. 2019).

Few research papers have placed significant emphasis on preventive maintenance leveraging the capabilities of digital twins. These studies delve into various aspects of how digital twins can enhance preventive maintenance strategies in different industrial contexts Table 2.

In our previous work (Abouzid and Saidi 2023), we've discovered the transformative power of digital twins, particularly when their predictive, monitoring, and diagnostic functionalities are harnessed effectively. By aligning the events observed in the physical system with the simulations generated by the digital twin, we can unlock its true potential. These simulations, depicted in Fig. 1, serve as invaluable tools for prediction, monitoring, and diagnosis.

Leveraging these functions allows us to implement a preventive maintenance approach with precision and foresight, enabling us to anticipate potential issues, monitor equipment health in real-time, and diagnose underlying problems promptly. This proactive stance not only mitigates the risk of unplanned downtime but also optimizes maintenance efforts, ensuring the continuous and efficient operation of critical systems.

3 Types of Maintenance Application with the Use of Digital Twin in Manufacturing Processes

3.1 Predictive Maintenance

Digital Twins play a crucial role in predictive maintenance by generating virtual replicas of physical assets or systems (Guo et al. 2021). These digital counterparts are dynamically enriched with real-time data obtained through sensors, facilitating continuous monitoring and simulation. This real-time integration enables predictive analysis and strategic maintenance planning.

In practical applications, the Digital Twin receives up-to-the-minute data streams from various sensors embedded in the physical assets. This constant influx of information empowers the Digital Twin to continuously simulate the state of the actual equipment. Through this process, manufacturers can scrutinize the virtual representation, gaining valuable insights into potential issues and deviations (Achouch et al. 2022).

The proactive nature of this approach allows manufacturers to foresee maintenance needs well in advance. By leveraging the insights garnered from the Digital Twin, they can strategically plan maintenance activities, preventing unexpected downtimes and optimizing the reliability of equipment. This foresighted strategy, driven by the Digital Twin's analytical capabilities, not only enhances equipment reliability but also contributes to significant reductions in downtime and more

Fig. 1 Digital twin functions

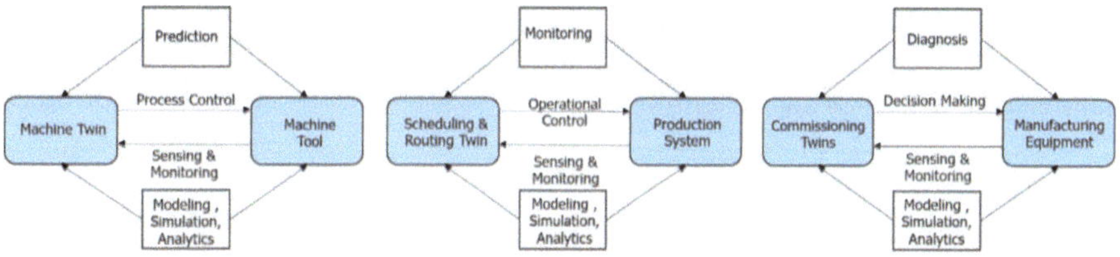

efficient allocation of maintenance resources, thereby optimizing overall system performance and minimizing maintenance costs.

3.2 Preventive Maintenance

Preventive maintenance is the cornerstone of ensuring operational reliability and efficiency in production lines, requiring a systematic approach that goes beyond reactive fixes to proactively address potential issues (Yan and Wu 2022). Central to this strategy is the creation of comprehensive Digital Twins for critical manufacturing equipment or systems. These digital replicas not only mirror the physical assets but also integrate real-time data from sensors and IoT devices, enabling continuous monitoring of key parameters like temperature, pressure, and performance metrics. By harnessing advanced analytics and predictive algorithms, these digital twins facilitate early detection of anomalies and potential failures, empowering maintenance teams to intervene preemptively. This proactive approach not only minimizes unplanned downtime but also optimizes maintenance schedules and resource allocation, ultimately enhancing overall maintenance efficiency and extending the lifespan of production line components. Through the implementation of digital twins and preventive maintenance strategies, organizations can foster a culture of reliability, agility, and continuous improvement in their manufacturing operations.

3.3 Commonalities Between Predictive and Preventive Maintenance

Both maintenance strategies, whether predictive or preventive, enhance uptime and diminish unplanned downtime, thereby enhancing the reliability and extending the lifecycle of assets.

Minimization of Downtime: Both approaches seek to minimize downtime by addressing potential issues before they escalate into critical failures, ensuring continuous operation in manufacturing processes.

Cost Savings: Both strategies contribute to cost savings by reducing the need for emergency repairs, avoiding production interruptions, and optimizing the use of maintenance resources.

Data Utilization: Both rely on the use of data, whether historical or real-time, to inform decision-making regarding maintenance activities and scheduling.

3.4 Differences Between Predictive and Preventive Maintenance

The primary distinction between preventive and predictive maintenance lies in their methods of triggering and scheduling maintenance tasks. Preventive maintenance adheres to a regular schedule triggered by factors such as time intervals or asset usage, whereas predictive maintenance relies on machine data to gauge the condition of the asset, scheduling maintenance accordingly.

Timing of Maintenance

Predictive Maintenance: Occurs based on real-time data analysis and predictions regarding when equipment is likely to fail.

Preventive Maintenance: Involves scheduled and routine maintenance activities regardless of the current condition of the equipment.

Condition Monitoring

Predictive Maintenance: Primarily relies on condition monitoring, utilizing sensors and data analytics to assess the current state of equipment.

Preventive Maintenance: Focuses on planned and routine tasks, often based on a predetermined schedule rather than real-time condition monitoring.

Resource Utilization

Predictive Maintenance: Often requires more sophisticated technology for real-time data analysis and monitoring and may demand a higher level of expertise.

Preventive Maintenance: Generally, involves routine tasks that are less resource intensive and may be performed by maintenance personnel following established procedures.

Flexibility

Predictive Maintenance: Offers more flexibility as maintenance activities are based on the actual condition of the equipment rather than a fixed schedule.

Preventive Maintenance: Follows a predetermined schedule, which may not always align with the actual condition of the equipment, potentially resulting in over maintenance.

Cost Structure

Predictive Maintenance: May have a higher initial investment due to the implementation of advanced technologies for real-time monitoring and analytics.

Table 3 Preventive maintenance and digital twin

Factors	Predictive	Preventive
Timing of maintenance	Based on real time data	Scheduled routines
Condition monitoring	Based on real time data	Scheduled routines
Resource utilization	Based on real time data	Scheduled routines
Flexibility	Based on real time data	Scheduled routines
Cost structure	Based on real time data	Scheduled routines

Preventive Maintenance: Tends to have more predictable and consistent costs as it is based on routine and planned activities (Table 3).

4 Preventive Maintenance Approach with Use of Digital Twin in Manufacturing Process

4.1 Overview of the Approach

In our proposed preventive maintenance approach for manufacturing processes, we harness the power of digital twin technology to enhance operational efficiency and equipment reliability. By creating digital replicas of critical manufacturing components or systems, we establish a virtual environment that mirrors the physical reality. This digital twin not only allows for real-time monitoring of equipment performance but also enables predictive analytics by continuously comparing simulated behavior with real-world data (Fig. 2).

4.2 Approach Key Steps

Identify Key Equipment

Description: Begin by identifying the machinery or systems that are integral to the manufacturing process.

Criteria for Selection: Consider the equipment's impact on production, its critical role in the supply chain, and the potential consequences of failure.

Involvement of Stakeholders: Collaborate with operations, maintenance, and production teams to ensure a comprehensive understanding of the critical components.

Build Digital Twins

Description: Develop accurate and detailed virtual replicas of identified critical manufacturing equipment.

Capture Geometry and Structure: Utilize Computer-Aided Design (CAD) models or 3D scanning technology to capture the physical geometry and structure of the equipment. And Include details such as dimensions, material properties, and connections to replicate the physical attributes faithfully.

Behavioral Modeling

Implement behavioral modeling to simulate the dynamic response of the equipment during various operational scenarios.

Consider factors such as load variations, temperature changes, and other environmental conditions affecting equipment behavior.

Operational Characteristics

Incorporate operational characteristics such as energy consumption, efficiency curves, and response times into the Digital Twin.

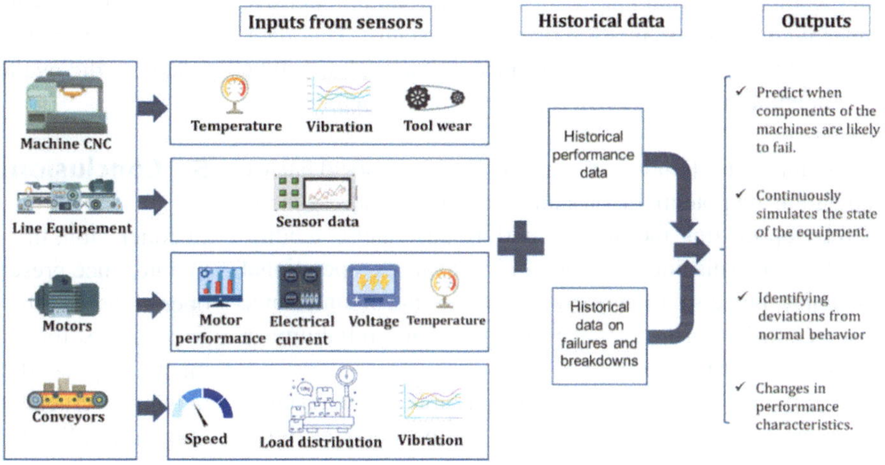

Fig. 2 Preventive maintenance approach with use of digital twin in manufacturing process

Ensure that the Digital Twin replicates the equipment's functionality and performance accurately.

Integrate with Real-Time Data

Description: Establish a seamless connection between the Digital Twin and the physical equipment's real-time data through sensor integration.

Sensor Selection and Placement: Identify and deploy sensors strategically on the critical equipment to capture relevant operational data. And common sensor types include temperature sensors, pressure sensors, accelerometers, and any other sensors pertinent to equipment performance.

Data Communication Protocols

Choose appropriate communication protocols (e.g., MQTT, OPC-UA) to facilitate the transfer of real-time data from sensors to the Digital Twin.

Ensure compatibility with existing industrial communication standards used in the manufacturing environment.

By diligently following these steps in the Digital Twin development process, we can create a robust and accurate representation of critical equipment. This lays the foundation for effective real-time monitoring, predictive analytics, and proactive maintenance strategies to ensure optimal performance and reliability in the manufacturing environment.

4.3 Motor Equipment Simulation with the Use of Digital Twin

In our specific use case, we're embarking on the creation of a digital twin for a motor, leveraging the capabilities of Amazon IoT Twin-Maker. Through this process, we aim to replicate the motor's behavior and characteristics in a virtual environment (Fig. 3). By integrating various metrics, including mainly speed and possibly other factors such as temperature and vibration, we'll establish a comprehensive understanding of the motor's operational status. This data, collected through IoT sensors installed on the motor, will be instrumental in formulating a robust preventive maintenance strategy.

By analyzing trends and patterns in the sensor data, we can anticipate potential issues and proactively address them, thereby optimizing the motor's performance and extending its lifespan. This initiative not only enhances operational efficiency but also minimizes downtime and maintenance costs, contributing to overall productivity and reliability in our manufacturing processes.

Within the Amazon platform, it's possible to create annotations and rules within the digital twin of the motor, enabling

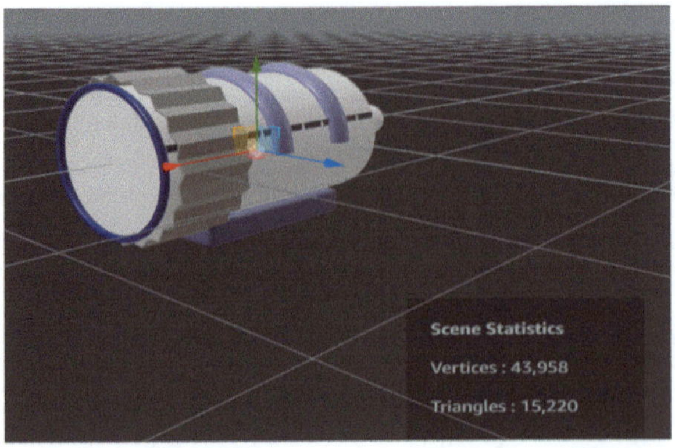

Fig. 3 Digital twin of motor

us to accurately assess its real-time state. As an illustration (Fig. 4), we can establish rules to monitor specific parameters, such as the motor's speed. For instance, we might configure a rule to indicate a red alert when the motor's speed exceeds 400, and conversely, display a green alert when the speed is equal to or below 400. This customized rule system allows for precise monitoring and immediate identification of any deviations from optimal operating conditions, facilitating proactive maintenance actions to uphold the motor's efficiency and reliability.

Additionally, performance analytics serve as another valuable tool in crafting an effective strategy for preventive maintenance. By delving into the performance data of the motor and analyzing key metrics such as energy consumption, temperature variations, and vibration levels, we gain deeper insights into its operational health (Fig. 5).

This comprehensive understanding enables us to anticipate potential issues before they escalate into critical failures, allowing for timely intervention and proactive maintenance. Moreover, performance analytics provide valuable feedback for refining and optimizing preventive maintenance schedules and strategies, ensuring maximum uptime and efficiency of the motor. Through the integration of performance analytics alongside digital twin technology, we can create a robust framework for preventive maintenance that enhances the overall reliability and longevity of the motor.

5 Conclusion and Perspectives

In conclusion, the integration of Digital Twins in automated maintenance presents a significant leap forward in industrial operations by providing virtual replicas of physical assets or processes for monitoring, analysis, and performance optimization. These dynamic models simulate real-world objects, playing a crucial role in data analysis and strategic planning within Smart Factories and Industry 4.0 environments.

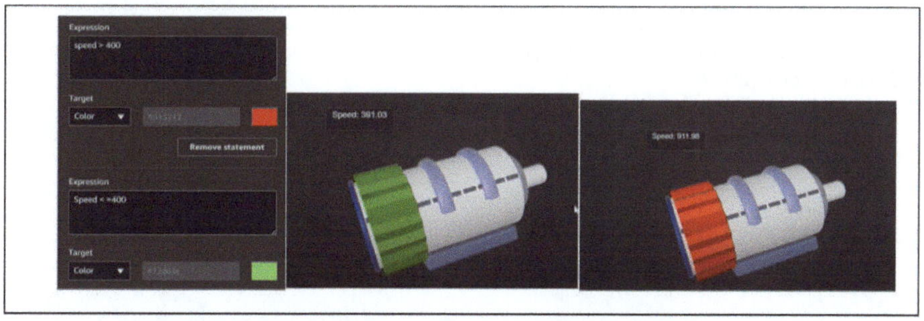

Fig. 4 Monitoring speed s motor

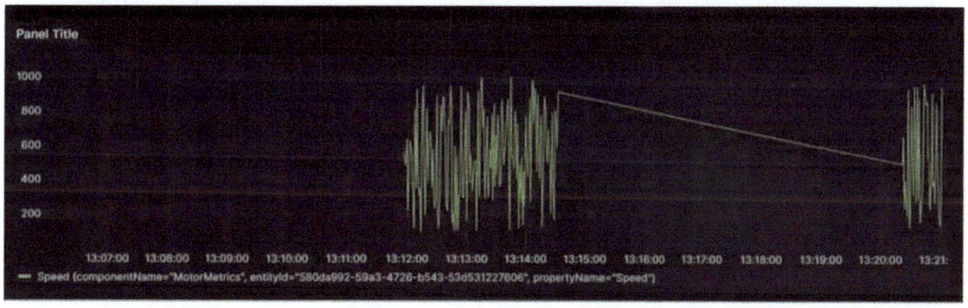

Fig. 5 Performance analytics of the motor

While predictive maintenance remains a key focus, preventive maintenance also contributes significantly to overall maintenance strategies. Leveraging Digital Twins enables remote monitoring and proactive maintenance actions, enhancing operational flexibility and efficiency.

The adoption of Digital Twins in maintenance marks a significant shift toward enhancing operational reliability and efficiency in industrial settings. It's crucial for organizations to tackle validity issues by refining CAD models, ensuring sensor data reliability, and using advanced data integration methods. These steps are essential for making Digital Twins more accurate and beneficial.

Future efforts may focus on refining predictive analytics capabilities, enhancing accuracy in predicting and preventing equipment failures. Advancements in IoT technologies and real-time data integration will further improve the effectiveness of Digital Twins in monitoring and maintaining industrial equipment.

References

Abouzid, I., Saidi, R.: Digital twin implementation approach in supply chain processes. Scientific African **21**, e01821 (2023)

Achouch, M., Dimitrova, M., Ziane, K.: On predictive maintenance in industry 4.0: Overview, models, and challenges. Appl. Sci. **12**(16), 8081 (2022)

Aivaliotis, P., Georgoulias, K., Arkouli, Z., Makris, S.: Methodology for enabling digital twin using advanced physics-based modelling in predictive maintenance. Procedia CIRP **81**, 417–422 (2019)

Aivaliotis, P., Georgoulias, P., Chryssolouris, G.: The use of digital twin for predictive maintenance in manufacturing. Int. J. Comput. Integr. Manuf. 1067–1080 (2019)

Bevilacqua, M., Bottani, E., Ciarapica, F.E., Costantino, F., Di Donato, L., Ferraro, A., Mazzuto, G.: Digital twin reference model development to prevent operators' risk in process plants. Sustainability **12**, 1088 (2020)

Centomo, S., Dall'Ora, N., Fummi, F.: The design of a digital-twin for predictive maintenance. In: 2020 25th IEEE International conference on emerging technologies and factory automation (ETFA) (2020)

Coraddu, A., Oneto, L., Baldi, F., Cipollini, F., Atlar, M.: Data-driven ship digital twin for estimating the speed loss caused by the marine fouling. Ocean Eng. **186**, 106063 (2019)

D'Amico, R.D., Erkoyuncu, J.A., Addepalli, S., Penver, S.: Cognitive digital twin: An approach to improve the maintenance management. CIRP J. Manuf. Sci. Technol. **38**, 613–630 (2022)

D'Amico, D., Ekoyuncu, J., Addepalli, S., Smith, C.: Conceptual framework of a digital twin to evaluate the degradation status of complex engineering systems. Procedia CIRP **86**, 61–67 (2019)

D'Amico, R.D., Erkoyuncu, J.A., Addepalli, S.: Cognitive digital twin: An approach to improve the maintenance management. CIRP J. Manuf. Sci. Technol. **38**, 613–630 (2022)

Errandonea, I., Beltrán, S., Arrizabalaga, S.: Digital twin for maintenance: A literature review **123**, 103316 (2020)

Guo, J., Yang, Z., Chen, C., Luo, W.: Real-time prediction of remaining useful life and preventive maintenance strategy based on digital twin. J. Comput. Inf. Sci. Eng. **21** (2021)

Khajavi S.H., Motlagh, N.H., Jaribion, A., Werner L.C., Holmström, J.: Digital twin: vision, benefits, boundaries, and creation for buildings. IEEE Access 147406–147419 (2019)

Kritzinger, W., Karner, M., Traar, G., Henjes, J., Sihn W.: Digital twin in manufacturing: A categorical literature review and classification **51**(11), 1016–1022 (2018)

Liu, Z., Meyendorf, N., Mrad, N.: The role of data fusion in predictive maintenance using digital twin. In: AIP conference proceedings, pp. 020023 (2018)

Rath, K.C., Chatterjee, P., Patro, P.M., Bojja, P., Kumar, A., Das, R.P.: 3D Printing technology in smart manufacturing systems for efficient production process (2023). https://doi.org/10.1002/9781119836780.ch4

Shao, G., Helu, M.: Framework for a digital twin in manufacturing: Scope and requirements. Manufacturing Letters **24**, 105–107 (2020)

Tao, F., Qi, Q., Wang, L., Nee, A.Y.C.: Digital twins and cyber–physical systems toward smart manufacturing and industry 4.0. Correl. Comp. **5**(4), 653–661 (2019)

Werner, A., Zimmermann, N., Lentes, J.: Approach for a holistic predictive maintenance strategy by incorporating a digital twin. Procedia Manufacturing **39**, 1743–1751 (2019)

Yan, Q., Wang, H., Wu, F.: Digital twin-enabled dynamic scheduling with preventive maintenance using a double-layer Q-learning algorithm. Comput. Oper. Res. **144**, 105823 (2022)

Streamlining the Usage of Digital Twin Technology in IoT Context

Aiman Benalla, Younes Karfa Bekali and Rajaa Saidi

Abstract

Recent advancements in communication, processing, storage, and sensing have led to a significant shift in business operations, known as digital transformation. This transformation aims to use data effectively through scalable digital solutions, especially in an era dominated by AI, machine learning, cloud computing, and IoT. A key concept in this transformation is digital twins, which refers to a digital representation of a tangible item or process which provides real-time insights for decision-making. This paper discusses the current literature on approaches, frameworks, and tools for representing a physical element's digital twin and traces its origins from a basic CAD description to a dynamic model in a continuous feedback loop with the real world. This paper proposes also an approach for streamlining the usage of digital twin technology based on Microsoft Azure tool.

Keywords

Digital twin • IoT • Industry 4.0 • Microsoft Azure

A. Benalla (✉) · Y. K. Bekali · R. Saidi
LRIT Laboratory, Faculty of Sciences, Mohammed V University, Rabat, Morocco
e-mail: aiman.benalla@um5r.ac.ma

Y. K. Bekali
e-mail: y.bekali@um5r.ac.ma

R. Saidi
e-mail: r.saidi@insea.ac.ma

R. Saidi
SI2M Laboratory, National Institute of Statistics and Applied Economics, Rabat, Morocco

1 Introduction

Through the use of **digital twin technologies**, **physical systems**, **objects**, or **processes** can be mapped in a **digital world**. It combines multiple technologies and varies in the scope of the physical entity being digitized. This ranges from creating digital models of objects, their processes, and environments. **Digitalization** has become a consensus; numerous businesses and industries already employ digital twins to identify difficulties and enhance productivity. Recent breakthroughs in **processing**, **sensing**, **communication**, and **storage** have contributed to the emergence of a new generation of information technologies, or "new ITs." This includes the **Internet of Things (IoT)** and sensor integration, **big data** analytics, **artificial intelligence (AI)**, **machine learning (ML)**, and **cloud computing**. These new ITs enable real-time monitoring, analysis, and simulation of the behavior and performance of the physical object by gradually merging the physical and virtual worlds. Thus, the Digital Twin came to be the key concept of Industry 4.0, serving as the foundation for the Industrial Internet of Things (IIoT).

This study outlines a method for implementing a digital-twin solution using Azure Digital Twins, which can be utilized for business optimization to achieve unprecedented levels of control and insight. The study covers the background and evolution of the digital twin concept, proposes an implementation approach, the development process, and includes a case study of a cold storage facility. Concluding remarks highlight future directions for research and development.

2 Digital Twin Definition and Evolution

2.1 Definition of Digital Twin

There are numerous definitions that have been put forth for the term "**digital twin.**"

Think of the **digital twin** as a virtual counterpart of a physical object, system, or process that enables monitoring, anal-

B. El Bhiri et al. (eds.), *AI and Digital Transformation: Innovations in Supply Chain, Education, and Energy Systems*, Sustainable Artificial Intelligence-Powered Applications, https://doi.org/10.1007/978-3-031-86837-5_18

ysis, and simulation of the real-world object's behavior and performance (Banerjee et al. 2017; Negri et al. 2017; Vachálek et al. 2017; Warmefjord et al. 2017).

The **digital twin** is referred to as virtual copy of a tangible object, a mechanism, or a procedure (HSDL 2013).

The **digital twin** is also denoted as a virtual model of a real-world entity, which could range from a physical item like a car to a conceptual idea like a business process (Mabkhot et al. 2018; Talkhestani et al. 2018).

According to the different given definitions, the definition of the "digital twin" concept varies depending on the field of study. To put it simply, the digital twin is a virtual representation of actual systems, processes, or objects at various stages of their lives.

2.2 Origins of Digital Twin

It all started with a basic CAD (Computer Aided Design) description, which characterized a physical object using real-world measurements. This was recognized as the **Geometry Model** which can be a two-dimensional object having length and width (2D) or a three-dimensional form extending the 2D with thickness or depth, as is more usual today, known as a 3D CAD model.

Over time, this Geometry Model evolved to become more detailed and actionable, transforming into a **Digital Model** illustrating the object's functional requirements, and specifications and can even mimic and simulate the object's physical behavior.

When a digital replica is linked to its real-world counterpart, an exchange pathway is formed between the virtual and real entities. This is referred to as a Digital Twin, which is a model that is constantly interacting and functioning in a continuous feedback loop with the real world. This connection enables the digital twin to properly mirror the current state of the physical thing, making it a powerful tool for real-time analysis, forecasting, visualization, and simulation, providing valuable insights into the object's behavior and performance (Qi et al. 2021; Tao et al. 2022).

Figure 1 encapsulates the evolution of digital modeling. Initially, CAD systems provided **Geometry Model**s, which were basic yet precise representations of objects in two dimensions-length and width-or in three dimensions with depth, known as 3D CAD models. These foundational models were essential for visualizing and constructing physical entities.

With the progression of technology, these models evolved into more intricate **Digital Model**s that not only represented an object's shape but also its functionality, requirements, and specifications. This evolution allowed for the simulation of the object's physical behavior, lessening the differences between its intended design and its functional implementation.

The culmination of this evolution is the Digital Twin, a dynamic model that forms a real-time, interactive link with its physical counterpart. This advanced stage enables continuous data exchange, enabling the digital twin to precisely reflect and assess the real object's present condition, thereby enhancing forecasting and decision-making processes.

2.3 Evolution and Progression of Digital Twin Concept

The digital twin concept has undergone tremendous evolution since its emergence. Table 1 illustrates this evolution.

3 Key Requirements for DT Implementation

3.1 Key Requirements for DT and It's Components

3.1.1 Key Requirements for DT

Outlined below are the fundamental requirements critical to the development of a digital twin, consisting of three integral components:

Model: In the context of digital twin technology, a "**model**" refers to a virtual model that is an exact, data-driven replica of a physical asset, system, or process.

Framework: The term describes the foundational software or system essential for creating a digital twin. It should have the ability to handle and integrate diverse data forms, including sensor outputs and contextual details, in order to build and sustain the digital twin.

Application: To ensure the digital twin meets its objectives in achieving business objectives, an application is necessary. This application operates within the framework to interpret the data and provide actionable insights or recommendations, which can lead to tangible business improvements like cost reductions or efficiency gains.

3.1.2 Digital Twin Components

Digital twins are composed of three distinct elements (Nath et al. 2021):

1. Information model
2. Algorithms and analytics
3. Action orchestration.

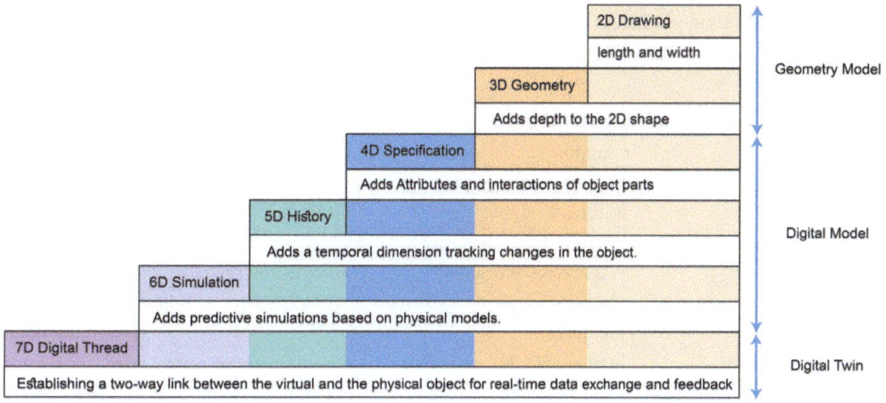

Fig. 1 Evolution of digital twins: a journey from 2D drawings to 7D interactive models

Table 1 Digital twin evolution

Evolution of DT concept	Year	Description
The digital twin introduced	2002	Michael Grieves was the first to introduce the concept of a "digital twin" in 2002, during his discourse on product lifecycle management (PLM) (Grieves 2016)
Early development	2003	Framling and his team suggested a system where each product has a digital twin, accurately mirroring the product's status and information (Främling et al. 2003)
Practical application	2010	In 2010, NASA began creating a virtual counterpart for its space assets (Shafto et al. 2012)
Concept expansion	2012	The practice of employing digital twins has been delved into for use disciplines such as in aircraft maintenance and air force product management (Glaessgen and Stargel 2012)
Formal definition	2014	In his 2014 white paper, Grieves articulated the digital twin concept, envisioning it as a system that includes physical entities, their virtual equivalents, and the data exchange that bridges the physical and virtual environments (Grieves 2014)
Recent applications	2024	Digital twin technology's deployment has proliferated across numerous sectors, notably within the manufacturing and construction industries (AlBalkhy et al. 2024; Hu et al. 2021; Yao et al. 2023)

These elements enable the digital twin to perform advanced functions such as simulation, optimization, analysis, and prediction.

3.2 Enabling Technologies for Digital Twin

The technological requirements for digital twins are (Nath et al. 2021; Tao et al. 2018):

- **Framework and Modeling**: this involves integrating software systems to manage Digital Twins-virtual models of physical entities-within Industrial Internet of Things (IIoT) platforms. These systems handle the metadata and contextual data necessary for the Digital Twin's operation.
- **Connectivity**: Seamless data flow is ensured from the device to the edge, from edge to cloud, and inter-cloud communications.
- **Data Management**: This includes the capture and secure storage of data.
- **Edge Computing**: Processing data at the edge to enhance efficiency and reduce latency.
- **Algorithms and analytics**: Utilizing algorithms for both real-time and centralized processing, and employing appropriate computing power to meet diverse data and analysis requirements.
- **Platforms and applications**: The software that drives digital twins, enabling various functionalities.
- **Visualization Tools**: User interfaces such as dashboards and alerts, along with immersive experiences provided by Augmented Reality (AR) and Virtual Reality (VR).
- **Insightful Actions**: Incorporating human oversight and field services to translate insights into actions.
- **Feedback Mechanisms**: Gathering feedback on products, operational processes, and training methods to inform improvements.
- **Agile Software Development:** Utilizing low-code for rapid deployment.

Table 2 Digital twin platforms benchmarking

Criteria	Azure Digital Twins [2020]	AWS IoT TwinMaker [2022]	Oracle IoT Digital Twin Framework [2019]	Eclipse Ditto [2019]
Framework and modeling	Rich modeling, DTDL support	Supports AWS CloudFormation templates	Custom models, supports PLM	JSON-based modeling
Data integration	IoT Hub, Event Hub, Time Series Insights	IoT Core, S3, DynamoDB, Grafana	Oracle IoT cloud service, ERP, SCM	MQTT, HTTP, WebSockets, Kafka
Simulation and analysis	Integrates with Azure Analytics services	Integrates with AWS analytics services	Integrates with Oracle analytics cloud	Limited, depends on integration
Security	Enterprise-grade security features	AWS security standards	Oracle Cloud security features	Open-source, security managed by user
Visualization tools	Offers 2D/3D scenes visualization	Provides an immersive 3D view of systems	Integration with Oracle products	Open standard, supports various protocols
Customization	High, using Azure functions, Logic Apps	High, using Lambda, other AWS services	High, using Oracle cloud services	High, open-source allows for customization
Developer tools	Visual Studio, Azure CLI	AWS CLI, SDKs, CloudFormation	Oracle developer tools	Eclipse IDE, REST APIs
Community	Strong, Microsoft community	Strong, AWS community	Oracle support	Strong, open-source community

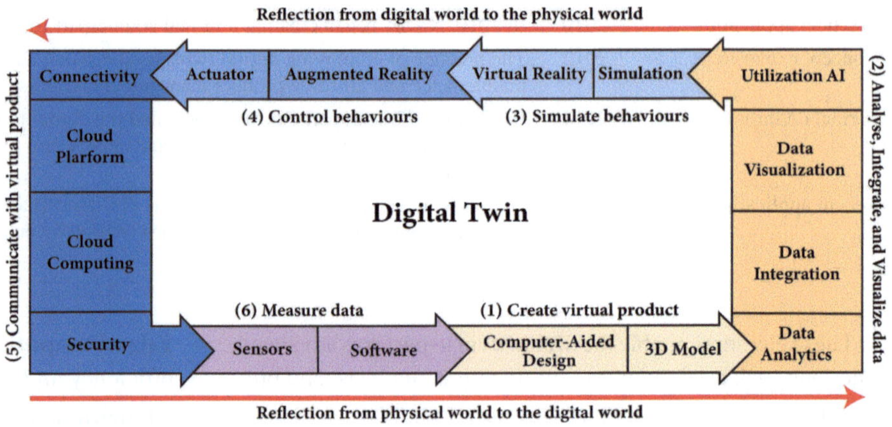

Fig. 2 Key technologies for the creation of digital twins

3.3 Review of Current Digital Twin Platforms

Drawing from the insights presented in Table 2 of the benchmarking study on digital twin platforms, as well as what is seen in Fig. 2 of Enabling technologies for digital twin, we recognize that digital twin technologies create a robust architecture. Within the context of this discourse. For the purposes of this article, we have chosen Azure Digital Twins (ADT) as a component of our technical solution. In the subsequent sections, we will outline a straightforward development process for a digital twin and provide an example of an implementation based on ADT.

4 Development Process and Implementation of a Digital Twin Prototype Using Azure

4.1 The Development Process of a Digital Twin

The development process for creating a digital twin is structured into five key stages, which we undertake using Azure Digital Twins platform (Nath et al. 2021; Yitmen et al. 2021):

1. Create DTDL models based on ontology.
2. Upload models to the Azure Digital Twin instance.
3. Create a Digital Twin instance for assets/objects based on the models.

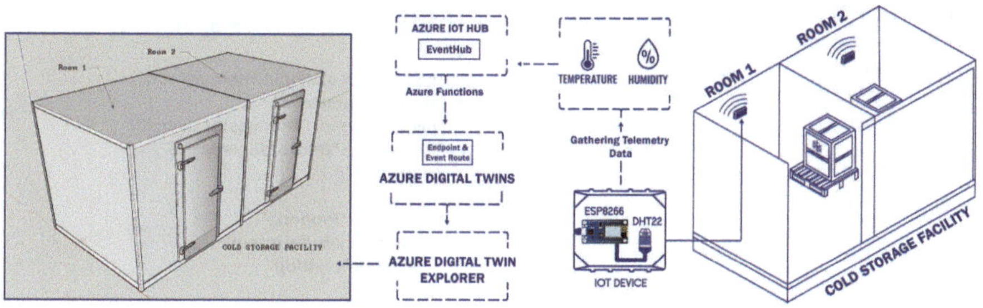

Fig. 3 Cold storage facility implementation employing digital twin technology

4. Develop an application to update the Digital Twin instance.
5. Develop an application tailored to the business use case.

4.2 Case Study

Digital twin technology emerges as a transformative force in business optimization, offering a dynamic approach to problem-solving across various industries. Businesses can achieve unprecedented levels of control and insight by developing digital twins for their physical assets, systems, or operations.

For instance, within the **healthcare sector**, the challenge of maintaining vaccine efficacy is paramount. This involves a meticulous balance of temperature and humidity during storage.

- **Business Problem:** The challenge of maintaining vaccine efficacy is paramount. This involves a meticulous balance of temperature and humidity during storage.
- **Objective:** The objective is to preserve the vaccines' potency and efficacy from production to delivery.
- **Solution:** The solution involves creating a digital twin of the storage environment, equipped with IoT sensors, could monitor and adjust temperature and humidity in real-time.

Figure 3 illustrates the proposed Cold Storage Facility implementation utilizing digital twin technology.

4.3 The Implementation of a Digital Twin

4.3.1 Create DTDL Models Based on Ontology

The first step is to create a DTDL model, where DTDL stands for Digital Twins Definition Language as defined by Microsoft. It enables the creation of a structured model file based on a specific ontology, defining the types and relationships of entities within the Digital counterpart space.

In our case, we will create a digital twin of a room equipped with a thermostat to measure temperature and humidity. This aligns with our business problem of maintaining vaccine temperature and humidity during storage. Thus, we will construct a digital twin of the storage environment, outfitted with IoT sensors, to monitor and regulate temperature and humidity in real-time.

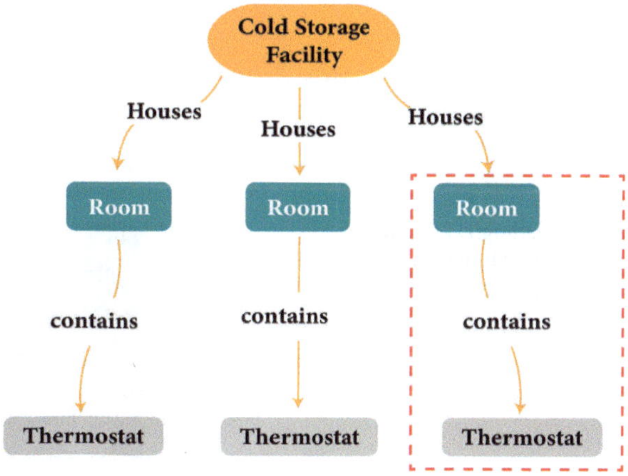

Fig. 4 The Proposed ontology for cold storage facility using DTDL model

The proposed ontology in Fig. 4 has clarified the relationships among models. The next step involves creating DTDL models with the help of metamodel classes to describe digital twin behaviors, concentrating on the highlighted section.

The DTDL comprises a set of meta-classes, which are:

1. **Interface**: A descriptive language for describing models.
2. **Property**: The simple types of metadata of a model.
3. **Telemetry**: The telemetry data that comes from a sensor. It can be used as metadata in a model.
4. **Relationship**: Relationship definitions between models.
5. **Component**: Modular parts of a model that encapsulate properties, telemetry, and commands.
6. **Command**: Actions that can be performed on the Digital Twin.

Figure 5 illustrates a digital twin representation of a Thermostat within its associated Room.

Fig. 5 DTDL model examples for thermostat and room components

Thermostat DTDL model

```
{
  "@id": "dtmi:example:Thermostat;1",
  "@type": "Interface",
  "displayName": "Thermostat model",
  "@context": "dtmi:dtdl:context;3",
  "contents": [
   {
    "@type": "Property",
    "name": "DisplayName",
    "schema": "string"
   },
   {
    "@type": "Property",
    "name": "Temperature",
    "schema": "double"
   },
   :
  ]
}
```

Room DTDL model

```
{
  "@id": "dtmi:example:Room;1",
  "@type": "Interface",
  "displayName": "Room model",
  "@context": "dtmi:dtdl:context;3",
  "contents": [
   {
    "@type": "Property",
    "name": "Temperature",
    "schema": "double"
   },
   {
    "@type": "Relationship",
    "name": "contains",
    "displayName": "contains",
    "target": "dtmi:example:Thermostat;1"
   },
   :
  ]
 }
]
```

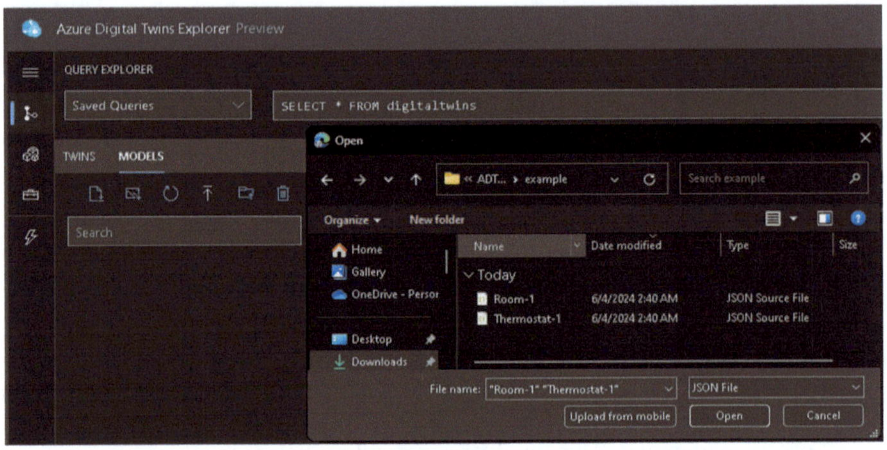

Fig. 6 Uploading DTDL model files via Azure Digital Twins Explorer interface

4.3.2 Upload Models to the Azure Digital Twin Instance

The created DTDL models can be deployed to a Digital Twin instance either through the Azure Digital Twins Explorer accessible via the Azure Portal's web interface or directly through the Azure Command-Line Interface (Azure CLI) (Fig. 6).

4.3.3 Create Digital Twin Instance Based on the DTDL Models

Next step is to create Digital Twin instances for Thermostat and Room. These instances will each be based on the Digital Twin models that we now have in the Azure Digital Twins service (Fig. 7).

4.3.4 Develop an App to Update the Digital Twin Instance

We can update the Azure Digital Twin instance using the REST API through Postman with custom collections for Data Plane and Control Plane, or we can do so via the Azure Digital Twins Explorer or the Azure CLI.

There are several ways to transmit telemetry data to the Azure Digital Twin:

- **Manual Telemetry Data Submission**: Send telemetry data to ADT manually using REST APIs.
- **Real-Time IoT Device Telemetry**: Receive telemetry from IoT devices in real-time or near-real-time.
- **External Telemetry Sources**: Employ any external source to provide telemetry data to ADT.

4.3.5 Develop an Application Tailored to the Business Use Case

The Fig. 8 will provide a high-level solution architecture using Microsoft Azure Digital Twins.

5 Conclusion and Future Work

This paper outlines the development process and methodology for implementing a digital twin solution using the Azure

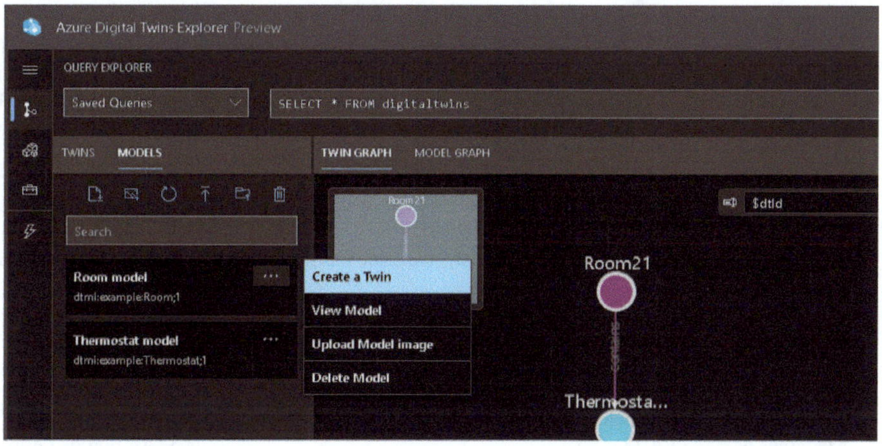

Fig. 7 Managing Digital Twins and model in Azure Digital Twins Explorer

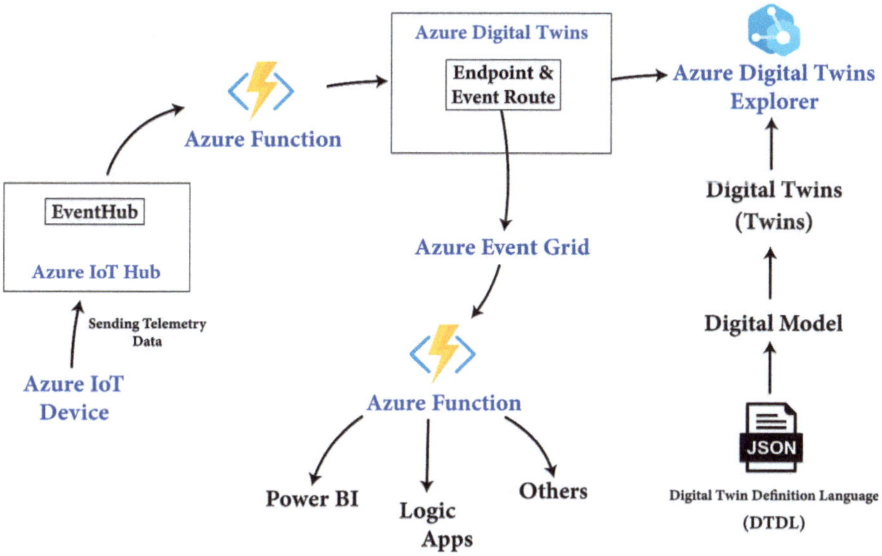

Fig. 8 High-level solution architecture using Microsoft Azure Digital Twins

Digital Twins platform, integrating with other services within the Microsoft ecosystem. It presents a specific application in the healthcare industry, emphasizing on how cold storage facilities can effectively manage the storage of sensitive materials such as vaccines. This involves the critical challenge of preserving vaccine effectiveness while maintaining precise regulation of temperature and humidity levels to ensure optimal storage conditions and uphold the potency and safety of vaccines for use.

In our forthcoming research, we intend to further examine and experiment with different frameworks and tools that can be integrated with digital twin technology. Given that digital twin technology has emerged as a vital aspect of digitization, our goal is to discover innovative methodologies and techniques to maximize its capabilities.

References

AlBalkhy, W., Karmaoui, D., Ducoulombier, L., Lafhaj, Z., Linner, T.: Digital twins in the built environment: definition, applications, and challenges. In: Automation in Construction, vol. 162, pp. 1–13. Elsevier (2024). https://doi.org/10.1016/j.autcon.2024.105368

Amazon Web Services.: AWS IoT TwinMaker (2022). https://aws.amazon.com/iot-twinmaker. Last accessed 05 June 2024

Banerjee, A., Dalal, R., Mittal, S., Joshi, K.P.: Generating digital twin models using knowledge graphs for industrial production lines. In: Proceedings of the Workshop on Industrial Knowledge Graphs (ACM Web Science Conference), p. 16 (2017)

Eclipse Foundation.: Eclipse Ditto: Digital twin framework for IoT (2019). https://eclipse.dev/ditto/index.html. Last accessed 05 June 2024

Främling, K., Holmström, J., Ala-Risku, T., Kärkkäinen, M.: Product agents for handling information about physical objects. Rep. Lab. Inf. Process. Sci. Ser. B **TKO-B**(153) (2003)

Glaessgen, E., Stargel, D.: The digital twin paradigm for future NASA and U.S. air force vehicles. In: Proceedings of the 53rd AIAA/ASME/ASCE/AHS/ASC Structures, Structural Dynamics, and Materials Conference, vol. 1818, pp. 1–13 (2012)

Grieves, M.: DT: Manufacturing Excellence through Virtual Factory Replication. Available at Research Gate (2014)

Grieves, M.: Origins of the digital twin concept. Available Online **2016**, 1–34 (2016)

Hu, W., Zhang, T., Deng, X., Liu, Z., Tan, J.: Digital twin: a state-of-the-art review of its enabling technologies, applications and challenges. J. Intell. Manuf. Spec. Equip. **2**(1), 1–34. Emerald Insight (2021)

Mabkhot, M.M., Al-Ahmari, A.M., Salah, B., Alkhalefah, H.: Requirements of the smart factory system: a survey and perspective. Machines **6**(2), 23 (2018)

Microsoft.: Azure Digital Twins (2020). https://azure.microsoft.com/en-us/products/digital-twins. Last accessed 05 June 2024

Nath, S.V., van Schalkwyk, P., Isaacs, D.: Building Industrial Digital Twins: Design, Develop, and Deploy Digital Twin Solutions for Real-World Industries Using Azure Digital Twins. Packt Publishing, Birmingham (2021)

Negri, E., Fumagalli, L., Macchi, M.: A review of the roles of digital twin in CPS-based production systems. In: Value Based and Intelligent Asset Management, Procedia Manuf., vol. 11, pp. 939–948. Springer, Heidelberg (2017). https://doi.org/10.1016/j.promfg.2017.07.198

Oracle.: IoT digital twin framework. Oracle Cloud Documentation (2019). https://docs.oracle.com/en/cloud/paas/iot-cloud/iotgs/iot-digital-twin-framework.html. Last accessed 05 June 2024

Qi, Q., Tao, F., Hu, T., Anwer, N., et al.: Enabling technologies and tools for digital twin. J. Manuf. Syst. **58**(B), 3–21. Elsevier (2021). https://doi.org/10.1016/j.jmsy.2019.10.001

Shafto, M., Conroy, M., Doyle, R., Glaessgen, E., Kemp, C., LeMoigne, J., Wang, L.: Modeling, simulation, information technology & processing roadmap. In: National Aeronautics and Space Administration (NASA), vol. 32, p. 138 (2012)

Talkhestani, B.A., Jazdi, N., Schlögl, W., Weyrich, M.: Consistency check to synchronize the digital twin of manufacturing automation based on anchor points. Procedia CIRP **72**, 159–164 (2018)

Tao, F., Sui, F., Liu, A., Qi, Q., Zhang, M., Song, B., Guo, Z., Lu, S.C.-Y., Nee, A.: Digital twin-driven product design framework. Int. J. Prod. Res. **56**(1–2), 688–701 (2018)

Tao, F., Xiao, B., Qi, Q., Cheng, J., Ji, P.: Digital twin modeling. J. Manuf. Syst. **64**, 372–389. Elsevier (2022). https://doi.org/10.1016/j.jmsy.2022.06.015

U.S. Air Force.: Global Horizons: Final Report. Homeland Security Digital Library (2013). https://www.hsdl.org/?view&did=741377. Last accessed 05 June 2024

Vachálek, J., Bartalský, L., Rovny, O., Šimiová, D., Morháč, M., Lokšík, M.: The digital twin of an industrial production line within the industry 4.0 concept. In: Proceedings of the PC, pp. 258–262. (2017). https://doi.org/10.1109/PC.2017.7976223

Warmefjord, K., Söderberg, R., Lindkvist, L., Lindau, B., Carlson, J. S.: Inspection data to support a digital twin for geometry assurance. In: Proceedings of the ASME International Mechanical Engineering Congress and Exposition, vol. 101, no. 2, Article no. V002T02A101. ASME (2017). https://doi.org/10.1115/IMECE2017-70398

Yao, J.F., Yang, Y., Wang, X.C., et al.: Systematic review of digital twin technology and applications. Vis. Comput. Ind. Biomed. **6**, 1–10 (2023). https://doi.org/10.1186/s42492-023-00137-4

Yitmen, I., Alizadehsalehi, S., Akıner, İ, Akıner, M.E.: An adapted model of cognitive digital twins for building lifecycle management. Appl. Sci. **11**(9), 4276 (2021)

A Comprehensive Framework for Predictive Maintenance in the Digital Era: The Road to an Intelligent Maintenance Strategy

Z. Mouhib, M. Gallab, S. Merzouk, M. Dinardo, A. Soulhi, and B. ElBhiri

Abstract

In the industrial sector, predictive maintenance (PdM) has become crucial with the advent of Industry 4.0 technologies, enhancing efficiency and equipment reliability. This paper introduces a comprehensive PdM framework for wide-ranging applications across various industries. It meticulously outlines the essential requirements, system architecture, and technological components needed to modernize maintenance practices. By integrating advanced technologies, the proposed framework aims to improve the efficiency and effectiveness of maintenance operations. This holistic approach ensures that companies can adopt tailored PdM strategies that meet their specific needs and constraints, ultimately leading to reduced downtime, optimized maintenance schedules, and improved overall productivity.

Z. Mouhib (✉) · M. Gallab
MIS-LISTD Laboratory, Computer Science Department, Mines-Rabat School (ENSMR), Rabat, Morocco
e-mail: zineb.mouhib@enim.ac.ma

M. Gallab
e-mail: gallab@enim.ac.ma

Z. Mouhib · S. Merzouk · B. ElBhiri
SMARTiLab Laboratory, Moroccan School of Engineering Sciences (EMSI), Rabat, Morocco
e-mail: s.merzouk@emsi.ma

B. ElBhiri
e-mail: b.elbhiri@emsi.ma

M. Dinardo
Department of Materials Engineering and Operations Management, University of Naples "Federico Iiii", Naples, Italy
e-mail: Mario.dinardo@unima.it

A. Soulhi
Mines-Rabat School (ENSMR), Rabat, Morocco
e-mail: soulhi@enim.ac.ma

Keywords

Redictive Maintenance (PdM) · Industry 4.0 · Framework · Requirements · Architecture

1 Introduction

Maintenance is a crucial field in the industrial sector, ensuring the smooth and efficient operation of machinery and systems. Its significance lies in its direct impact on production efficiency, safety, and overall profitability (Eddarhri et al. 2022). Industrial maintenance involves a variety of activities aimed at keeping equipment in optimal working condition, preventing unexpected breakdowns, and extending the lifespan of assets (Silvestri et al. 2020). These activities include routine inspections, repairs, and servicing, all focused on minimizing downtime and maintaining continuous production flow (Gackowiec 2019).

Moreover, the emergence of Industry 4.0 technologies has revolutionized the industrial landscape, significantly changing maintenance approaches. Technologies such as the Internet of Things (IoT), Big Data analytics, and artificial intelligence (AI) have introduced new capabilities for monitoring and managing industrial assets (Achouch et al. 2022). These advancements have led to the development of more sophisticated maintenance strategies that leverage real-time data and predictive analytics to foresee potential issues before they occur (Cachada et al. 2018). In particular, the integration of IoT devices enables continuous data collection from various sensors, providing a wealth of information that can be analyzed to predict equipment failures (Soori et al. 2024).

Consequently, predictive maintenance (PdM) has emerged as a proactive approach, offering substantial benefits over traditional maintenance methods (Groba et al. 2007). PdM uses advanced analytics and machine learning models to analyze data from equipment sensors, allowing for the early

B. El Bhiri et al. (eds.), *AI and Digital Transformation: Innovations in Supply Chain, Education, and Energy Systems*, Sustainable Artificial Intelligence-Powered Applications, https://doi.org/10.1007/978-3-031-86837-5_19

detection of anomalies and potential failures (Krupitzer et al. 2020). This method not only prevents unexpected downtime but also optimizes maintenance schedules, reducing unnecessary maintenance activities and their associated costs (Krupitzer et al. 2020). Therefore, structuring an efficient predictive maintenance system is essential to maximize the benefits of these advanced technologies. A well-designed PdM system ensures that high-quality data is collected, accurately analyzed, and effectively utilized to make informed maintenance decisions.

In this context, the paper proposes a PdM framework designed to be universally applicable, ensuring that companies across various industries can adopt predictive maintenance strategies tailored to their specific needs and constraints. Through a detailed exploration of the requirements, system architecture, and technological components, this study aims to provide a robust blueprint for transforming maintenance practices in the modern industrial era.

The paper is structured as follows: First, the Introduction is followed by Sect. 2, which delves into the theoretical background of Predictive Maintenance. Next, Sect. 3 details the various components of the proposed system. Subsequently, Sect. 4 reviews different architectures from existing literature and highlights how our framework addresses their limitations. Finally, the Conclusion summarizes the contributions of our proposed system and explores potential directions for future research.

2 Predictive Maintenance (PdM): Theorical Background

Predictive maintenance (PdM) represents a prospective maintenance strategy rooted in data-based methodologies, particularly machine learning models, to proactively predict and avoid equipment or system failures. This approach prevents costly downtime and optimizes the availability of engineering systems (Silvestrin et al. 2019). It relies on continuous monitoring of various operational indicators, such as vibrations in rotating machinery or thermal images of electrical equipment, to predict emerging problems and mitigate the risk of catastrophic failures (p. 760–767. IEEE, Xiamen, China(2019). Indeed, the application of predictive maintenance is based on techniques that can be classified into six categories which are: process parameter measurements,,oil analysis, thermographic analysis, acoustic analysis, vibration analysis, and other PdM techniques (Mobley 2002).

This approach fundamentally transforms maintenance into a state-based preventive effort, using the actual operating state of the equipment in conjunction with historical data to predict future states (Selcuk 2017). In addition, relying on real-time data and advanced analytics, predictive maintenance can revolutionize maintenance decision-making, providing factual data to plan tasks, reducing unplanned failures, and preventing major mechanical failures (Motaghare et al. 2018).

In fact, the concept of PdM has been around for a long time (Mobley 2002), but it is only in recent times, with the emergence of advanced technologies, that PdM has become accessible on a larger scale (Ran et al. 2019). The expansion of Internet of Things (IoT) devices has revolutionized data collection, enabling real-time equipment and systems monitoring. These devices can collect vast amounts of data on various parameters such as temperature, vibration, and pressure, which are crucial for predicting potential failures (Motaghare et al. 2018).

Moreover, advancements in Big Data analytics and Machine Learning (ML) have transformed how this data is processed and utilized (Motaghare et al. 2018). Machine Learning models can analyze both historical and real-time data to identify patterns and accurately predict future failures. This capability allows for more proactive and precise maintenance schedules, reducing unplanned downtime and extending the lifespan of equipment (Nguyen et al. 2015). In addition, deep learning, a subset of machine learning, has significantly enhanced. predictive maintenance capabilities. Deep learning algorithms can handle vast amounts of complex data and uncover intricate patterns that traditional algorithms might miss (Motaghare et al. 2018). This results in even more accurate predictions and insights, facilitating better decision-making in maintenance planning.

These technological innovations, and others, have collectively democratized PdM, making it accessible to a broader range of industries and companies.

3 The Proposed Framework

3.1 Methodology

In the context of the new industrial era, characterized by rapid technological advancements and increasing complexity, effective maintenance actions have become more critical than ever. Predictive maintenance is crucial for ensuring the reliability and efficiency of industrial operations. By anticipating potential failures before they occur, it minimizes downtime, reduces costs, and enhances overall productivity. As industries adopt smarter and more interconnected systems, the ability to seamlessly integrate predictive maintenance solutions with cutting-edge technologies and adhere to evolving industrial standards is essential. This proactive approach not only addresses the constraints of modern industrial environments but also paves the way for more sustainable and resilient operations.

As mentioned by Yongyi et al. (Motaghare et al. 2018), to design an efficient PdM universal system, three key requirements must be satisfied: First, the proposed system must be in harmony with different industrial fields. Secondly, a smooth integration with new and future technologies is a must. And finally, the system must fulfill the fundamental tasks of PdM.

In this context, a framework for predictive maintenance is proposed, taking into account the different requirements cited above. The primary objective of this framework is to present a comprehensive and detailed model of predictive maintenance. This model includes a curated selection of technologies tailored to each step of the process. By doing so, the framework aims to streamline the implementation of predictive maintenance systems within industrial companies. Therefore, the framework provides a holistic plan that is versatile and adaptable, ensuring that companies from various sectors or fields of activity can benefit from it without encountering specific limitations. This approach not only enhances the efficiency and effectiveness of maintenance operations but also supports a wide range of industries in adopting advanced maintenance strategies seamlessly.

Given the complex nature of the system, it is essential to employ a structured and detailed approach to its design and implementation. To achieve this, we will utilize SysML (Systems Modeling Language) diagrams specifically suited for modeling complex systems.

Among the various types of SysML diagrams, we will focus on the requirement diagram, to capture and manage the system's requirements, ensuring that all necessary functionalities and constraints are clearly defined and traceable throughout the development process. And the block definition diagram, to present the technological selection for each step of the system.

The two elements share two crucial factors that drive the desired improvement: reducing unnecessary time and costs (Fig. 1).

4 The Framework Requirements

An efficient predictive maintenance system must predict anomalies before they occur effectively. To achieve this, three elements are required: first, having high-quality data; second, analyzing and processing this data using an appropriate algorithm; and finally, ensuring the results are accessible through real-time alerts and reports.

To gather further information and insights on the required system specifications, the main criteria can be outlined in the requirements diagram as follows.

We begin with ensuring High-Quality Data, which is fundamental for the system's accurate anomaly prediction. Firstly, Data Accuracy is paramount. The data must be highly accurate and error-free, which is essential for reliable

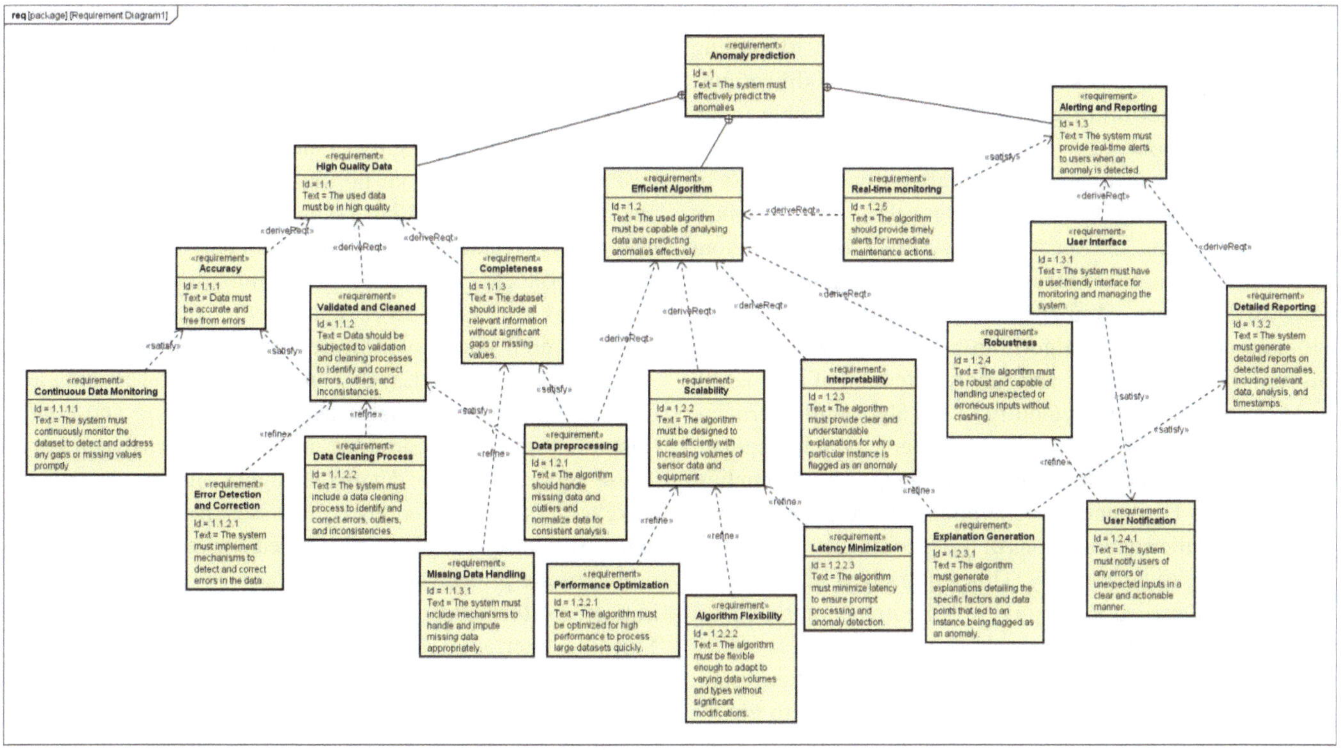

Fig. 1 Predictive maintenance system requirements

predictions. To achieve this, the system should implement robust validation and cleaning mechanisms. And, continuously monitor the dataset to promptly detect and address any gaps or missing values.

Subsequently, in the Validated and Cleaned phase, the system must undergo rigorous validation and data cleaning processes. These processes are critical for identifying and rectifying errors, outliers, and inconsistencies. Specifically, Error Detection and Correction mechanisms must be in place to ensure that any anomalies in the data are promptly identified and corrected. Additionally, a comprehensive Data Cleaning Process is required to address and rectify data issues systematically.

Furthermore, the Completeness of the dataset is crucial. The dataset should encompass all relevant information without significant gaps or missing values. If missing data is encountered, the system must employ effective Missing Data Handling techniques to impute and manage the missing values appropriately. This ensures that the data remains comprehensive and reliable for analysis.

Next, we focus on the Efficient Algorithm. The algorithm should be able to analyze data and predict anomalies effectively. Initially, this starts with Data Preprocessing, where the algorithm handles missing data and outliers, ensuring data is normalized for consistent analysis.

Scalability is another critical requirement. The algorithm might be designed to scale efficiently with increasing volumes of sensor data and equipment inputs. This involves Performance Optimization to enable the algorithm to process large datasets swiftly and effectively. Additionally, Algorithm Flexibility is necessary to adapt to varying data volumes and types without significant modifications, ensuring the algorithm remains robust and adaptable under different conditions.

To further enhance the system's reliability, Robustness of the algorithm is emphasized. The algorithm must be robust enough to handle unexpected or incorrect inputs without crashing, maintaining system stability and reliability. Simultaneously, Latency Minimization is essential for real-time anomaly detection. The algorithm must minimize latency to ensure prompt processing and timely detection of anomalies.

Moreover, the Interpretability of the algorithm is crucial. The system must provide clear and understandable explanations for why particular instances are flagged as anomalies. This includes Explanation Generation, where the algorithm generates detailed explanations of the specific factors and data points leading to an anomaly. This requirement also satisfies the need for Detailed Reporting by providing comprehensive insights into the detected anomalies.

Additionally, the system requires Real-time Monitoring. The algorithm should provide timely information to enable proactive maintenance actions, ensuring continuous and effective monitoring of the system.

When an anomaly is detected, Alerting and Reporting is activated. The system must provide real-time alerts to users, ensuring they are immediately informed of any detected anomalies. This involves a User Interface that is user-friendly for monitoring and managing the system effectively. Concurrently, the system must generate Detailed Reporting on detected anomalies, including relevant data, analysis, and timestamps. Moreover, User Notification is essential to inform users of any errors or unexpected inputs in a clear and actionable manner.

5 Predictive Maintenance Architecture

The architecture of the PdM system, as depicted in Fig. 2 below, is organized into three main layers: the Data Layer, the Processing Layer, and the Action Layer. Each layer contains specific components that work together to ensure the system operates effectively (Fig. 2).

Firstly, the Data Layer is essential for the acquisition and storage of information. Data collection begins with sensors attached to machines, which gather real-time data such as temperature, vibration, acoustic emissions, ultrasound signals, and infrared thermography readings. Additionally, inspection data and historical data are integrated to provide a comprehensive dataset for analysis. Consequently, all this collected data is then transferred to a central database, which serves as a repository where the information is stored for subsequent processing and analysis Next, the Processing Layer is responsible for real-time condition monitoring and anomaly detection. The collected data undergoes real-time condition monitoring using techniques such as oil analysis, vibration analysis, acoustic emissions, ultrasound, and infrared thermography. When anomalies are detected based on deviations from normal patterns, alerts are generated to highlight potential issues. Subsequently, the system uses advanced data analysis and modeling techniques to interpret the collected data. Machine learning and deep learning algorithms are applied to enhance predictive capabilities and identify trends that may indicate potential failures.

Finally, the Action Layer focuses on decision-making and interventions. The results of anomaly detection and condition monitoring are compiled into comprehensive reports. These reports provide detailed insights and summaries for stakeholders. The system then processes the anomaly-related data to identify problems that need correction, and these problems are prioritized based on their severity and potential impact. Furthermore, data visualization tools and Human–Machine Interface (HMI) systems allow operators to visualize the data. This interface enables operators to interact with the data and understand the issues that need attention. Based on the prioritized alerts, maintenance schedules are planned. Operators use the HMI to schedule and execute

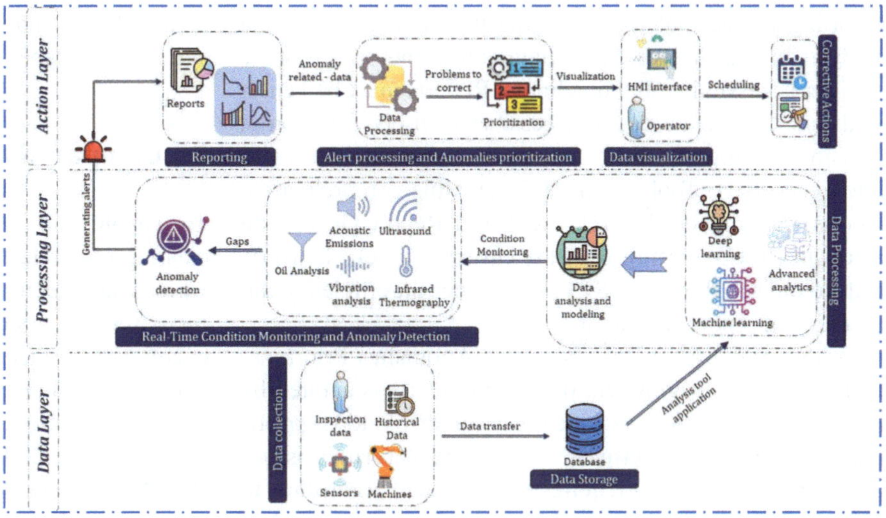

Fig. 2 Predictive maintenance architecture

maintenance tasks, ensuring timely intervention to prevent equipment failures.

6 Predictive Maintenance: Technological Selection

To select the appropriate technologies for each step, it is crucial to consider the various constraints associated with each stage. To illustrate these constraints and the chosen technologies, we present the Block Definition Diagram shown below.

The first stage, Data Acquisition, involves collecting data from various sources. This stage faces constraints such as environmental conditions, data volume and variety, power supply, latency, and integration complexity. These challenges arise because data needs to be collected from a diverse range of sensors and devices, often operating in harsh or variable environments. Additionally, the data must be transmitted reliably and quickly, requiring stable power sources and efficient integration with existing systems. To address these constraints, a variety of technologies can be employed. For example, IoT Sensors and Edge Computing enable local data processing, thereby reducing latency and power consumption (Carvalho et al. 2019). Besides, Wireless Sensor Networks (WSN) and Low-Power Sensors ensure data collection from remote or challenging environments. Hafeez et al. 2021; Akyildiz et al. 2002). Furthermore, Radio-Frequency Identification (RFID) and other advanced sensor technologies enhance data acquisition efficiency and accuracy (Markiewicz et al. 2019).

Once the data is acquired, it needs to be stored securely and efficiently. The Data Storage stage faces constraints such as scalability, data security and privacy, data integrity, and redundancy and backup requirements. These constraints emphasize the necessity for systems capable of managing large data volumes while ensuring data protection and availability for future use. To overcome these challenges, technologies like Cloud Computing provide scalable storage solutions (Adgar et al. 2007), Big Data technologies facilitate the management of large datasets (Wang et al. 2017), and Blockchain ensures data integrity and enhances security through decentralized data storage (Daily and Peterson xxxx; K. et Walther, J. K. et Walther, J. 2017).

Next, in the Data Processing stage, the main constraints are the complexity of data, timeliness, integration with other systems, and data preprocessing requirements. Processing complex datasets in real-time requires sophisticated algorithms and high computational power. Additionally, seamless integration with other systems is crucial for the smooth operation of the predictive maintenance system. To tackle these issues, advanced analytics and Artificial intelligence (AI) / Machine Learning (ML) algorithms are utilized for complex data analysis (Nguyen et al. 2015; Chen et al. 2020). Edge Analytics and High-Performance Computing (HPC) support real-time data processing (Sang et al. 2020; Cardoso and Ferreira 2020), while Fog Computing enhances processing power by distributing computational tasks closer to the data source (Illa and Padhi 2018).

Moreover, the Real-time Condition Monitoring and Anomaly Detection stage is critical for detecting possible issues before they result in system failures. This stage faces constraints such as processing power, algorithm accuracy, data latency, resource availability, and latency in detection. Effective real-time monitoring requires robust algorithms and sufficient computational resources to ensure timely and accurate detection of anomalies. Technologies such as Machine Learning and Deep Learning for advanced anomaly detection (Borghesi et al. 2019), Artificial Intelligence to enhance the system's ability to learn and improve over time (Chen et al. 2020), and Digital Twin technology, which creates virtual replicas of physical systems for detailed supervising and analysis (Teoh et al. 2023), are employed. Subsequently, reporting is essential for conveying the findings from data analysis to stakeholders. The main constraints in this stage are user accessibility and timeliness. Reports must be generated quickly and be easily accessible to users for prompt decision-making. To facilitate effective reporting, Business Intelligence (BI) Tools are used for creating comprehensive reports (Al-amri et al. 2021), AI-based Report Generation automates the reporting process, ensuring timely delivery, and Automated Reporting Systems enhance the efficiency and accuracy of report generation.

Additionally, Alert Processing and Anomalies Prioritization involves managing and prioritizing alerts generated by the system. Constraints include the accuracy of alerts, prioritization logic, and responsiveness. Ensuring that critical issues are addressed promptly requires sophisticated algorithms to filter and prioritize alerts based on their severity. AI and ML algorithms are used for intelligent alert processing, and Rule-Based Systems ensure consistent and logical prioritization of alerts (Klathae and Ruangchoengchum 2019).

Furthermore, Data Visualization is vital for making complex data understandable. Constraints in this stage include clarity, interactivity, real-time updating, and user interaction. Visualizations must be clear and interactive, allowing users to explore data and gain insights quickly. Technologies such as Augmented Reality (AR) provide immersive and interactive visualizations (Babu 2012), and Human–Machine Interface (HMI) enables intuitive user interaction with the system (Meriem et al. 2023).

Following this, the Corrective Actions stage focuses on responding to identified issues. Constraints here include timeliness, resource management, integration with maintenance systems, and documentation and tracking. Effective corrective actions require timely responses, efficient resource management, and thorough documentation. Technologies supporting corrective actions include Robotic Process Automation (RPA) to automate repetitive tasks (Natephra and Motamedi 2019), Maintenance Management Software (CMMS) to help in tracking and managing maintenance activities (Ulrich et al. 2019), Mobile Devices and Apps for real-time communication and documentation (Dechamma and Shobha 2020), and Autonomous Maintenance Robots to perform maintenance tasks autonomously, reducing downtime (Almomani and Aldaihani 2021).

Finally, the Connectivity stage ensures that all components of the system are interconnected. Constraints include reliability, latency, scalability, and security. A robust and secure connectivity infrastructure is essential for the smooth operation of the predictive maintenance system. To ensure reliable and fast communication, technologies such as Industrial Internet of Things (IIoT) and 5G Networks are used (Mourtzis et al. 2021; Schiffer et al. 2010). Cyber-Physical Systems (CPS) merge physical and digital components (Zahariadis et al. 2017), and Cybersecurity Technologies protect the system from cyber threats.

7 Comparative Study and Discussion

To situate the proposed model within the context of existing research, we provide a comprehensive overview of various architectures that have been proposed in the past for describing predictive maintenance. These architectures encompass a range of methodologies and technologies, each contributing unique insights and approaches to the field.

First, the model introduced by Nguyen and Medjaher (Civerchia et al. 2017) employs Long Short-Term Memory (LSTM) networks to predict equipment failures. It starts with data acquisition from sensors monitoring parameters like temperature, vibration, pressure, and operational speed. This real-time data is then used to train LSTM classifiers with historical data of past operational states and failure events. Consequently, LSTM networks capture temporal dependencies in time-series data, identifying trends and anomalies that precede failures. During operation, the trained models estimate the probabilities of potential failures by continuously monitoring sensor data, thus enabling timely interventions. The use of LSTM networks is particularly advantageous as they handle sequences of data and retain information over long periods, which is essential for predicting failures caused by complex, cumulative factors.

The second model, proposed by Caliskan et al. (Meesublak and Klinsukont 2020), focuses on enhancing predictive maintenance in manufacturing environments using a Siamese network architecture combined with transfer learning. Initially, data is collected from IoT sensors on manufacturing equipment, gathering parameters such as temperature, vibration, and pressure. After collecting the data, feature extraction techniques transform raw sensor data into meaningful features for the Siamese network. Subsequently, using transfer learning, the model leverages pre-existing data from similar environments to accelerate learning and enhance performance. The Siamese network then identifies similarities and differences in data pairs, detecting potential failures or anomalies.

Next, the model proposed by Groba et al. (Groba et al. 2007) integrates diverse predictive maintenance techniques, emphasizing high automation and seamless interoperability with enterprise systems. The process begins with data acquisition from shop floors and enterprise systems, including sensors and maintenance logs, ensuring a holistic view of the operational environment. Following data preprocessing and storage in a centralized database, key performance indicators are modeled to reflect equipment health. Predictive algorithms are then used to forecast equipment conditions, enabling early warnings and proactive maintenance. Finally, the system supports maintenance decision-making through dashboards and reports, optimizing maintenance schedules and resource allocation.

In addition, Cheng et al.'s model (Nguyen and Medjaher 2019) leverages Building Information Modeling (BIM) and IoT data for predictive maintenance of MEP (Mechanical, Electrical, and Plumbing) components. Data is collected from BIM and IoT sensors, offering a comprehensive view of the building's infrastructure. Once the data is preprocessed, it trains predictive models using ANNs (Artificial Neural Networks) and SVMs (Support Vector Machines). These models continuously analyze real-time data, predicting component failures and enabling proactive maintenance actions. Maintenance. managers subsequently receive alerts and recommendations through an integrated dashboard, optimizing maintenance schedules and preventing unexpected breakdowns. Conversely, Kaparthi and Bumblauskas (Caliskan et al. 2023) introduce a model that uses decision tree-based machine learning techniques to classify and predict equipment states. Data from sensors undergoes feature extraction, transforming it into meaningful features for decision tree algorithms. Furthermore, random forests improve prediction accuracy by combining multiple decision trees, providing reliable classifications of equipment operational states. This model's systematic approach ensures accurate and robust predictions, facilitating timely maintenance interventions.

In other hand, Gallab al.'s model (Cheng et al. 2020) outlines a comprehensive framework for transforming maintenance practices through digital technologies. The process begins with system identification, followed by determining organizational needs and objectives. Subsequently, appropriate digital technologies, such as IoT sensors and AI, are implemented. Maintenance X.0 is then deployed at strategic, tactical, and operational levels, integrating digital tools into daily maintenance practices. This framework ensures seamless communication and data flow, optimizing maintenance operations across the organization.

Moreover, the model of Li et al. (Kaparthi and Bumblauskas 2020) focuses on optimizing maintenance schedules using advanced data mining and ML techniques. The process begins with data preprocessing, ensuring high-quality input for model training, where historical data trains machine learning models to recognize patterns and anomalies. Once the models are trained, they provide real-time fault diagnosis, enabling immediate responses to potential issues. Maintenance decision support subsequently generates actionable insights and recommendations, ensuring efficient resource allocation and optimal timing for maintenance tasks.

Additionally, Cachada et al.'s architecture (Cachada et al. 2018) aligns predictive maintenance with Industry 4.0 principles, leveraging advanced data analytics and augmented reality (AR). IoT sensors collect real-time data, which undergoes offline analysis to detect patterns and predict failures. Consequently, dynamic monitoring continuously analyzes equipment conditions, triggering alerts for anomalies. Intelligent decision support provides maintenance technicians with actionable insights through AR (Augmented Reality), improving accuracy and efficiency in maintenance tasks.

Furthermore, Sundaram and Zeid's framework (Gallab et al. 2024) offers an interoperable solution for smart manufacturing, using data-driven and hybrid models for predictive maintenance. Initially, data is collected from IoT sensors, ERP (Enterprise Resource Planning), and MES (Manufacturing Execution Systems), then analyzed using machine learning and physics-based models. Real-time condition monitoring subsequently detects anomalies, while maintenance applications guide teams with actionable recommendations. This framework ensures comprehensive and effective maintenance operations in diverse manufacturing environments.

Finally, Abdullahi et al.'s model (Li et al. 2017) proposes a distributed digital twin framework for proactive maintenance. Data from sensors is used to create and update digital twins, virtual replicas of physical assets. Real-time anomaly detection uses advanced analytics and machine learning to trigger alerts for deviations from normal conditions. Consequently, this approach enhances the accuracy and responsiveness of maintenance activities, though implementing and maintaining digital twins in distributed environments remains complex.

Despite their advanced capabilities, these models encounter significant limitations. Each model addresses predictive maintenance using specific technologies and algorithms, often tailored to a particular sector or enterprise. This specialization results in several challenges, particularly high initial implementation costs, the need for substantial investment and training, and the lack of flexibility in choosing the appropriate models or technologies.

By comparing these models with the one proposed in this paper, it becomes evident that our proposed model is a comprehensive and standardized solution that can adapt to various industrial sectors for several compelling reasons.

Firstly, the requirements diagram plays a crucial role in providing a clear overview of the diverse requirements essential for designing an effective predictive maintenance system. This diagram illustrates the interconnections between different requirements, highlighting how certain requirements can impact others. By doing so, it ensures that all necessary considerations are accounted for, facilitating a thorough and well- informed design process.

Furthermore, the architecture presented in Fig. 3 delves into the intricate stages of the predictive maintenance system. This detailed architectural view breaks down the process into manageable steps, offering enhanced visibility into each phase. By detailing the sequential stages, stakeholders gain a deeper understanding of the entire process, from initial data

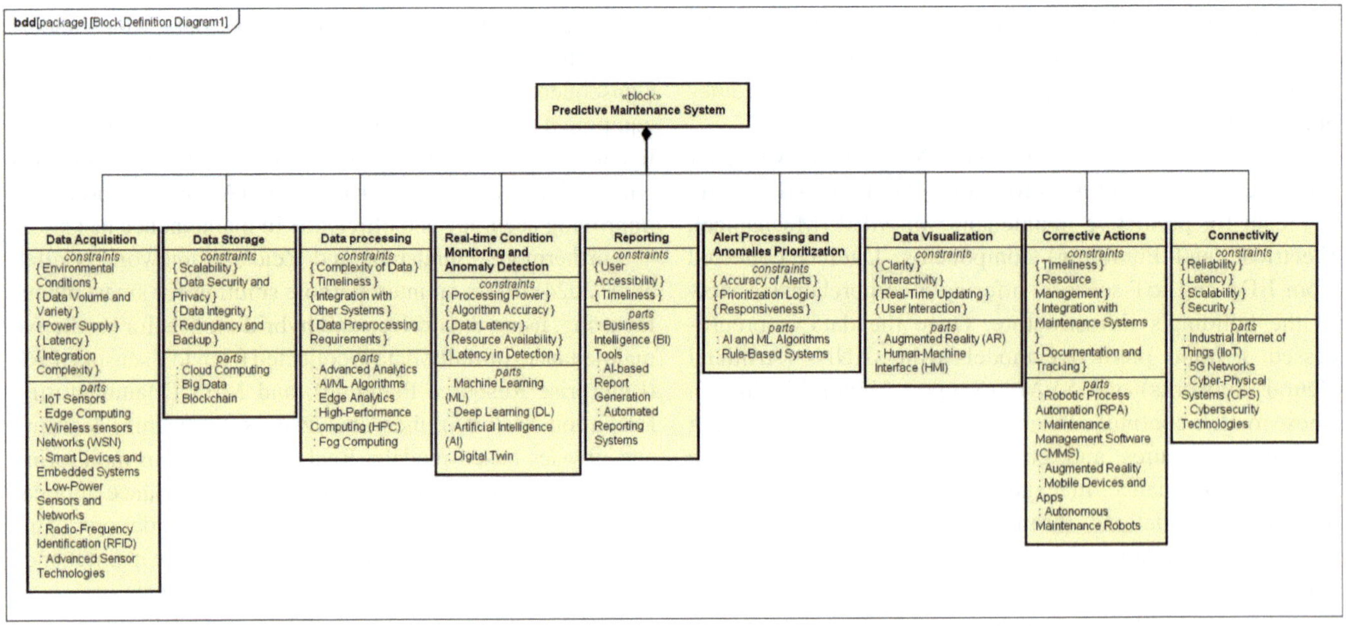

Fig. 3 Technological selection of PdM system inspired by Carvalho et al. (2019)

collection to the final maintenance actions. Consequently, this clarity aids in identifying potential bottlenecks and optimizing each stage for better overall performance.

In addition, the block definition diagram further refines the system by outlining specific constraints and corresponding technologies for each stage. This detailed exposition includes an examination of the various constraints associated with each phase, such as data processing requirements, computational limitations, and integration challenges. Moreover, it identifies suitable technologies that can be employed at each stage to address these constraints effectively. This granular level of detail ensures that the system is robust, efficient, and tailored to fulfill the specific demands of various industrial contexts.

Moreover, the primary advantage of this model lies in its flexibility and adaptability. It is designed to be customizable according to the unique needs and resources of different enterprises. Whether a company has limited resources or requires a highly specialized maintenance system, this model can be tailored to fit their specific circumstances. The standardized nature of the model means it can be implemented across various sectors without losing its effectiveness or requiring significant modifications.

8 Conclusion and Perspectives

The adoption of Predictive Maintenance (PdM) represents a significant advancement in maintenance strategies, leveraging modern technologies such as IoT, machine learning, and data analytics to foresee and prevent equipment failures. This approach not only mitigates costly downtimes but also improves the overall efficiency and lifespan of industrial machinery. Through continuous monitoring of operational indicators and utilizing sophisticated algorithms, PdM transforms maintenance from a reactive to a proactive endeavor, aligning maintenance schedules with the actual condition of the equipment.

The proposed framework detailed in this paper outlines a comprehensive model for implementing PdM across various industries. This model is designed to be adaptable and scalable, ensuring compatibility with different industrial environments and future technologies. Key requirements for a successful PdM system include the acquisition of high-quality data, efficient data processing algorithms, and robust real-time monitoring and alert systems. The integration of SysML diagrams aids in managing system requirements and technological selections, ensuring clarity and coherence throughout the system development process.

By comparing existing PdM models, it is evident that while many offers specialized solutions for particular sectors, our proposed framework provides a standardized yet flexible approach. This model's adaptability allows it to meet the diverse needs of various industries, making it a versatile tool for enhancing maintenance operations.

Looking ahead, there is considerable potential for further research to explore implementing the proposed system with various technologies customized to the specific needs, resources, and characteristics of different industries. This investigation should. assess the compatibility and effectiveness of diverse technologies, taking into account elements such as costs, scalability, and simplicity of integration.

Furthermore, performing case studies or experimental evaluations across different industrial sectors can provide valuable understanding into the system's efficiency and flexibility.

References

Abdullahi, I., Longo, S., Samie, M.: Towards a distributed digital twin framework for predictive maintenance in manufacturing systems (2024)

Achouch, M., Dimitrova, M., Ziane, K., Sattarpanah Karganroudi, S., Dhouib, R., Ibrahim, H., Adda, M.: On predictive maintenance in industry 4.0: Overview, models, and challenges. Appl. Sci. **12**, 8081 (2022). https://doi.org/10.3390/app12168081

Adgar, A., Addison, J.F.D., Yau, C.Y.: Applications of RFID technology in maintenance systems. In: Proceedings of the second world congress on engineering asset management (WCEAM) (2007)

Akyildiz, I.F., Su, W., Sankarasubramaniam, Y., Cayirci, E.: Wireless sensor networks: a survey. Comput. Netw. **38**, 393–422 (2002). https://doi.org/10.1016/S1389-1286(01)00302-4

Al-amri, R., Murugesan, R.K., Man, M., Abdulateef, A.F., Al-Sharafi, M.A., Alkahtani, A.A.: A review of machine learning and deep learning techniques for anomaly detection in IoT data. Appl. Sci. **11**, 5320 (2021). https://doi.org/10.3390/app11125320

Almomani, H., Aldaihani, A.H.: Using computerized maintenance management system (CMMS) in roads maintenance operations. **6** (2021)

Babu, K.V.S.N.J.: Business intelligence: concepts, components, techniques and benefits. SSRN J. (2012). https://doi.org/10.2139/ssrn.2150581

Borghesi, A., Bartolini, A., Lombardi, M., Milano, M., Benini, L.: Anomaly detection using autoencoders in high performance computing systems. AAAI. **33**, 9428–9433 (2019). https://doi.org/10.1609/aaai.v33i01.33019428

Cachada, A., Moreira, P.M., Romero, L., Barbosa, J., Leitno, P., Geraldes, C.A.S., Deusdado, L., Costa, J., Teixeira, C., Teixeira, J., Moreira, A.H.J.: Maintenance 4.0: Intelligent and predictive maintenance system architecture. In: 2018 IEEE 23rd International conference on emerging technologies and factory automation (ETFA). IEEE, p. 139-146, Turin (2018)

Caliskan, A., O'Brien, C., Panduru, K., Walsh, J., Riordan, D.: An efficient siamese network and transfer learning-based predictive maintenance system for more sustainable manufacturing. Sustainability. **15**, 9272 (2023). https://doi.org/10.3390/su15129272

Cardoso, D., Ferreira, L.: Application of predictive maintenance concepts using artificial intelligence tools. Appl. Sci. **11**, 18 (2020). https://doi.org/10.3390/app11010018

Carvalho, T.P., Soares, F.A.A.M.N., Vita, R., Francisco, R.D.P., Basto, J.P., Alcalá, S.G.S.: A systematic literature review of machine learning methods applied to predictive maintenance. Comput. Ind. Eng. **137**, 106024 (2019). https://doi.org/10.1016/j.cie.2019.106024

Chen, Q., Zhu, Z., Si, S., Cai, Z.: Intelligent maintenance of complex equipment based on blockchain and digital twin technologies. In: 2020 IEEE International conference on industrial engineering and engineering management (IEEM). IEEE, p 908–912, Singapore, Singapore (2020)

Cheng, J.C.P., Chen, W., Chen, K., Wang, Q.: Data-driven predictive maintenance planning framework for MEP components based on BIM and IoT using machine learning algorithms. Autom. Constr. **112**, 103087 (2020). https://doi.org/10.1016/j.autcon.2020.103087

Civerchia, F., Bocchino, S., Salvadori, C., Rossi, E., Maggiani, L., Petracca, M.: Industrial internet of things monitoring solution for advanced predictive maintenance applications. J. Ind. Inf. Integr. **7**, 4–12 (2017). https://doi.org/10.1016/j.jii.2017.02.003

Daily, J., Peterson, J.: Predictive maintenance: how big data analysis can improve maintenance. In: Richter, K., Walther, J. (éd.) Supply chain integration challenges in commercial aerospace. p. 267–278. Springer International Publishing, Cham (2017)

Dechamma, P.R., Shobha, N.S.: A review on robotic process automation. Int. J. Res. Eng. Sci. Manag. **3**, 237–244 (2020)

Eddarhri, M., Adib, J., Hain, M., Marzak, A.: Towards predictive maintenance: the case of the aeronautical industry. Procedia Computer Science. **203**, 769–774 (2022). https://doi.org/10.1016/j.procs.2022.07.115

Gackowiec, P.: General overview of maintenance strategies – concepts and approaches. Multidiscip. Asp. Prod. Eng.. **2**, 126–139 (2019). https://doi.org/10.2478/mape-2019-0013

Gallab, M., Ahidar, I., Zrira, N., Ngote, N.: Towards a digital predictive maintenance (DPM): healthcare case study. Procedia Computer Science. **232**, 3183–3194 (2024). https://doi.org/10.1016/j.procs.2024.02.134

Groba, C., Cech, S., Rosenthal, F., Gossling, A.: Architecture of a predictive maintenance framework. In: 6th International conference on computer information systems and industrial management applications (CISIM'07). IEEE, p. 59–-64, Elk, Poland (2007)

Hafeez, T., Xu, L., Mcardle, G.: Edge intelligence for data handling and predictive maintenance in IIOT. IEEE Access. **9**, 49355–49371 (2021). https://doi.org/10.1109/ACCESS.2021.3069137

Illa, P.K., Padhi, N.: Practical guide to smart factory transition using iot, big data and edge analytics. IEEE Access. **6**, 55162–55170 (2018). https://doi.org/10.1109/ACCESS.2018.2872799

Kaparthi, S., Bumblauskas, D.: Designing predictive maintenance systems using decision tree-based machine learning techniques. IJQRM. **37**, 659–686 (2020). https://doi.org/10.1108/IJQRM-04-2019-0131

Klathae, V., Ruangchoengchum, P.: The predictable maintenance 4.0 by applying digital technology: a case study of heavy construction machinery. Rev. Integr. Bus. Econ. Res. **8**, 34–46 (2019)

Krupitzer, C., Wagenhals, T., Züfle, M., Lesch, V., Schäfer, D., Mozaffarin, A., Edinger, J., Becker, C., Kounev, S.: A survey on predictive maintenance for industry 4.0 (2020). http://arxiv.org/abs/2002.08224

Li, Z., Wang, Y., Wang, K.-S.: Intelligent predictive maintenance for fault diagnosis and prognosis in machine centers: Industry 4.0 scenario. Adv. Manuf. **5**, 377–387 (2017). https://doi.org/10.1007/s40436-017-0203-8

Markiewicz, M., Wielgosz, M., Bochenski, M., Tabaczynski, W., Konieczny, T., Kowalczyk, L.: Predictive maintenance of induction motors using ultra-low power wireless sensors and compressed recurrent neural networks. IEEE Access. **7**, 178891–178902 (2019). https://doi.org/10.1109/ACCESS.2019.2953019

Meesublak, K., Klinsukont, T.: A cyber-physical system approach for predictive maintenance. In: 2020 IEEE International conference on smart internet of things (SmartIoT). IEEE, pp. 337–341, (2020)

Meriem, H., Nora, H., Samir, O.: Predictive maintenance for smart industrial systems: a roadmap. Procedia Computer Science. **220**, 645–650 (2023). https://doi.org/10.1016/j.procs.2023.03.082

Mobley, R.K.: An introduction to predictive maintenance. Butterworth-Heinemann, Amsterdam; New York (2002)

Motaghare, O., Pillai, A.S., Ramachandran, K.I.: Predictive maintenance architecture. In: 2018 IEEE International conference on computational intelligence and computing research (ICCIC). IEEE, p. 1–4, Madurai, India (2018)

Mourtzis, D., Angelopoulos, J., Panopoulos, N.: Design and development of an IoT enabled platform for remote monitoring and predictive maintenance of industrial equipment. Procedia Manufacturing. **54**, 166–171 (2021). https://doi.org/10.1016/j.promfg.2021.07.025

Natephra, W., Motamedi, A.: Live data visualization of IoT sensors using augmented reality (AR) and BIM. In: Présenté à 36th International symposium on automation and robotics in construction, Banff, AB, Canada mai 24 (2019)

Nguyen, K.T.P., Medjaher, K.: A new dynamic predictive maintenance framework using deep learning for failure prognostics. Reliab. Eng. Syst. Saf. **188**, 251–262 (2019). https://doi.org/10.1016/j.ress.2019.03.018

Nguyen, K.-A., Do, P., Grall, A.: Multi-level predictive maintenance for multi-component systems. Reliab. Eng. Syst. Saf. **144**, 83–94 (2015). https://doi.org/10.1016/j.ress.2015.07.017

Ran, Y., Zhou, X., Lin, P., Wen, Y., Deng, R.: A survey of predictive maintenance: systems, purposes and approaches (2019). http://arxiv.org/abs/1912.07383

Sang, G.M., Xu, L., De Vrieze, P., Bai, Y., Pan, F.: Predictive maintenance in Industry 4.0. In: Proceedings of the 10th International conference on information systems and technologies. pp. 1–11 (2020)

Schiffer, S., Wortmann, A., Lakemeyer, G.: Self-maintenance for autonomous robots controlled by readyLog. In: Proceedings of the 7th IARP Workshop on technical challenges for dependable robots in human environments. pp. 101–107 (2010)

Selcuk, S.: Predictive maintenance, its implementation and latest trends. Proceedings of the institution of mechanical engineers Part B. J. Eng. Manuf. **231**, 1670–1679 (2017). https://doi.org/10.1177/0954405415601640

Silvestri, L., Forcina, A., Introna, V., Santolamazza, A., Cesarotti, V.: Maintenance transformation through Industry 4.0 technologies: A systematic literature review. Comput. Ind. **123**, 103335 (2020). https://doi.org/10.1016/j.compind.2020.103335

Silvestrin, L.P., Hoogendoorn, M., Koole, G.: A comparative study of state-of-the-art machine learning algorithms for predictive maintenance. In: 2019 IEEE Symposium series on computational intelligence (SSCI). IEEE, p. 760–767, Xiamen, China (2019)

Soori, M., Karimi, F., Jough, G., Dastres, R., Arezoo, B.: Internet of things and data analytics for predictive maintenance in industry 4.0. A Review (2024). https://doi.org/10.13140/RG.2.2.30521.79207

Sundaram, S., Zeid, A.: Smart prognostics and health management (SPHM) in smart manufacturing: an interoperable framework. Sensors. **21**, 5994 (2021). https://doi.org/10.3390/s21185994

Teoh, Y.K., Gill, S.S., Parlikad, A.K.: IoT and Fog-Computing-Based predictive maintenance model for effective asset management in industry 4.0 using machine learning. IEEE Internet Things J. **10**, 2087–2094 (2023). https://doi.org/10.1109/JIOT.2021.3050441

Ulrich, T.A., Boring, R.L., Rashdan, A.A.: Visualization strategy and human-machine interface development of a data driven condition monitoring system for maintenance. (2019)

Wang, J., Zhang, L., Duan, L., Gao, R.X.: A new paradigm of cloud-based predictive maintenance for intelligent manufacturing. J. Intell. Manuf. **28**, 1125–1137 (2017). https://doi.org/10.1007/s10845-015-1066-0

Zahariadis, T., Voulkidis, A., Karkazis, P., Trakadas, P.: Preventive maintenance of critical infrastructures using 5G networks & drones. In: 2017 14th IEEE International conference on advanced video and signal based surveillance (AVSS). IEEE, p. 1–4, Lecce (2017)

MIX
Papier aus verantwortungsvollen Quellen
Paper from responsible sources
FSC® C105338

If you have any concerns about our products,
you can contact us on
ProductSafety@springernature.com

In case Publisher is established outside the EU,
the EU authorized representative is:
Springer Nature Customer Service Center GmbH
Europaplatz 3, 69115 Heidelberg, Germany

Printed by Libri Plureos GmbH
in Hamburg, Germany